Metabolism and Functions of Bioactive Ether Lipids in the Brain

Akhlaq A. Farooqui • Tahira Farooqui
Lloyd A. Horrocks

Metabolism and Functions of Bioactive Ether Lipids in the Brain

Springer

Akhlaq A. Farooqui
The Ohio State University
Department of Molecular and
Cellular Biochemistry
Columbus, OH
USA
farooqui.1@osu.edu

Tahira Farooqui
The Ohio State University
Department of Molecular and
Cellular Biochemistry
Columbus, OH
USA
farooqui.2@osu.edu

Lloyd A. Horrocks (late)
The Ohio State University
Department of Molecular and
Cellular Biochemistry
Columbus, OH
USA

ISBN: 978-1-4419-2652-4 e-ISBN: 978-0-387-77401-5
DOI: 10.1007/978-0-387-77401-5

Printed on acid-free paper

9 8 7 6 5 4 3 2 1

springer.com

Lloyd A. Horrocks (1932-2007); A mentor, colleague, and friend.

We dedicate this monograph to the memory of Professor Lloyd A. Horrocks, an outstanding neurochemist and a world-renowned authority on plasmalogens metabolism and phospholipases A_2 in brain.

Akhlaq A. Farooqui
Tahira Farooqui

Preface

Neurochemical importance of ether lipids is becoming increasingly evident. These lipids include plasmalogen and platelet activating factor. They play crucial roles in membrane fusion, ion transport, inflammation, oxidative stress, and learning and memory. Significant advances have been made in our understanding of structure and functions of plasmalogens and platelet activating factor in neural and non-neural tissues. Increased degradation of plasmalogens is associated with neurochemical and neuropathological changes associated with acute neural trauma (stroke, spinal cord trauma, and head injury) and neurodegenerative diseases (such as Alzheimer disease). The decrease in activity of plasmalogen-synthesizing enzymes is involved in peroxisomal disorders (such as Zellweger syndrome and Rhizomelic chondrodysplasia punctata). The increase in platelet activating factor levels has been reported to occur in ischemic injury, bacterial meningitis, AIDS, prion diseases, and multiple sclerosis. Miller-Dieker lissencephaly is caused by a mutation in PAF-acetyl hydrolase. In the past decade, there has been considerable development not only in our knowledge of the biochemistry of ether lipids but also in our understanding of signal transduction processes associated with their metabolism in the brain. The molecular mechanism that governs the transfer of the death signal from neural cell surface to the nucleus depends upon levels of lipid mediators generated by the degradation of ether lipids and crosstalk among ether lipid, diacyl glycerophospholipid, and glycosphingolipid-derived lipid mediators. This cross-talk modulates the intensity of oxidative stress and neuroinflammation. Thus, interactions among ether lipid, diacyl glycerophospholipid, and glycosphingolipid-derived lipid mediators play a major role in neuronal cell injury and death following acute neural trauma and neurodegenerative diseases. At present, it is unclear whether these processes are primary initiating points in neurodegeneration, or if they are the end result of the neurodegenerative process itself. In recent years, we have been empowered by technological advances in lipidomics, proteomics, and genomics. Investigators are using these techniques not only to identify and determine levels of lipid mediators but also for developing diagnostic test for neurodegenerative diseases associated with altered ether lipid metabolism.

The purpose of this monograph is to present the readers with a coherent overview and cutting edge information in a manner that is useful not only to students and teachers but also to researchers and physicians. This monograph has 12 chapters. The first chapter describes the occurrence and importance of ether lipids in brain. Chapters 2 and 3 cover cutting-edge information on the biosynthesis and degradation of plasmalogens in the central nervous system. Chapter 4 describes the determination and purification of plasmalogen-selective phospholipase A_2 and lysoplasmalogenase from brain. Chapters 5 and 6 are devoted to roles and involvement of plasmalogens in neurological disorders. Chapters 7 and 8 describe the biosynthesis and degradation of platelet- activating factor in the central nervous system. Chapters 9 and 10 describe the functions and association of platelet activating factors with neurological disorders. Chapter 11 deals with neurochemical effects of antitumor ether lipids. Finally, Chapter 12 provides readers and researchers with perspective that will be important for future research work on bioactive ether lipids. Our writing style and demonstrated ability to present complicated material on bioactive ether lipid metabolism makes this book particularly accessible to neuroscience graduate students, teachers, and fellow researchers. It can be used as a supplement for a range of neuroscience courses. This monograph is essential reading for the busy physician or pathologist who wants to be up to date with the latest developments on plasmalogens and platelet-activating factor metabolism.

Clinicians will find this book useful for understanding the molecular aspects of neurodegeneration in stroke and Alzheimer disease that are mediated by plasmalogen-selective phospholipase A_2 and PAF acetyl hydrolase. To our knowledge, no one has written a monograph on bioactive ether lipids and so this monograph is the first to provide students, teachers, researchers, and clinicians a comprehensive description of metabolism and role of plasmalogen and platelet-activating factor along with abnormal signal transduction processes in neurological disorders.

The choices of topics presented in this monograph are personal. They are not only based on our interest in ether lipid metabolism in neurological disorders, but also in an area where major progress has been made. We have tried to ensure uniformity in mode of presentation as well as a logical progression from one topic to another, and have provided extensive referencing. For the sake of simplicity and uniformity, a large number of figures and line diagrams of signal transduction pathways are also included. We hope that our attempt to integrate and consolidate the knowledge of signal transduction processes associated with ether lipid metabolism in brain will provide the basis of more dramatic advances and developments on the involvement of bioactive ether lipids in neurological disorders.

Akhlaq A. Farooqui
Tahira Farooqui

Acknowledgments

We thank Sage Science Press, USA, and Portland Press Ltd., UK, for granting permission to reproduce figures from our earlier papers published by them. We also thank Siraj A. Farooqui for drawing chemical structures and signal transduction pathways associated with the synthesis and degradation of bioactive ether lipids. We would also like to express our gratitude to Ann H. Avouris of Springer, New York for her able and professional handing of the manuscript.

Contents

About the Authors

Dr. Akhlaq A. Farooqui is a leader in the field of bioactive ether lipids, glutamate-mediated neurotoxicity, and brain phospholipases A_2. In collaboration with the late Dr. Lloyd A. Horrocks, he discovered a plasmalogen-selective phospholipase A_2 in brain and showed its stimulation in kainate-mediated neurotoxicity and brain tissue from patients with Alzheimer disease. He has also found a decrease in plasmalogen levels in brain from Alzheimer disease patients. This decrease in plasmalogens is due to the stimulation of phospholipases A_2. Dr. Akhlaq A. Farooqui has authored two monographs, one entitled *Glycerophospholipids in Brain: Phospholipase A_2 in Neurological Disorders* (Springer, 2007) and *Neurochemical Aspects of Excitotoxicity* (Springer, 2008). Dr. Tahira Farooqui is an expert on glycerophospholipid and sphingolipid metabolism and neural plasticitiy. She has published extensively on the molecular mechanism of neuroinflammation, interactions between glycerophospholipid and sphingolipid-derived lipid mediators, and neural plasticity in brain.

List of Abbreviations for Glycerophospholipids

Phosphatidylcholine	PtdCho
Phosphatidylethanolamine	PtdEtn
Choline plasmalogen	Plscho
Ethanolamine plasmalogen	PlsEtn
Ethanolamine plasmalogens	PlsEtns
Phosphatidylinositol	PtdIns
Phohatidylinositol 4-phosphate	PtdIns4*P*
Phosphatidylinositol 4,5- bisphosphate	PtdIns(4,5)P_2
Inositol-1,4,5-trisphosphate	Ins-1,4,5- P_3
Phosphatidic acid	PtdH
Phosphatidylserine	PtdSer
Ceramide	Cer
Sphingosine	Sph

Chapter 1
Occurrence and Importance of Ether Lipids in Brain

1.1 Introduction

Ether glycerophospholipids are major constituents of neural cell membranes. The overall physicochemical characteristics of ether glycerophospholipids are similar to those of ester-bonded glycerophospholipids except for differences in the phase-transition temperature from gel to liquid crystalline and from lamellar to hexagonal phases. These differences may be responsible for determining physical properties of neural membranes, such as bilayer thickness, area per molecule, side-chain packing, free volume, and lateral domains (Paltauf, 1994; Lohner, 1996). The replacement of one or both acyl ester bonds with an alkenyl or alkyl ether bond produces changes in membrane properties (Lohner, 1996), such as a decrease in membrane dipole potential and alterations in thermotropic phase behavior, ion permeability, and side-chain mobility (Paltauf, 1994). Although the occurrence of ether glycerophospholipid species with inositol or serine as a head group has been reported, the most abundant glycerophospholipid species in brain are those with ethanolamine and choline as head groups. Artificial model membranes composed of ether lipids show markedly different molecular dynamics than membranes consisting of diacyl phospholipids (Lohner, 1996). Studies on model membranes indicate that high ether lipid content provides membranes with an unique microenvironment that is necessary for their optimal function. This includes maintenance of activities of membrane-bound enzymes, regulation of permeability, and optimal function of receptors and ion channels. Perturbation of an ether lipid-rich microenvironment in membranes produces significantly more derangements in membrane dynamics than the perturbation of model membranes composed of diacyl glycerophospholipids. Some neutral lipids also contain ether bonds (Foglia et al., 1988; Bordier et al., 1996). They include 1-*O*-alkyl-2,3-*O*-diacyl-*sn*-glycerols, 1-*O*-alk-1′-enyl-2,3-*O*-diacyl-*sn*-glycerols, and 1-*O*-alkyl-2-*O*-acyl-*sn*-glycerols that are analogs of triacylglycerol and diacylglycerols, respectively (Snyder, 1996). These lipids protect against radiation damage and possess antitumor properties. 1-*O*-alk-1′-enyl-2-*O*-acyl-*sn*-glycerols and 1-*O*-alkyl-2-acyl-*sn*-glycerols are natural constituents of myocardium. These ether lipids stimulate protein kinase C activity suggesting that ether lipids may play an important role in regulating protein kinase C-mediated cellular differentiation (Ford et al., 1989).

1.2 Classification of Ether Lipids Found in Brain

Depending on the substituents at carbon-1, most ether glycerophospholipids are divided into two groups: (a) alkenylacyl glycerophospholipids and (b) alkylacyl glycerophospholipids. In mammalian tissues, the alkenylacyl glycerophospholipids are represented by plasmalogens, whereas the alkylacyl glycerophospholipids are represented by platelet-activating factor and its analogs. Plasmalogens contain a vinyl ether (enol ether) linkage at the *sn*-1 position with 16:0, 18:0, and 18:1 (n-7 and n-9) side-chains (alk-1-enyl groups), an ester bond linking arachidonic acid or docosahexaenoic acid or another unsaturated fatty acid at the *sn*-2 position, and a phosphoethanolamine or phosphocholine group at the *sn*-3 position of the glycerol moiety (Fig. 1.1). High levels of ethanolamine plasmalogens occur in brain, lungs, kidney, heart, skeletal muscles, and testes, whereas high levels of choline plasmalogens are found in heart and skeletal muscle. Macrophages and neutrophils also have high levels of ethanolamine plasmalogens as well as plasmanylcholine (1-*O*-alkyl-2-acyl-*sn*-glycerol-3-phosphocholine). Plasmalogens are a reservoir for arachidonate and docosahexaenoate. They not only serve as endogenous antioxidants but also play an important role in neural membrane fusion (Farooqui and Horrocks, 2001). In contrast, platelet-activating factor (PAF) has an *O*-alkyl ether linkage at the *sn*-1 position (fatty alcohol side-chain), a short acyl chain (acetyl moiety) at the *sn*-2 position, and a phosphocholine group at the *sn*-3 position of the glycerol moiety. PAF stimulates a wide range of biological responses ranging from aggregation and degranulation of platelets and neutrophils to a variety of other cellular effects, such as the stimulation of chemotaxis, chemokinesis, superoxide formation, protein phosphorylation, activation of protein kinase C, glycogenolysis, and tumor necrosis

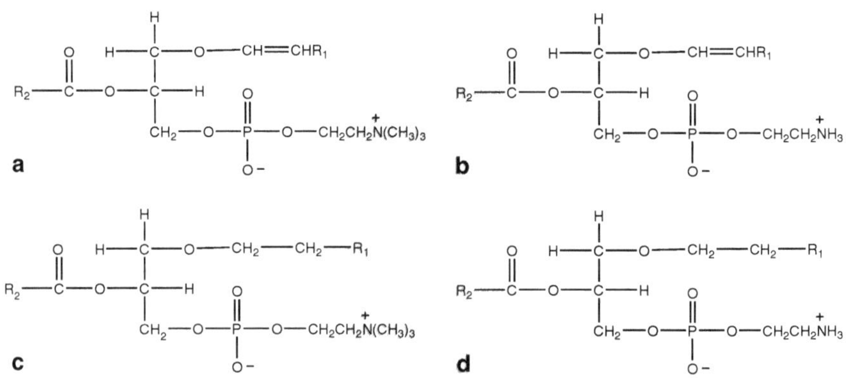

Fig. 1.1 Structures of glycerophospholipids containing choline, ethanolamine, and ether glycerophospholipids. Plasmenylcholine, 1-alk-1′-enyl-2-acyl-*sn*-glycero-3-phosphocholine (**a**); plasmenylethanolamine, 1-alk-1′-enyl-2-acyl-*sn*-glycero-3-phosphoethanolamine (**b**); phosphatidylcholine, 1, 2-diacyl-*sn*-glycero-3-phosphocholine (**c**); and phosphatidylethanolamine, 1, 2-diacyl-*sn*-glycero-3-phosphoethanolamine (**d**). Current nomenclature uses 1Z-alkenyl instead of alk-1′enyl (Fahy et al., 2005).

factor production (Snyder, 1995). PAF acts by binding to a unique G-protein-receptor with seven transmembrane segments. These receptors are linked to intracellular signal transduction pathways, including turnover of phosphatidylinositol, elevation of intracellular Ca^{2+} concentration, and activation of kinases. All these processes are associated with signal transduction processes involved in modulation of neural cell function (Farooqui and Horrocks, 2004).

It is not possible to separate ether lipids from other glycerophospholipids by conventional procedures. They must be chemically modified before using chromatographic procedures for their separation and characterization. The alkenyl and alkyl chains at the *sn*-1 position and acyl chains at the *sn*-2 position may have different number of carbon atoms. Thus, many molecular species exist within each type of ether lipid (Creer and Gross, 1985; Blank et al., 1994; Blank et al., 1995; Guan et al., 2001). Because each molecular species may have a different turnover and role in brain tissue, it is important to isolate and characterize each molecular specie and investigate its role in neuronal and glial cell membranes under normal and pathological conditions. In spite of the rapid development on glycerophospholipid separations in recent years, there are no methods available for separating the molecular species of naturally occurring bioactive ether lipids (Maeba and Ueta, 2004b). In addition, ether lipids in neural membranes include more complex ether lipids. The analogs of seminolipids are 2-eicosa-5′,8′,11′,14′-tetraenylglycerol, 1-*O*-alkyl-2-acyl-*sn*-glycero-3-phosphoinositol (GPI anchor), and phosphatidylinositol ether lipid analogs (PIAs).

1.3 Physicochemical Properties of Ether Lipids

The overall physicochemical properties of ether glycerophospholipids are similar to those of ester-bonded glycerophospholipids, except for differences in the phase-transition temperature from gel to liquid crystalline and from lamellar to hexagonal phases. These differences may be responsible for determining physical properties of neural membranes, such as bilayer thickness, area per molecule, side-chain packing, free volume, and lateral domains (Paltauf, 1994; Lohner, 1996). In artificial membrane systems, there are close similarities between the molecular arrangement of the ether glycerophospholipids and the corresponding ester-bonded analogs. The replacement of one or both acyl ester bonds with an alkenyl or alkyl ether bond produces changes in membrane properties (Lohner, 1996), such as a decrease in membrane dipole potential and alterations in thermotropic phase behavior, ion permeability, and side-chain mobility (Paltauf, 1994). NMR and crystallographic studies indicate that in ether lipids (plasmalogens), the *sn*-2 acyl chain of arachidonate or docosahexaenoate is oriented perpendicular to the membrane surface. In contrast, diacyl glycerophospholipids acyl chains are bent at the *sn*-2 position of glycerol moiety (Pearson and Pascher, 1979; Seelig and Waespe-Sarcevic, 1978; Han and Gross, 1990; Han and Gross, 1991). There are differences in the orientation of the polar head group between plasmalogens and diacyl phosphatidylcholines.

The head groups in choline plasmalogens are more perpendicular to the membrane surface than the head groups in diacyl phosphatidylcholines. Collectively, these studies suggest that the presence of ether lipids provides membranes with conformations necessary for optimal function.

Among ether lipids, plasmalogens are much more susceptible to oxidative reaction than their fatty acid ester analogs. This is due to the presence of the vinyl ether (enol ether) linkage. The vinyl ether bond of plasmalogens interacts with metal ions such as copper and iron with 1:1 stoichiometry and forms a complex (Hahnel et al., 1999). The formation of the metal ion and plasmalogen complex may be partially responsible for the retardation of metal ion-mediated lipid peroxidation in neural membranes. The vinyl ether bond of plasmalogens is also a target of reactive chlorinating species in activated neutrophils. This process results in the generation of α-chloro fatty aldehyde and *sn*-1 lyso-glycerophospholipid (Thukkani et al., 2003). Both these metabolites are cytotoxic. α-Chloro fatty aldehyde forms stable adducts with lysine of protein and primary amines of ethanolamine glycerophospholipids (Wildsmith et al., 2006). α-Chloro fatty aldehyde not only acts as a neutrophil chemoattractant, but also downregulates endothelial nitric oxide synthase expression (Thukkani et al., 2003; Kim et al., 2004). Collectively, these studies suggest that reactive chlorinating species target the vinyl ether bond of plasmalogens during neutrophil activation generating α-chloro fatty aldehyde. This aldehyde forms a Schiff base with membrane proteins and ethanolamine glycerophospholipids resulting in alterations in membrane dynamics and inducing the recruitment of neutrophils to areas involved in active inflammation (Thukkani et al., 2002). Thus by quenching hypochlorous acid, plasmalogens serve as protective agents for the host cells.

Plasmalogens also contain arachidonic or docosahexaenoic acids at the *sn*-2 position of the glycerol moiety. These fatty acids have many double bonds in their structure. During oxidant attack, there may be an intramolecular competition between the vinyl ether double bond at the *sn*-1 position and the fatty acid double bond at the *sn*-2 position for reaction with the oxidants (Engelmann, 2004). During this reaction, the oxygen radical may abstract a hydrogen atom from the vinyl ether linkage, resulting in the decomposition of plasmalogens into aldehydes and 1-lysophospholipid. The vinyl ether bond at the *sn*-1 position reacts with oxidant more rapidly than the double bonds of polyunsaturated fatty acids at the *sn*-2 position. Collective evidence suggests that plasmalogens are a first line of defense against oxidative stress and may have a decisive role in protecting lipoproteins and lowering the oxidizability of cellular membranes from oxidative stress (Maeba and Ueta, 2004a; Engelmann, 2004).

1.4 Fecapentaenes: The Novel Plasmalogens

Fecapentaene-12 and fecapentaene-14 (Fig. 1.2) are polyunsaturated ether lipids generated by colonic microflora in humans and pigs (Van Tassell et al., 1989). The fecapentaenes are potent direct-acting genotoxins that are detected in the feces of

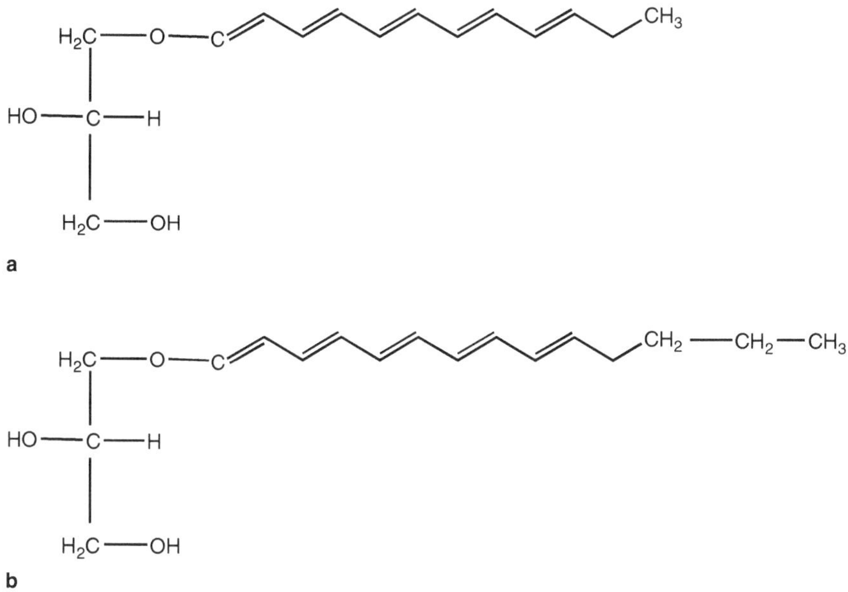

Fig. 1.2 Structures of fecapentaenes: fecapentaene-12 (**a**) and fecapentaene-14 (**b**).

most individuals on normal western diets. Fecapentaenes may play an essential role in the initiation of colorectal cancer. Increased fecal mutagenicity has been observed in populations eating a high fat and low fiber diet (Mower et al., 1982). Although the exact mechanism associated with fecapentaene-mediated carcinogenicity remains unclear, fecapentaenes may act through their interactions with DNA. Fecapentaene-12-mediated DNA damage includes DNA cross-linking, DNA protein cross-linking, and single-strand breaks (Povey et al., 1990). Fecapentaene-12 induces direct oxidative DNA damage via production of the reactive oxygen species O_2^-, H_2O_2, and ($^{\bullet}OH$). Prostaglandin H synthase (PGHS) plays an important role in the generation of reactive oxygen species. Nonsteroidal antiinflammatory agents (NSAIDs) block the induction of oxidative DNA base damage by fecapentaene-12 in HeLa cells (Plummer et al., 1995). This may occur in two ways: first, by formation of highly reactive fecapentaene-12 hydroperoxides, which would generate oxygen radicals through Fenton-like reactions, and second, by the generation of oxygen radicals through peroxidase-mediated cooxidation of fecapentaene-12. Fecapentaene-12 also induces the formation of 7,8-dihydro-8-oxo-2′-deoxyguanosine (8-oxodG), a dose-dependent marker for oxidative DNA damage. Furthermore, fecapentaene-12 can also associate noncovalently with duplex DNA. Collective evidence suggests that fecapentaene-12 provides an interesting new example highlighting the potential for hydrophobic long chain hydrocarbons to associate noncovalently with duplex DNA (Szekely and Gates, 2006). Fecapentaene-12 decreases the content of cellular free low-molecular-weight thiols including glutathione (Dypbukt et al., 1989). Fecapentaene-12 reacts directly with glutathione causing both decreased levels of free thiol and some concomitant formation of oxidized glutathione, indicating that the thiol depletion is a result of both alkylation and oxidative reactions. Collectively, these studies suggest that fecapentaene-12 is a potent cytotoxic and genotoxic agent that reacts with cellular thiols and cause several types of DNA damage.

1.5 Other Ether Lipids Found in Mammalian Tissues

1-*O*-alkyl-2-acyl-*sn*-glycero-3-phosphocholines are found in platelets, neutrophils, macrophages, monocytes, endothelial cells, and mast cells. This ether lipid is a precursor of platelet-activating factor (PAF). Dialkyl glycerophosphocholines are minor constituents of bovine heart and spermatozoa. 1-*O*-alkyl-2-acyl-*sn*-glycero-3-phosphoinositols are a component of glycosylphosphatidylinositol anchors (GPI-anchors) (Fig. 1.3) that are involved in the attachment of several proteins to the cell surface (Paltauf, 1994). For example, the presence of alkyl ether lipids in GPI anchors has been reported for many mammalian enzymes such as human and bovine erythrocyte acetylcholinesterase, placental alkaline phosphatase, membrane dipeptidase, and 5′-nucleotidase (Roberts et al., 1988; Hooper, 1997; Chatterjee and Mayor, 2001). The release of GPI-anchored enzymes from the cell surface by specific phospholipases may play a key role in modulation of functional properties of neural membranes. Reconstitution of GPI-anchored proteins into bilayers of defined ether lipids provides a powerful tool to explore the interactions of these proteins with the membrane and investigate how bilayer properties modulate their structure, function, and cleavage by phospholipases (Chatterjee and Mayor, 2001). The 1-*O*-alkyl-2-*O*-acyl-*sn*-2-glycerol residue in glycosylphosphatidylinositols

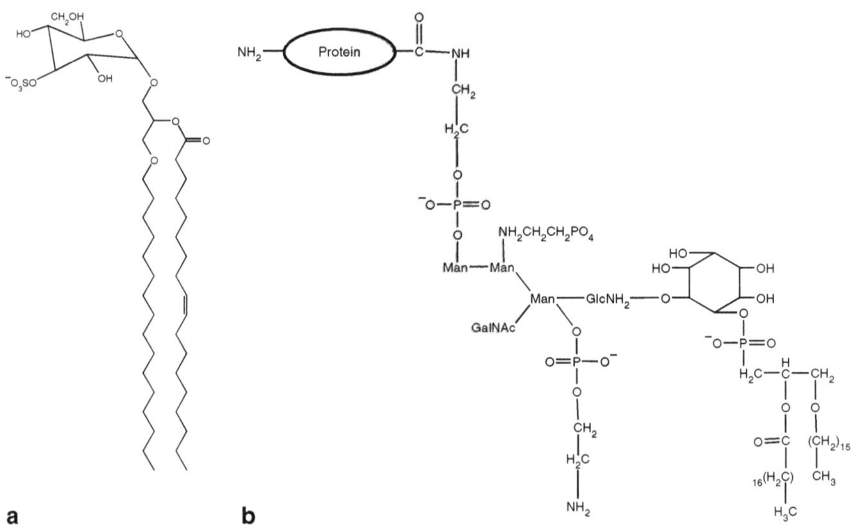

Fig. 1.3 Structures of complex ether lipids: seminolipids (**a**) and GPI-anchor (**b**).

modulates the activity of glucocorticoid receptors (Schulman et al., 1992). GPI-anchoring of proteins is a posttranslational modification, which takes place in the endoplasmic reticulum. It not only attaches proteins to the luminal side of the membrane but also facilitates transport of certain solutes through membranes.

Other types of minor ether lipids in mammalian tissues include cholesterol ethers and vinyl ethers, glycerol thio-ethers, and dialkyl glycerophosphocholines, which have been found in bovine heart. 1-*O*-alkyl-1′-enyl-2-acyl-*sn*-glycerols and 1-*O*-alkyl-2-acyl-*sn*-glycerols are naturally occurring constituents of rabbit myocardium. The levels of these ether-linked diradylglycerols are markedly increased in rabbit heart tissue following ischemia (Ford and Gross, 1988; Ford and Gross, 1989). Like 1,2-diacyl-*sn*-glycerols, 1-*O*-alkyl-1′-enyl-2-acyl-*sn*-glycerols, and 1-*O*-alkyl-2-acyl-*sn*-glycerols stimulate protein kinase C and contribute to signal transduction processes during cell stimulation and trauma (Ford et al., 1989).

1-*O*-alk-1′-enyl-2-lyso-*sn*-glycero-3-phosphates (alkenyl-G*P* or lyso PlsH) occur in some commercially available sphingolipid preparations because they are difficult to remove. Alkenyl-G*P* markedly stimulates mitogen-activated protein kinases and elicits mitogenic responses in 3T3 fibroblasts. This ether lipid may function as a component of a growth factor (Liliom et al., 1998a). Its presence in sphingolipid preparations may explain some results attributed to a sphingolipid.

Minor ether phospholipids not only facilitate the attachment of proteins with biomembranes, but also stabilize and modulate receptors localized in biomembranes. Alkenyl-G*P* was recently detected in fluid bathing the cornea. Corneal injury in rabbits produces a marked increase in the levels of this ether lipid along with lyso-phosphatidic acid (Liliom et al., 1998b). These lipids may be involved not only in maintaining the integrity of the normal cornea, but also in promoting cellular regeneration of the injured cornea (Liliom et al., 1998b).

Rat brain contains acyl and alkyl lyso-phosphatidic acids (Sugiura et al., 1999). Acyl lyso-phosphatidic acids are converted to phosphatidic acid, a common precursor for diacyl glycerophospholipids, whereas alkyl lyso-phosphatidic acids are a precursor for ether-linked glycerophospholipids (Sugiura et al., 1999). The predominant molecular species of acyl lyso-phosphatidic acids contain 18:1, 18:0, and 16:0 (46.9%, 22.5%, and 18.8%, respectively). A significant amount of 20:4-containing species (7.2%) is also present in the acyl lyso-phosphatidic acids fraction. In contrast, rat brain alkyl lyso-phosphatidic acids consist of species with 16:0, 18:0, and 18:1. Acyl and alkyl lyso-phosphatidic acids not only induce the rounding of neuroblastomaxglioma hybrid NG108–15 cells, but also elicit a transient increase in cellular Ca^{2+} with equal potency. Collectively, these studies suggest that in brain tissue acyl and alkyl lyso-phosphatidic acids play important physiological roles as intercellular signaling molecules, as well as metabolic intermediates for diacyl glycerophospholipids and ether lipids.

2-Arachidonoylglycerols (2-AG) are endogenous cannabinoid receptor ligands that exert its effects by binding to central and peripheral cannabinoid receptors (Mechoulam et al., 1995; Sugiura et al., 1995). An ether-linked analog of 2-AG, 2-eicosa-5′,8′,11′,14′-tetraenylglycerol (HU310, noladin ether) (Fig. 1.4) was isolated from pig brain (Hanuš et al., 2001). Noladin ether is also found in rat brain (Fezza

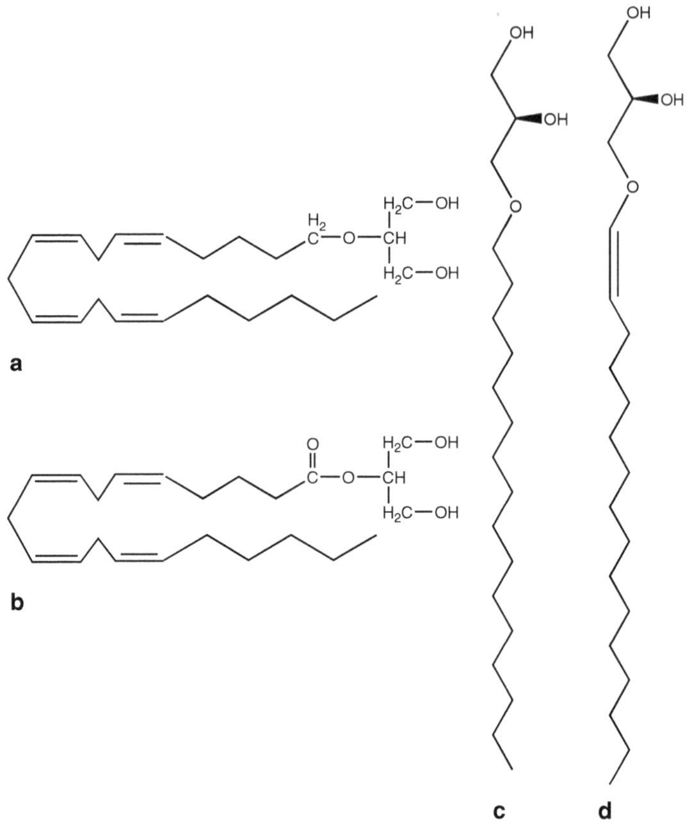

Fig. 1.4 Structures of other ether lipids. Noladin ether lipid (2-eicosa-5′,8′,11′,14′-tetraenyl-glycerol) (**a**); 2-arachidonoylglycerol (**b**); alkylglycerol (AG) (**c**); and sn-1-alkenylglycerol (2-AEG) (**d**).

et al., 2002). In contrast, recent mass spectrographic studies indicate that noladin ether is not found in brain tissue from rat, mouse, hamster, guinea pig, and pig (Oka et al., 2003). The reason for this discrepancy is not fully understood. The role of noladin ether in mammalian brain is not known. However, noladin ether may be another endogenous cannabinoid receptor ligand involved in the modulation of several cannabimimetic activities, such as inhibition of lymphocyte proliferation, hypothermia, reduced locomotor activity, and analgesia (Hanuš et al., 2001; Fezza et al., 2002). Noladin ether is not hydrolyzed by monoacylglycerol lipase, the enzyme that hydrolyzes 2-arachidonoylglycerol.

Topical administration of the novel putative endogenous cannabinoid noladin ether reduces intraocular pressure in rabbits. This intraocular pressure reduction is most probably mediated through the CB1 receptor. The effect on intraocular pressure of noladin ether differs from those of the known endogenous cannabinoids *N*-arachidonoylethanolamide and 2-arachidonoylglycerols, probably because of its more stable chemical structure (Laine et al., 2002). Noladin ether also affects the aqueous humor outflow facility in a porcine anterior-segment-perfused organ culture model. The effect of noladin is dose-dependent. Pretreatment with SR141716A, a selective CB1 antagonist, blocks it. The molecular mechanism involved in noladin-mediated processes is not fully understood. However, noladin ether-induced enhancement of the outflow facility may be mediated through the trabecular meshwork CB1 receptor that involves p42/44 MAP kinase signaling pathway and changes in actin cytoskeletons (Njie et al., 2006).

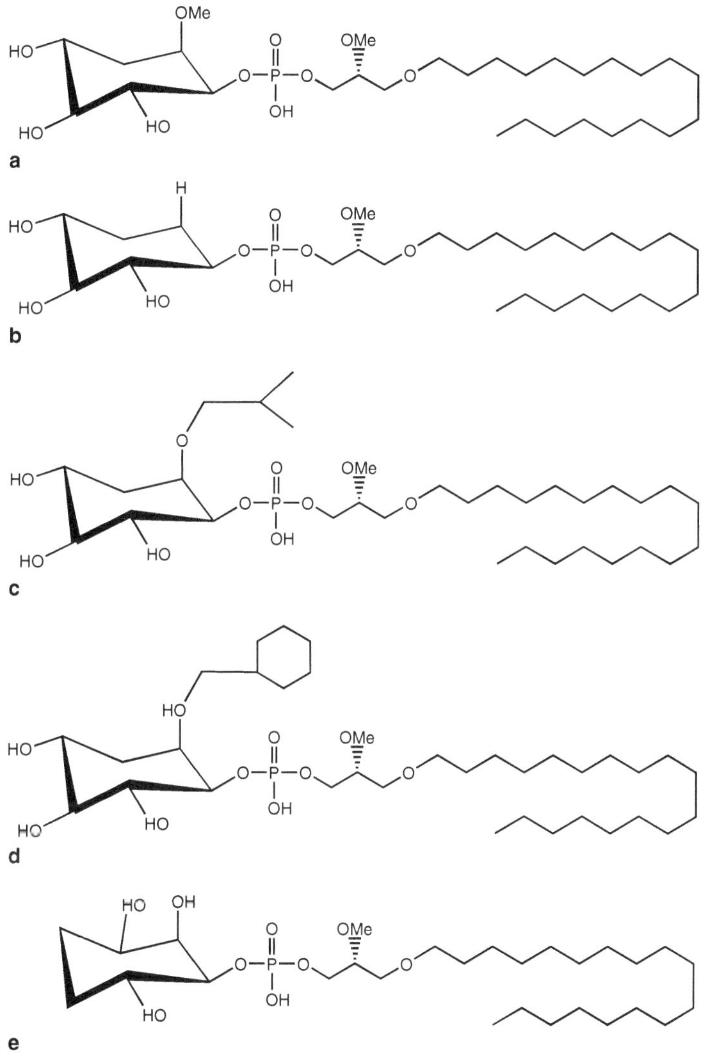

Fig. 1.5 Structure of phosphatidylinositol ether lipid analogs (PIAs): PIA5 (**a**), PIA6 (**b**), PIA23 (**c**), PIA24 (**d**), and PIA25 (**e**). Phosphatidylinositol ether lipid analogs (PIAs) inhibit serine/threonine Akt (protein kinase B) translocation, phosphorylation, and kinase activity and preferentially induce apoptosis in breast and lung cancer cell lines with high levels of active Akt (Gill et al., 2006).

Synthesis of phosphatidylinositol ether lipid analogs (PIAs) has also been reported (Gills and Dennis, 2004) (Fig. 1.5). PIAs inhibit Akt translocation, phosphorylation, and kinase activity. Akt is an attractive therapeutic target in cancer because it contributes to tumorigenesis and therapeutic resistance. Furthermore, PIAs selectively induce apoptosis in cancer cell lines that depend on Akt for survival (Gills and Dennis, 2004). Although PIAs are widely active in cancer cells, which correlate with the presence of its intended target, the activity of PIAs is biologically distinct from other known inhibitors of the PI3K/Akt/mTOR pathway (Gills et al., 2006).

A modified galactosylceramide, with a long-chain cyclic acetal at the sugar moiety, plasmalogalactosylceramide (3-*O*-(4′,6′-plasmalogalactosyl) 1-*O*-alkylglycerol), was isolated from equine brain (Yachida et al., 1998). The cyclic acetal linkage, its linked position, and the length of the acetal chain of the natural plasmalo lipid were determined by proton NMR spectroscopy and fast-atom bombardment-mass spectrometry, as well as by gas chromatography-mass spectrometry and gas–liquid chromatography (Yachida et al., 1999). At present no information is available about the metabolism and role of this ether lipid in other mammalian brains.

1.6 Lipid Metabolism in Ether Lipid-Deficient Mice

Ether lipid-deficient mice (Gorgas et al., 2006) display severe phenotypic alterations such as arrest of spermatogenesis, development of cataracts, defects in peroxisomal function, and abnormalities in myelinogenesis. Peroxisomes are involved in anabolic and catabolic processes associated with lipid metabolism. Peroxisomes perform a number of essential metabolic functions including β-oxidation of straight and branched chain very long and long chain fatty acids, ether phospholipid biosynthesis, fatty acid α-oxidation, dolichol synthesis, oxidation of D-amino acids and polyamines, glyoxylate detoxification, and inactivation of hydrogen peroxide (Wanders and Waterham, 2006). The involvement of peroxisomes in these metabolic pathways necessitates the transport of metabolites in and out of peroxisomes. Recently, considerable progress has been made in the characterization of several metabolite transport systems across the peroxisomal membrane (Wanders and Waterham, 2006; Visser et al., 2007). This is exemplified by the identification of a specific transporter for adenine nucleotides and several half-ABC (ATP-binding cassette) transporters that may be present as hetero- and homo-dimers.

Peroxisome-mediated processes also include the synthesis of cholesterol and bile acids (Verhoeven et al., 1998). Long-chain acyl-CoA synthetase occurs in the peroxisome. Earlier reports indicated that this enzyme is similar to the enzymes in the outer mitochondrial membrane and the endoplasmic reticulum. However, the occurrence of two highly homologous but different cDNAs encoding rat liver and brain long-chain acyl-CoA synthetases has been reported (Causeret et al., 1993). Evidence is now accumulating for a distinct synthetase that specifically activates very-long chain fatty acids in peroxisomes, outer mitochondrial membrane, and the endoplasmic reticulum

(Causeret et al., 1993). Long-chain fatty alcohols, obligate precursors of ether lipids, are biosynthesized by the reduction of the corresponding acyl-CoAs by an acyl-CoA reductase. Peroxisome proliferators do not stimulate these enzymes. However, feno- and ciprofibrate treatments results in a sixfold increase in the palmitoyl-CoA synthetase mRNA level in rat liver (Causeret et al., 1993).

In ether lipid-deficient mice, mutations in dihydroxyacetone phosphate acyltransferase and alkyl dihydroxyacetone phosphate synthase result in the deficiency of ether lipids, especially ethanolamine plasmalogens in neural membranes (Gorgas et al., 2006). In addition to plasmalogens, alterations in cholesterol synthesis and efflux were also observed. In normal mice and cell cultures, both plasmalogen synthesis and very long-chain fatty acid β-oxidation can be accelerated by increasing the number of peroxisomes with peroxisome proliferators such as fibrate derivatives and 4-phenylbutyrate (Hayashi and Takahata, 1991; Hayashi and Oohashi, 1995; Wei et al., 2000). Collective evidence suggests that the normal metabolism of plasmalogens and cholesterol is essential for optimal neural membrane function. Deficiencies of plasmalogen and cholesterol may result in impaired membrane trafficking in ether-deficient mice. No changes were observed in PtdCho and sphingomyelin levels, but significant changes in the distribution of cholesterol and disorganization of myelin membrane lipid raft microdomains were observed in ether lipid-deficient mice (Gorgas et al., 2006).

In normal mice, dihydroxyacetone phosphate acyltransferase is highly expressed in the inner segment of photoreceptors and in the retinal pigment epithelium (RPE), suggesting two distinct sites for plasmalogen biosynthesis (Acar et al., 2007). A deficiency of dihydroxyacetone phosphate acyltransferase results in abnormal ocular and visual function. Major ocular abnormalities in dihydroxyacetone phosphate acyltransferase deficient mice include microphthalmia, bilateral cataract, alterations in the retinal pigment epithelium, optic nerve hypoplasia, and irido-corneal adhesion. Ethanolamine plasmalogens are enriched in the epithelial cell layer and are abundantly present in the outer cortical zone, but significantly lesser in lens (Gorgas et al., 2006). Collective evidence suggests that mammalian ether lipids may participate in (a) the assembly and function of lipid rafts, (b) the assembly and disassembly of different types of cell–cell contracts, and (c) the intracellular transport of cholesterol (Gorgas et al., 2006).

The testis and epididymis of normal mice contain the protein and mRNA for dihydroxyacetonephosphate acyltransferase and alkyl-dihydroxyacetonephosphate synthase. In the testis, peroxisomes are localized exclusively in Leydig cells and not in cells of the seminiferous tubules, implying that the latter do not contribute to the biosynthesis of plasmalogens of the sperm membrane. In contrast, peroxisomes are clearly visualized in the epithelial cells of the epididymis. The results suggest that peroxisomes in epithelial cells of the rat epididymis play a pivotal role in the biosynthesis of plasmalogens destined for delivery to the sperm plasma membrane (Reisse et al., 2001). Feeding a DHA-enriched diet led to significant improvements in sperm quality in horses (Brinsko et al., 2005). Plasmalogen deficiency is associated with male infertility (Rodemer et al., 2003). DHA-enriched diets also stimulate the formation of peroxisomes, and thus of plasmalogen synthesis (Maldergem et al., 2005; Deckelbaum et al., 2006; André et al., 2006).

High concentrations of plasmalogens in male reproductive tissues suggest that these lipids may have a role in spermatogenesis and fertilization. Choline plasmalogens may contribute to nondiffusible membrane regions that confer stability to acrosomal membranes (Wolf et al., 1988). In ether lipid-deficient mice, the disruption of the dihydroxyacetone phosphate acyltransferase gene expression causes infertility and arrest of the spermatogenic cycle (Rodemer et al., 2003; Gorgas et al., 2006). These studies support the view that plasmalogens play an important role during spermatogenesis. Collectively, studies on lipid metabolism in ether lipid-deficient mice may provide important information on the metabolism and role of various lipids in neural and nonneural tissues in in vivo settings (Rodemer et al., 2003; Gorgas et al., 2006).

Testis and spermatozoa contain seminolipid or glycerolipid 3-*O*-sulphogalactosyl-1-alkyl-2-acyl-glycerol (sulfogalactosyl-alkylacylglycerol) (Ishizuka et al., 1978) (Fig. 1.3). During spermatogenesis, seminolipid is synthesized rapidly in early phase of spermatocyte development and maintained on the cell surface of germ cells. It plays a crucial role in formation of macro and microdomains, during sperm maturation and sperm capacitation (Ishizuka et al., 1978). The corresponding sulfated diacyl glycerophospholipid is found in brain, and is closely associated with myelin synthesis (Pieringer et al., 1977). In rat brain, the relative molar concentrations of the diacyl and alkylacyl types of sulfogalactosylglycerols change with age. Until the 19th day, the diacyl type prevails, but later, at 68 days, 85% are of the alkylacyl type. Among the ether moieties, the hexadecyl chain predominated (80%), whereas palmitoyl, stearoyl, and oleoyl moieties are present as the main acyl groups.

1.7 Conclusion

Collective evidence suggests that in comparison to the ester-bonded glycerophospholipids, ether lipids provide neural membranes with specialized physical and physicochemical properties such as increased fluidity and membrane fusion capability. Many proteins interact differently with ether glycerophospholipids than with diacyl glycerophospholipids. For example, the selective binding of ethanolamine plasmalogen, which has predominantly unsaturated acyl groups, is essential for Ca^{2+} transport (Ford and Hale, 1996). Some ether lipids, such as 1-*O*-hexadec-1′-enyl-2-arachidonoyl-*sn*-glycero-3-phosphoethanolamine, have a stimulatory effect on the Ca^{2+} pump (Duhm et al., 1993; Ford and Hale, 1996). Ether lipids also participate in neural cell differentiation. They may act as potential antioxidants in which the alkenyl bond absorbs oxygen radicals to protect arachidonic and docosahexaenoic acids in the *sn*-2 position, with the production of harmless products that can be recycled (Zoeller et al., 1988). Most information described earlier has been obtained during studies in vitro. Most known functions of ether lipids have been discovered using systems in vitro. It is not known whether the behavior in vivo of ether lipids is similar to their behavior in vitro. More studies are required on the metabolism in vivo of ether lipids in brain.

References

Acar N., Gregoire S., Andre A., Juaneda P., Joffre C., Bron A. M., Creuzot-Garcher C. P., and Bretillon L. (2007). Plasmalogens in the retina: In situ hybridization of dihydroxyacetone phosphate acyltransferase (DHAP-AT) – the first enzyme involved in their biosynthesis – and comparative study of retinal and retinal pigment epithelial lipid composition. Exp. Eye Res. 84:143–151.

André A., Cabaret S., Berdeaux O., Juanéda P., Sébédio J. L., and Chardigny J. M. (2006). Bioequivalence of docosahexaenoic acid and α-linolenic acid supplementations on plasmalogen, long-chain aldehyde, and docosahexaenoic acid levels in the brain of very old rats. Nutr. Res. 26:214–220.

Blank M. L., Smith Z. L., Cress E. A., and Snyder F. (1994). Molecular species of ethanolamine plasmalogens and transacylase activity in rat tissues are altered by fish oil diets. Biochim. Biophys. Acta Lipids Lipid Metab. 1214:295–302.

Blank M. L., Smith Z. L., Fitzgerald V., and Snyder F. (1995). The CoA-independent transacylase in PAF biosynthesis: Tissue distribution and molecular species selectivity. Biochim. Biophys. Acta Lipids Lipid Metab. 1254:295–301.

Bordier C. G., Sellier N., Foucault A. P., and Le Goffic F. (1996). Purification and characterization of deep sea shark *Centrophorus squamosus* liver oil 1-O-alkylglycerol ether lipids. Lipids 31:521–528.

Brinsko S. P., Varner D. D., Love C. C., Blanchard T. L., Day B. C., and Wilson M. E. (2005). Effect of feeding a DHA-enriched nutriceutical on the quality of fresh, cooled and frozen stallion semen. Theriogenology 63:1519–1527.

Causeret C. C., Bentejac M. M., Bugaut M. M. (1993). Proteins and enzymes of the peroxisomal membrane in mammals. Biol. Cell 77:89–104.

Chatterjee S. and Mayor S. (2001). The GPI-anchor and protein sorting. Cell Mol. Life Sci. 58:1969–1987.

Creer M. H. and Gross R. W. (1985). Reversed-phase high-performance liquid chromatographic separation of molecular species of alkyl ether, vinyl ether, and monoacyl lysophospholipids. J. Chromatogr. 338:61–69.

Deckelbaum R. J., Worgall T. S., and Seo T. (2006). n-3 Fatty acids and gene expression. Am. J. Clin. Nutr. 83:1520–1525.

Duhm J., Engelmann B., Schönthier U. M., and Streich S. (1993). Accelerated maximal velocity of the red blood cell Na^+/K^+ pump in hyperlipidemia is related to increase in 1-palmitoyl-2-arachidonoyl-plasmalogen phosphatidylethanolamine. Biochim. Biophys. Acta Biomembr. 1149:185–188.

Dypbukt J. M., Edman C. C., Sundqvist K., Kakefuda T., Plummer S. M., Harris C. C., and Grafström R. C. (1989). Reactivity of fecapentaene-12 toward thiols, DNA, and these constituents in human fibroblasts. Cancer Res. 49:6058–6063.

Engelmann B. (2004). Plasmalogens: Targets for oxidants and major lipophilic antioxidants. Biochem. Soc. Trans. 32:147–150.

Fahy E., Subramaniam S., Brown H. A., Glass C. K., Merrill A. H. J., Murphy R. C., Raetz C. R. H., Russell D. W., Seyama Y., Shaw W., Shimizu T., Spener F., Van Meer G., VanNieuwenhze M. S., White S. H., Witztum J. L., and Dennis E. A. (2005). A comprehensive classification system for lipids. J. Lipid Res. 46:839–861.

Farooqui A. A., and Horrocks L. A. (2001). Plasmalogens: workhorse lipids of membranes in normal and injured neurons and glia. Neuroscientist. 7:232–245.

Farooqui A. A., and Horrocks L. A. (2004). Plasmalogens, platelet-activating factor, and other ether lipids. In: Nicolaou A. and Kokotos G. (eds.), *Bioactive Lipids*. Oily Press, Bridgwater, England, pp. 107–134.

Fezza F., Bisogno T., Minassi A., Appendino G., Mechoulam R., and Di Marzo V. (2002). Noladin ether, a putative novel endocannabinoid: inactivation mechanisms and a sensitive method for its quantification in rat tissues. FEBS Lett. 513:294–298.

Foglia T. A., Nungesser E., and Marmer W. N. (1988). Oxidation of 1-*O*-(alk-1-enyl)-2,3-di-*O*-acylglycerols: Models for plasmalogen oxidation. Lipids 23:430–434.
Ford D. A. and Gross R. W. (1988). Identification of endogenous 1-O-alk-1′-enyl-2-acyl-*sn*-glycerol in myocardium and its effective utilization by choline phosphotransferase. J. Biol. Chem. 263:2644–2650.
Ford D. A. and Gross R. W. (1989). Differential accumulation of diacyl and plasmalogenic diglycerides during myocardial ischemia. Circ. Res. 64:173–177.
Ford D. A. and Hale C. C. (1996). Plasmalogen and anionic phospholipid dependence of the cardiac sarcolemmal sodium-calcium exchanger. FEBS Lett. 394:99–102.
Ford D. A., Miyake R., Glaser P. E., and Gross R. W. (1989). Activation of protein kinase C by naturally occurring ether-linked diglycerides. J. Biol. Chem. 264:13818–13824.
Gills J. J. and Dennis P. A. (2004). The development of phosphatidylinositol ether lipid analogues as inhibitors of the serine/threonine kinase, Akt. Expert Opin. Invest. Drugs. 13:787–797.
Gills J. J., Holbeck S., Hollingshead M., Hewitt S. M., Kozikowski A. P., and Dennis P. A. (2006). Spectrum of activity and molecular correlates of response to phosphatidylinositol ether lipid analogues, novel lipid-based inhibitors of Akt. Mol. Cancer Ther. 5:713–722.
Gorgas K., Teigler A., Komljenovic D., and Just W. W. (2006). The ether lipid-deficient mouse: Tracking down plasmalogen functions. Biochim. Biophys. Acta Mol. Cell Res. 1763:1511–1526.
Guan Z. Z., Grunler J., Piao S. F., and Sindelar P. J. (2001). Separation and quantitation of phospholipids and their ether analogues by high-performance liquid chromatography. Anal. Biochem. 297:137–143.
Hahnel D., Huber T., Kurze V., Beyer K., and Engelmann B. (1999). Contribution of copper binding to the inhibition of lipid oxidation by plasmalogen phospholipids. Biochem. J. 340:377–383.
Han X. and Gross R. W. (1991). Alterations in membrane dynamics elicited by amphiphilic compounds are augmented in plasmenylcholine bilayers. Biochim. Biophys. Acta Biomembr. 1069:37–45.
Han X. L. and Gross R. W. (1990). Plasmenylcholine and phosphatidylcholine membrane bilayers possess distinct conformational motifs. Biochemistry 29:4992–4996.
Hanuš L., Abu-Lafi S., Fride E., Breuer A., Vogel Z., Shalev D. E., Kustanovich I., and Mechoulam R. (2001). 2-Arachidonyl glyceryl ether, an endogenous agonist of the cannabinoid CB1 receptor. Proc. Natl. Acad. Sci. USA 98:3662–3665.
Hayashi H. and Oohashi M. (1995). Incorporation of acetyl-CoA generated from peroxisomal β-oxidation into ethanolamine plasmalogen of rat liver. Biochim. Biophys. Acta Lipids Lipid Metab. 1254:319–325.
Hayashi H. and Takahata S. (1991). Role of peroxisomal fatty acyl-CoA β-oxidation in phospholipid biosynthesis. Arch. Biochem. Biophys. 284:326–331.
Hooper N. M. (1997). Glycosyl-phosphatidylinositol anchored membrane enzymes. Clin. Chim. Acta. 266:3–12.
Ishizuka I., Inomata M., Ueno K., and Yamakawa T. (1978). Sulfated glyceroglycolipids in rat brain. Structure sulfation in vivo, and accumulation in whole brain during development. J. Biol. Chem. 253:898–907.
Kim S. Y., Min D. S., Choi J. S., Choi Y. S., Park H. J., Sung K. W., Kim J., and Lee M. Y. (2004). Differential expression of phospholipase D isozymes in the hippocampus following kainic acid-induced seizures. J. Neuropathol. Exp. Neurol. 63:812–820.
Laine K., Jarvinen K., Mechoulam R., Breuer A., and Jarvinen T. (2002). Comparison of the enzymatic stability and intraocular pressure effects of 2-arachidonylglycerol and noladin ether, a novel putative endocannabinoid. Invest. Ophthalmol. Vis. Sci. 43:3216–3222.
Liliom K., Fischer D. J., Virág T., Sun G., Miller D. D., Tseng J. L., Desiderio D. M., Seidel M. C., Erickson J. R., and Tigyi G. (1998a). Identification of a novel growth factor-like lipid, 1-*O*-*cis*-alk-1′-enyl-2-lyso-*sn*-glycero-3-phosphate (alkenyl-GP) that is present in commercial sphingolipid preparations. J. Biol. Chem. 273:13461–13468.

Liliom K., Guan Z., Tseng J. L., Desiderio D. M., Tigyi G., and Watsky M. A. (1998b). Growth factor-like phospholipids generated after corneal injury. Am. J. Physiol. 274:C1065–C1074.

Lohner K. (1996). Is the high propensity of ethanolamine plasmalogens to form non-lamellar lipid structures manifested in the properties of biomembranes? Chem. Phys. Lipids. 81:167–184.

Maeba R. and Ueta N. (2004a). A novel antioxidant action of ethanolamine plasmalogens in lowering the oxidizability of membranes. Biochem. Soc. Trans. 32:141–143.

Maeba R. and Ueta N. (2004b). Determination of choline and ethanolamine plasmalogens in human plasma by HPLC using radioactive triiodide (1–) ion ($^{125}I_3^-$). Anal. Biochem. 331:169–176.

Maldergem L., Moser A., Vincent M. F., Roland D., Reding R., Otte J. B., Wanders R., and Sokal E. (2005). Orthotopic liver transplantation from a living-related donor in an infant with a peroxisome biogenesis defect of the infantile Refsum disease type. J. Inherited Metab. Dis. 28:593–600.

Mechoulam R., Ben Shabat S., Hanuš L., Ligumsky M., Kaminski N. E., Schatz A. R., Gopher A., Almog S., Martin B. R., and Compton D. R. (1995). Identification of an endogenous 2-monoglyceride, present in canine gut, that binds to cannabinoid receptors. Biochem. Pharmacol. 50:83–90.

Mower H. F., Ichinotsubo D., Wang L. W., Mandel M., Stemmermann G., Nomura A., Heilbrun L., Kamiyama S., and Shimada A. (1982). Fecal mutagens in two Japanese populations with different colon cancer risks. Cancer Res. 42:1164–1169.

Njie Y. F., Kumar A., Qiao Z., Zhong L., and Song Z. H. (2006). Noladin ether acts on trabecular meshwork cannabinoid (CB1) receptors to enhance aqueous humor outflow facility. Invest Ophthalmol. Vis. Sci. 47:1999–2005.

Oka S., Tsuchie A., Tokumura A., Muramatsu M., Suhara Y., Takayama H., Waku K., and Sugiura T. (2003). Ether-linked analogue of 2-arachidonoylglycerol (noladin ether) was not detected in the brains of various mammalian species. J. Neurochem. 85:1374–1381.

Paltauf F. (1994). Ether lipids in biomembranes. Chem. Phys. Lipids. 74:101–139.

Pearson R. H. and Pascher I. (1979). The molecular structure of lecithin dihydrate. Nature 281:499–501.

Pieringer J., Rao G. S., Mandel P., and Pieringer R. A. (1977). The association of the sulphogalactosylglycerolipid of rat brain with myelination. Biochem. J. 166:421–428.

Plummer S. M., Hall M., and Faux S. P. (1995). Oxidation and genotoxicity of fecapentaene-12 are potentiated by prostaglandin H synthase. Carcinogenesis. 16:1023–1028.

Povey A. C., Plummer S. M., Grafstrom R. C., and Harris C. C. (1990). Genotoxic mechanisms of fecapentaene-12 in human cells. Prog. Clin. Biol. Res. 347:155–166.

Reisse S., Rothardt G., Volkl A., and Beier K. (2001). Peroxisomes and ether lipid biosynthesis in rat testis and epididymis. Biol. Reprod. 64:1689–1694.

Roberts W. L., Myher J. J., Kuksis A., and Rosenberry T. L. (1988). Alkylacylglycerol molecular species in the glycosylinositol phospholipid membrane anchor of bovine erythrocyte acetylcholinesterase. Biochem. Biophys. Res. Commun. 150:271–277.

Rodemer C., Thai T. P., Brugger B., Kaercher T., Werner H., Nave K. A., Wieland F., Gorgas K., and Just W. W. (2003). Inactivation of ether lipid biosynthesis causes male infertility, defects in eye development and optic nerve hypoplasia in mice. Hum. Mol. Genet. 12:1881–1895.

Schulman G., Bodine P. V., and Litwack G. (1992). Modulators of the glucocorticoid receptor also regulate mineralocorticoid receptor function. Biochemistry 31:1734–1741.

Seelig J. and Waespe-Sarcevic N. (1978). Molecular order in cis and trans unsaturated phospholipid bilayers. Biochemistry 17:3310–3315.

Snyder F. (1995). Platelet-activating factor: The biosynthetic and catabolic enzymes. Biochem. J. 305:689–705.

Snyder F. (1996). Ether-linked lipids and their bioactive species: Occurrence, chemistry, metabolism, regulation, and function. In: Vance D. E. and Vance J. E. (eds.), *Biochemistry of Lipids, Lipoproteins and Membranes*. Elsevier Science, The Netherlands, pp. 183–209.

Sugiura T., Kondo S., Sukagawa A., Nakane S., Shinoda A., Itoh K., Yamashita A., and Waku K. (1995). 2-arachidonoylglycerol: A possible endogenous cannabinoid receptor ligand in brain. Biochem. Biophys. Res. Commun. 215:89–97.

Sugiura T., Nakane S., Kishimoto S., Waku K., Yoshioka Y., Tokumura A., and Hanahan D. J. (1999). Occurrence of lysophosphatidic acid and its alkyl ether-linked analog in rat brain and comparison of their biological activities toward cultured neural cells. Biochim. Biophys. Acta Mol. Cell Biol. Lipids 1440:194–204.

Szekely J. and Gates K. S. (2006). Noncovalent DNA binding and the mechanism of oxidative DNA damage by fecapentaene-12. Chem. Res. Toxicol. 19:117–121.

Thukkani A. K., Hsu F. F., Crowley J. R., Wysolmerski R. B., Albert C. J., and Ford D. A. (2002). Reactive chlorinating species produced during neutrophil activation target tissue plasmalogens – Production of the chemoattractant, 2-chlorohexadecanal. J. Biol. Chem. 277:3842–3849.

Thukkani A. K., McHowat J., Hsu F. F., Brennan M. L., Hazen S. L., and Ford D. A. (2003). Identification of α-chloro fatty aldehydes and unsaturated lysophosphatidylcholine molecular species in human atherosclerotic lesions. Circulation 108:3128–3133.

Van Tassell R. L., Piccariello T., Kingston D. G., and Wilkins T. D. (1989). The precursors of fecapentaenes: purification and properties of a novel plasmalogen. Lipids 24:454–459.

Verhoeven N. M., Roe D. S., Kok R. M., Wanders R. J., Jakobs C., and Roe C. R. (1998). Phytanic acid and pristanic acid are oxidized by sequential peroxisomal and mitochondrial reactions in cultured fibroblasts. J. Lipid Res. 39:66–74.

Visser W. F., van Roermund C. W., IJlst L., Waterham H. R., and Wanders R. J. (2007). Metabolite transport across the peroxisomal membrane. Biochem. J. 401:365–375.

Wanders R. J. A. and Waterham H. R. (2006). Peroxisomal disorders: The single peroxisomal enzyme deficiencies. Biochim. Biophys. Acta Mol. Cell Res. 1763:1707–1720.

Wei H., Kemp S., McGuinness M. C., Moser A. B., and Smith K. D. (2000). Pharmacological induction of peroxisomes in peroxisome biogenesis disorders. Ann. Neurol. 47:286–296.

Wildsmith K. R., Albert C. J., Hsu F. F., Kao J. L. F., and Ford D. A. (2006). Myeloperoxidase-derived 2-chlorohexadecanal forms Schiff bases with primary amines of ethanolamine glycerophospholipids and lysine. Chem. Phys. Lipids. 139:157–170.

Wolf D. E., Lipscomb A. C., and Maynard V. M. (1988). Causes of nondiffusing lipid in the plasma membrane of mammalian spermatozoa. Biochemistry. 27:860–865.

Yachida Y., Kashiwagi M., Mikami T., Tsuchihashi K., Daino T., Akino T., and Gasa S. (1998). Stereochemical structures of synthesized and natural plasmalogalactosylceramides from equine brain. J. Lipid Res. 39:1039–1045.

Yachida Y., Kashiwagi M., Mikami T., Tsuchihashi K., Daino T., Akino T., and Gasa S. (1999). Novel plasmalogalactosylalkylglycerol from equine brain. J. Lipid Res. 40:2271–2278.

Zoeller R. A., Morand O. H., and Raetz C. R. H. (1988). A possible role for plasmalogens in protecting animal cells against photosensitized killing. J. Biol. Chem. 263:11590–11596.

Chapter 2
Biosynthesis of Plasmalogens in Brain

2.1 General Considerations and Distribution of Plasmalogens in Brain

Plasmalogens account for the major portion of the ethanolamine glycerophospholipids in the adult human brain (50%), but the brain of newborn babies has low levels (7% of total phospholipids mass) (Horrocks and Sharma, 1982). Levels of ethanolamine plasmalogen (PlsEtn) increase rapidly during the intense period of myelination and ethanolamine glycerophospholipids of myelin sheath contain up to 70% PlsEtn. An eight-fold increase in PlsEtn levels per gram of brain tissue occurs in white matter during first year of life so that PlsEtn accounts for 20% of the glycerophospholipid mass and 70% of the ethanolamine glycerophospholipids (Balakrishnan et al., 1961). At that time, myelination is rapid. The highest level of myelin is between 30 and 40 years of age (Toews and Horrocks, 1976). In human brain, there is a steep rise in PlsEtn content, followed by a further rise up to 30–40 years of age. This is followed by a decline of PlsEtn levels during normal aging. At 70 years of age, the levels of PlsEtn are 18% less than at 40 years of age (Rouser and Yamamoto, 1968; Horrocks et al., 1981). In chicks, there is a marked increase in plasmalogen levels in synaptosomes during the first 3 days after hatching (Getz et al., 1968). Collectively, these studies suggest that plasmalogens are major glycerophospholipids in brain tissue. Their metabolism may be involved in signal transduction processes associated with neural cell functions such as synaptogenesis, myelination, and ion transport (Farooqui and Horrocks, 2001).

Plasmalogens impart membranes with different biophysical properties such as phase transition temperature, bilayer thickness, acyl chain packing free volume, and lateral domain. The perpendicular orientation of the *sn*-2 acyl chain at the membrane surface and the lack of a carbonyl group at the *sn*-1 position in plasmalogens affect the hydrophilicity of the head group, resulting in stronger intermolecular hydrogen bonding between the head groups (Lohner, 1996). These properties allow PlsEtns to adopt the inverse hexagonal phase and may be responsible for a different membrane potential compared with other glycerophospholipids (Lohner, 1996). This property affects lipid packing, fluidity, and interaction with neural membrane receptors and ion channels. In cellular membranes and lipoproteins, plasmalogens account for 15–20% and 5% of all phospholipids, respectively (Nagan and Zoeller, 2001; Engelmann et al., 1994). PlsEtn and PlsCho are the two major plasmalogen species found in mammalian cell membranes. In most cells, PlsEtns exceed the choline plasmalogens by 10-fold, with the exception of cardiac and skeletal muscle where choline plasmalogen dominates. The level of plasmalogens in brain tissue depends on the degree of myelination and increases rapidly during myelinogenesis (Horrocks, 1972; Horrocks and Sharma, 1982). Factors that modulate the levels of plasmalogens in neurons, astrocytes, and oligodendrocytes during myelination and aging remain unknown.

2.2 Biosynthesis of Plasmalogens

The enzymes for plasmalogen biosynthesis have not been purified and characterized from brain tissue. The reasons for this lack of information on the purification and characterization of plasmalogen biosynthesizing enzymes from brain are not known. However, the low activity of plasmalogen synthesizing enzymes, complex, laborious, and time consuming assays for determining activities, and the heterogeneity and complex organization of brain tissue may be responsible for the lack of information. Several investigators have reviewed the biosynthesis of plasmalogens in nonneural tissues (Fig. 2.1) (Horrocks and Sharma, 1982; Lee, 1998; Nagan and Zoeller, 2001; Murphy, 2001; Brites et al., 2004). The starting metabolite for plasmalogen biosynthesis is dihydroxyacetone phosphate from glycolysis, which is used to form the glycerol backbone of the plasmalogen. The biosynthesis of plasmalogens is initiated in peroxisomes and completed in the endoplasmic reticulum. Thus, the first three enzymes of plasmalogen biosynthesis, dihydroxyacetone phosphate acyltransferase, alkyl dihydroxyacetone phosphate synthase, and acyl/alkyl dihydroxyacetone reductase, are located in peroxisomes. The endoplasmic reticulum contains the other enzymes, namely 1-alkyl-*sn*-Gro*P* acyltransferase, 1-alkyl-2-acyl-*sn*-Gro*P* phosphohydrolase, and 1-alkyl-2-acyl-*sn*-Gro: CDP-choline (CDP-ethanolamine) choline (ethanolamine) phosphotransferase.

The rate-limiting step for plasmalogen biosynthesis has not been identified. However, it is proposed that regulation point lies downstream from first three steps (Nagan and Zoeller, 2001). This suggestion is based on the incorporation of 1-*O*-[9′-(1″-pyrenyl)]nonyl-*sn*-glycerol (pAG), a fluorescent ether lipid with a pyrene moiety covalently attached at the alkyl chain terminus (Zheng et al., 2006) (Fig. 2.2). CHO-K1 and NRel-4 cells take up this ω-pyrene-labeled 1-*O*-alkyl-*sn*-glycerol. NRel-4 cells are a variant defective in dihydroxyacetone phosphate acyltransferase. Treatment of CHO-K1 and NRel-4 cells results in the incorporation of pAG into ethanolamine and choline phospholipids as well as into a neutral lipid fraction tentatively identified as alkyldiacylglycerols. NRel-4 cells incorporate more fluorescence in the phospholipid fraction than CHO-K1, specifically in the ethanolamine phospholipids. Analysis of the fluorescent lipids demonstrates that 93% of the pAG is taken up by glycerolipids with the intact ether bond. Although

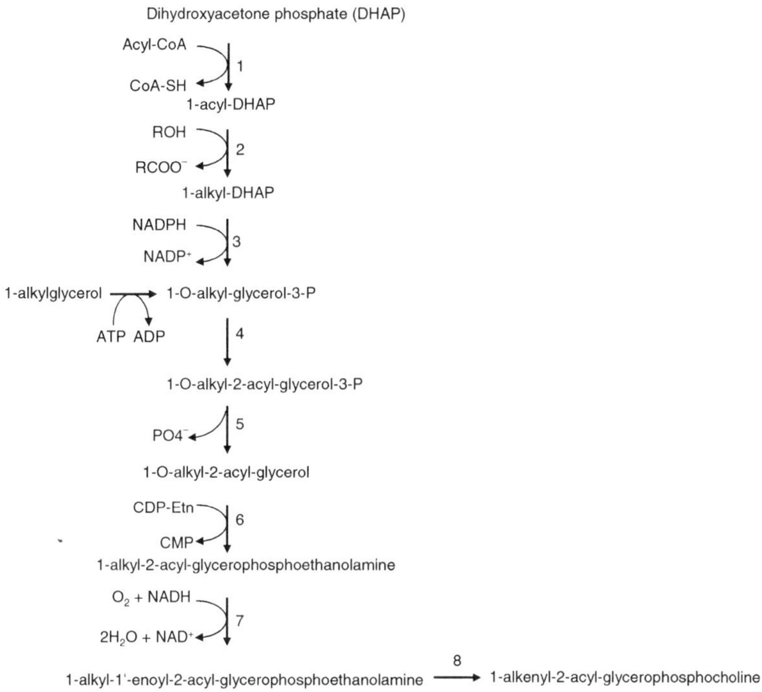

Fig. 2.1 Biosynthesis of plasmalogens in mammalian tissues. (1) Dihydroxyacetone phosphate acyltransferase, (2) 1-acyl dihydroxyacetone phosphate synthase, (3) 1-acyl/alkyl dihydroxyacetone phosphate reductase, (4) 1-alkyl-*sn*-glycerophosphate acyltransferase, (5) 1-alkyl-2-acylglycerophosphate phosphohydrolase, (6) CDP-ethanolamine transferase, (7) 1-alkyl-2-acyl-*sn*-glycerophosphoethanolamine desaturase, and methyltransferases and base-exchange enzymes.

the addition of 20 μM 1-*O*-hexadecyl-*sn*-glycerol (HG) (Fig. 2.2) to the medium fully restores PlsEtn biosynthesis in NRel-4 cells, pAG only partially restores PlsEtn synthesis (Zheng et al., 2006). Both pAG and HG inhibit CHO-K1 and NRel-4 cell growth. The molecular mechanism associated with this inhibitory process remains unknown (Zheng et al., 2006). Incubation of cells with pAG followed by irradiation with long-wavelength (>300 nm) ultraviolet light produces cytotoxicity due to the generation of reactive oxygen species such as singlet oxygen. NRel-4 cells exhibit an increase in sensitivity to UV light compared with CHO-K1 cells. This photodynamic cytotoxicity approach can be used to select for mutants that are defective in downstream steps in ether lipid biosynthesis (Zheng et al., 2006).

Dihydroxyacetone phosphate acyltransferase may be a crucial enzyme for plasmalogen biosynthesis, but it is not a rate-limiting step for plasmalogen synthesis (Nagan and Zoeller, 2001). Sphinganine is an effective donor of the 1-alkenyl chain of plasmalogens (Stoffel et al., 1970). Most fatty alcohols from sphinganine predominantly incorporate into choline plasmalogens and only a little into PlsEtns. The reason

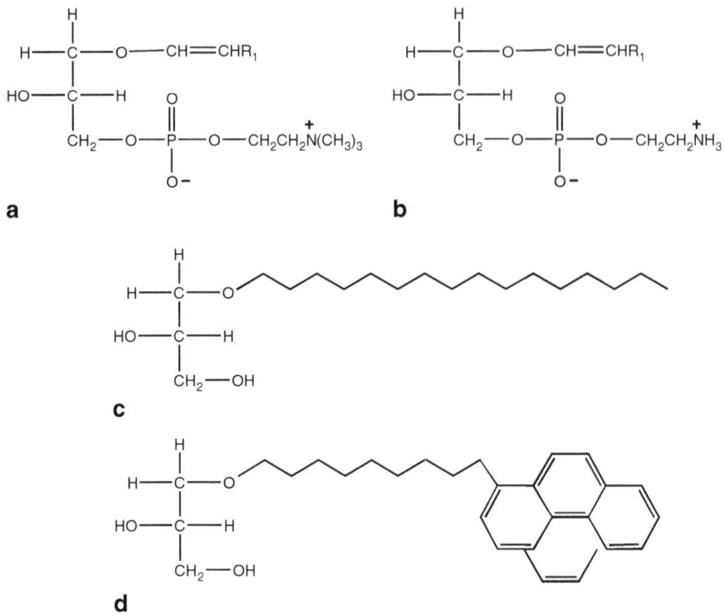

Fig. 2.2 Chemical structures of lyso-choline plasmalogen (**a**), lyso-ethanolamine plasmalogen (**b**), 1-*O*-hexadecyl-*sn*-glycerol (**c**), and 1-*O*-[9′-(1″-pyrenyl)]nonyl-*sn*-glycerol (pAG) (**d**).

for the preferential incorporation of fatty alcohols from sphinganine into choline plasmalogens remains unknown. Detailed investigations are required on this topic.

2.2.1 *Dihydroxyacetone Phosphate Acyltransferase*

This enzyme catalyzes the esterification of the free hydroxyl group of dihydroxyacetone phosphate by utilizing long chain (>C10) acyl CoA to form acyl dihydroxyacetone phosphate. This enzyme is associated with peroxisomal membranes. It was purified from rat and guinea pig livers and human placenta with multiple column chromatographic procedures (Jones and Hajra, 1983; Webber and Hajra, 1993; Ofman and Wanders, 1994; Causeret et al., 1997). The purified enzyme has molecular mass of 69 kDa and migrates as a single band on SDS-polyacrylamide gel electrophoresis, but gel filtration studies indicate a molecular mass of 90 kDa (Webber and Hajra, 1993). The purified enzyme has a K_m value of 70 μM and a V_{max} value of 4 μmol acyl dihydroxyacetone phosphate formed per minute per milligram of protein (Table 2.1) Free coenzyme A inhibits the acyltransferase reaction with an inhibition constant (K_i) of approximately 0.76 mM. Dihydroxyacetone phosphate acyltransferase is resistant to inhibition

Table 2.1 Kinetic properties of enzymes associated with plasmalogen biosynthesis.

Enzyme	pH optimum	K_m value (μM)	V_{max} (nmol/min/mg)	Reference
DHAP-AT	7.6	70	4,000	(Webber and Hajra, 1993)
DHAP-S	7.5	68	42.80	(Zomer et al., 1993)
Acyl/alkyl dihydroxyacetone reductase	7.5	21	67,000	(Datta et al., 1990)
1-Alkyl-2-acyl-*sn*-Gro*P* phosphohydrolase	6.0–6.5	–	813	(Jamal et al., 1991)
CDP-ethanolamine:DAG ethanolaminephospho-transferase	8.0	0.57	252	(Mancini et al., 1999)
Plasmanylethanolamine desaturase	7.1			(Paltauf, 1994; Zheng et al., 2006)
Alkylglycerol phosphotransferase				(Zheng et al., 2006)

Table 2.2 Localization and molecular weights of enzymes associated with plasmalogen biosynthesis.

Enzyme	Localization	Molecular mass (kDa)	Reference
Dihydroxyacetone phosphate acyltrans-ferase	Peroxisome	77	(Ofman et al., 1998)
Alkyl dihydroxyacetone phosphate synthase	Peroxisome	65	(Zomer et al., 1993)
Acyl/alkyl dihydroxyacetone reductase	Peroxisome	60	(Datta et al., 1990)
Alkylglycerophosphate acyltransferase	Endoplasmic reticulum	–	(Stamps et al., 1997)
Alkylacylglycerophosphate phospho-hydrolase	Endoplasmic reticulum	–	–
CDP-ethanolamine:diacylglycerol ethanolaminephosphotransferase	Endoplasmic reticulum	38	(Mancini et al., 1999)
Cyanide-sensitive Δ1-alkyl desaturase	–	–	(Paltauf, 1994)

by *N*-ethylmaleimide, a sulfhydryl group blocking agent, and utilizes only dihydroxyacetone phosphate as the acyl group acceptor.

Human placental dihydroxyacetone phosphate acyltransferase was purified using octyl-Sepharose CL-4B, Hydroxyapatite HTP, and CM-Sepharose CL-6B column chromatographic procedures along with PBE 94 chromatofocusing and TSK G3000 SW size exclusion chromatography. The purified enzyme has an isoelectric point of 5.1–5.3. The molecular mass of native enzyme is 60–80 kDa as calculated from HPLC size exclusion chromatography (Table 2.2). SDS-PAGE indicates a molecular mass of 65 kDa.

The cDNA for dihydroxyacetone phosphate acyltransferase was cloned (Thai et al., 1997). The nucleotide-derived amino acid sequence revealed a protein consisting of 680 amino acid residues of molecular mass, 77 kDa containing a C-terminal type 1 peroxisomal targeting signal (PTS). Monospecific antibodies prepared against this polypeptide efficiently immunoprecipitate dihydroxyacetone phosphate acyltransferase activity from solubilized peroxisomal preparations, thus confirming that the cloned cDNA codes for dihydroxyacetone phosphate acyltransferase (Thai et al., 1997). Using the amino acid sequence of human acyl-CoA:dihydroxyacetone phosphate acyltransferase as bait to screen the database of expressed sequence tags (dbEST), several partial mouse cDNA clones showing high identity have been identified (Ofman and Wanders, 1994; Ofman et al., 1999). Primers were selected based on the dbEST sequences and used for amplification of this transcript from cDNA prepared from mouse skin fibroblasts. The complete nucleotide sequence has revealed an open reading frame of 2,034 bp encoding a protein consisting of 678 amino acids with a molecular mass of 77 kDa. The deduced amino acid sequence shows high identity (80%) with human dihydroxyacetone phosphate acyltransferase and also suggests a typical peroxisomal targeting signal type 1 (PTS1) at its extreme carboxy-terminus (alanine–lysine–leucine). Definitive evidence that this cDNA indeed codes for dihydroxyacetone phosphate acyltransferase is obtained by heterologous expression in the yeast *Saccharomyces cerevisiae* (Ofman et al., 1999). Northern blot analysis indicates high expression of dihydroxyacetone phosphate acyltransferase especially in mouse heart, liver, and testis.

Dihydroxyacetone phosphate acyltransferase is absolutely required for the synthesis of plasmalogen but its activity is not a limiting factor for plasmalogen synthesis in CHO cells (Liu et al., 2005). Earlier studies indicate that acylation of dihydroxyacetone phosphate is important for the biosynthesis of nonether glycerolipids (Hajra et al., 2000), but recent studies have clearly shown that dihydroxyacetone phosphate acyltransferase does not contribute to the synthesis of diacyl glycerolipids. Acylated dihydroxyacetone phosphate is also synthesized by a NEM-sensitive microsomal dihydroxyacetone phosphate acyltransferase activity (Schlossman and Bell, 1977). This enzyme does not contribute to plasmalogen synthesis but may be involved in the synthesis of nonether glycerolipids (Liu et al., 2005). These authors isolated a fibroblast-like cell line CHO-K1 that is deficient in plasmalogens due to the loss of various steps of plasmalogen biosynthesis pathways (Nagan et al., 1997; Nagan et al., 1998). The mutant CHO cell line NRel-4 has markedly reduced plasmalogen levels because of decreased dihydroxyacetone phosphate acyltransferase activity. The lower activity of this enzyme is due to reduced levels of the message for dihydroxyacetone phosphate acyltransferase. Expression of the dihydroxyacetone phosphate acyltransferase gene in NRel-4 cells results in the restoration of plasmalogen biosynthesis suggesting that this enzyme is essential for plasmalogen synthesis (Liu et al., 2005).

The gene for dihydroxyacetone phosphate acyltransferase is located on chromosome 1q42.12–43. It spans approximately 28 kb and consists of 16 exons and 15 introns (Ofman et al., 2001). In brain, the mRNA for dihydroxyacetone phosphate acyltransferase is mainly localized in white matter peroxisomes. This expression is weak compared to intense expression in liver peroxisomes (André et al., 2005b). The blood–brain barrier prevents the transport of plasmalogens to the brain. The expression of dihydroxyacetone phosphate acyltransferase may be lower than in liver, but is sufficient to maintain a plasmalogen pool necessary for structural and metabolic activities in brain tissue (André et al., 2005b).

Dihydroxyacetone phosphate acyltransferase is highly expressed in the inner segment of photoreceptors and in the retinal pigment epithelium (RPE), suggesting two distinct sites for plasmalogen biosynthesis (Acar et al., 2007). PlsEtn is the main class of plasmalogens in both neural retina and RPE (28–29% of the total ethanolamine glycerophospholipids). Compared to other tissues, photoreceptors and the RPE monolayer contain a greater proportion of octadecanal in the *sn*-1 position of plasmenylethanolamine. The RPE monolayer is located in a highly oxygenated environment and is exposed to high levels of visible light. Therefore, it is at risk for oxidative damage (Cai et al., 2000; Acar et al., 2007). In the RPE monolayer, PlsEtns may be involved in protection against oxidative stress (Acar et al., 2007).

2.2.2 Alkyl Dihydroxyacetone Phosphate Synthase

This enzyme replaces the acyl chain in acyl dihydroxyacetone phosphate with a long-chain fatty alcohol to form 1-alkyl-*sn*-glycero-3-phosphate (Paltauf, 1994). The fatty alcohol may be supplied either from dietary intake or by the reduction of long-chain acyl-CoA through the action of an acyl-CoA reductase. Alkyl dihydroxyacetone phosphate synthase is localized in peroxisomes. It was solubilized from an enriched peroxisome fraction with Triton X-100 and potassium chloride. The solubilized enzyme was purified by chromatography on QAE-Sephadex, Matrex Red, phosphocellulose, and Concanavalin A. SDS-polyacrylamide gel electrophoresis of alkyl dihydroxyacetone phosphate synthase indicates a molecular mass of 65 kDa. Chromatofocusing studies indicate an isoelectric point of pH 5.9. The pH optimum of alkyl dihydroxyacetone phosphate synthase is between pH 7 and 8. The purified enzyme has a specific activity of 350 nmol/min/mg protein corresponding to a purification of at least 13,000-fold (Zomer et al., 1993). Recombinant alkyl dihydroxyacetone phosphate synthase from guinea pig liver follows ping–pong rather than a sequential reaction mechanism (de Vet et al., 1999).

N-ethylmaleimide, *p*-bromophenacylbromide, and 2,4-dinitrofluorobenzene irreversibly inhibit the alkyl dihydroxyacetone phosphate synthase activity. Saturating concentrations of palmitoyl dihydroxyacetone phosphate protect the enzyme from inactivation. The rate of inactivation of the enzyme by *p*-bromophenacylbromide depends upon pH and is highest under alkaline conditions. Collectively, these results suggest the involvement of cysteine, histidine, and lysine residues in the reaction catalyzed by alkyl dihydroxyacetone phosphate synthase. The divalent cations Mg^{2+}, Zn^{2+}, and Mn^{2+} inhibit the enzymic activity, whereas Ca^{2+} has no effect. Mutational analysis indicates that histidine 617 is an essential amino acid for catalytic activity: replacement of this residue by alanine results in complete loss of enzymic activity (de Vet et al., 1999). A recombinant enzyme after the deletion of five C-terminal amino acids shows no activity, indicating the importance of the C-terminus for catalytic activity.

Although the reaction catalyzed by alkyl dihydroxyacetone phosphate synthase is not a net redox sensitive reaction, the amino acid sequence of the enzyme indicates the presence of a flavin adenine dinucleotide (FAD)-binding domain (de Vet et al., 2000). On the basis of fluorescence properties and UV–visible absorption spectra, alkyl dihydroxyacetone phosphate synthase contains an essential FAD molecule that acts as a cofactor. The FAD participates directly in catalysis. During incubation of the enzyme with the substrate, palmitoyl dihydroxyacetone phosphate, the flavin moiety is reduced, indicating that in this initial step the substrate is oxidized (de Vet et al., 2000). Stopped flow assay studies show that the reduction of the flavin moiety is a monophasic process yielding an oxygen-stable, reduced-enzyme species. Upon addition of hexadecanol to the reduced enzyme species, the flavin moiety is efficiently reoxidized. Thus the collective evidence suggests that FAD participates in the reaction catalyzed by alkyl dihydroxyacetone phosphate synthase (de Vet et al., 2000).

From amino acid sequence information, cDNAs encoding alkyl dihydroxyacetone phosphate synthase have been cloned from both guinea pig and human liver. In both cases, the enzyme is synthesized as a precursor protein with a N-terminal cleavable presequence containing a PTS type 2 (de Vet and van den Bosch, 2000). Human fibroblasts derived from Zellweger syndrome and rhizomelic chondrodysplasia punctata patients contain much lower levels of the enzyme protein (see Chap. 6).

Radiation inactivation experiments were used to determine the in situ functional size of dihydroxyacetone phosphate acyltransferase and alkyl dihydroxyacetone phosphate synthase. Alkyl dihydroxyacetone phosphate synthase displays single exponential decay when enzymic activity and immunoreactive protein levels are measured with target sizes of 79 kDa and 78 kDa, respectively. Dihydroxyacetone phosphate acyltransferase activity is increased at lower doses and decays upon further irradiation with an apparent target size of 62 kDa. These data indicate that the functional unit sizes for both enzymes in situ are represented by single polypeptide chains (Biermann et al., 1998). After cross-linking, alkyl dihydroxyacetone phosphate synthase can be detected in a 210-kDa complex together with dihydroxyacetone phosphate acyltransferase. Both enzymes are located entirely on the luminal side of the peroxisomal membrane (Biermann et al., 1999). Coimmunoprecipitation studies confirm that the two enzymes interact with each other in a heterotrimeric complex. Furthermore, alkyl dihydroxyacetone phosphate synthase also forms a homotrimeric complex in the absence of dihydroxyacetone phosphate acyltransferase as observed by immunoblot analysis after cross-linking experiments with either dihydroxyacetone phosphate acyltransferase deficient human fibroblast homogenates or recombinant (His)6-tagged alkyl dihydroxyacetone phosphate synthase. In summary, alkyl dihydroxyacetone phosphate synthase interacts selectively with dihydroxyacetone phosphate acyltransferase in a heterotrimeric complex and in the absence of dihydroxyacetone phosphate acyltransferase can also form a homotrimeric complex (Biermann et al., 1999).

2.2.3 Acyl/alkyl Dihydroxyacetone Phosphate Reductase

In the presence of NADPH, acyl/alkyl dihydroxyacetone phosphate reductase reduces alkyl dihydroxyacetone phosphate at the *sn*-2 position to generate 1-alkyl 2-lyso-*sn*-glycero-3-phosphate, the ether-linked analog of lyso-phosphatidic acid. In guinea pig and rat liver, acyl/alkyl dihydroxyacetone phosphate reductase is localized in peroxisomal and microsomal fractions (Ghosh and Hajra, 1986). From the distribution of marker enzymes, about two-thirds of the acyl/alkyl dihydroxyacetone phosphate reductase activity is present in peroxisomes with the rest in microsomes. The properties of this enzyme in peroxisomes and microsomes are similar with respect to heat inactivation, pH optima, sensitivity to trypsin, and inhibition by $NADP^+$ and acyl CoA. The enzymic activity in peroxisomes and microsomes from mouse liver is increased to the same extent by chronically feeding clofibrate, a hypolipidemic drug. The kinetic properties of this enzyme in these two different organelles are also similar. From these results, the same enzyme is present in two different subcellular compartments of liver.

Acyl/alkyl dihydroxyacetone phosphate reductase was purified from pig liver peroxisomes using multiple column chromatographic procedures (Datta et al., 1990). The purified enzyme migrates as a single band on SDS-polyacrylamide gel electrophoresis with an apparent molecular weight of 60 kDa. The molecular weight of the native enzyme is estimated to be 75 kDa by size exclusion chromatography. The protein is very hydrophobic. It contains 27% hydrophobic amino acids and so it requires strong detergents to solubilize the enzymic activity. The K_m value of the purified enzyme for hexadecyl dihydroxyacetone phosphate is 21 μM, and the V_{max} value in the presence of 0.07 mM NADPH is 67 μmol/min/mg protein (Datta et al., 1990). The turnover number (K_{cat}), after correcting for the isotope effect of the cosubstrate $NADP^3H$, was 6,000 mol/min/mol of enzyme, assuming the enzyme has a molecular weight of 60 kDa. The purified enzyme also uses palmitoyl dihydroxyacetone phosphate as a substrate with a K_m value of 15.4 μM and a V_{max} of 75 μmol/min/mg protein. Palmitoyl dihydroxyacetone phosphate competitively inhibits the reduction of hexadecyl dihydroxyacetone phosphate, indicating that the same enzyme catalyzes the reduction of both acyl dihydroxyacetone phosphate and alkyl dihydroxyacetone phosphate. NADH can substitute for NADPH, but the K_m of the enzyme for NADH (1.7 mM) is much higher than that for NADPH (20 μM). The purified enzyme is competitively (against NADPH) inhibited by $NADP^+$ and palmitoyl-CoA. The enzyme is stable on storage at 4°C in the presence of NADPH and dithiothreitol (Datta et al., 1990).

Enzymes catalyzing subsequent steps of plasmalogen biosynthesis are localized in the endoplasmic reticulum and may not be identical to those involved in diacyl glycerophospholipid synthesis (Lee, 1998). These enzymes include alkyl-GP acyltransferase (1-alkyl-*sn*-Gro*P* acyltransferase), alkylacyl-GP phosphohydrolase I (1-alkyl-2-acyl-*sn*-Gro*P* phosphohydrolase), 1-alkyl-2-acyl-*sn*-Gro:CDP-choline (CDP-ethanolamine) choline (ethanolamine) phosphotransferase, and plasmanylethanolamine desaturase.

2.2.4 Alkylglycerophosphate Acyltransferase

This enzyme is also known as lysophosphatidate acyltransferase (LPAAT). It catalyzes the transfer of acyl group from acyl-CoA to alkyl-lysophosphatidate. Triton X-100 and bovine serum albumin stimulate it. Alkyl-GP acyltransferase is stereospecific. Thus only the 1-alkyl-*sn*-glycero-3-phosphate isomer, and not other optical enantiomers, is active with brain alkyl-GP acyltransferase. Substrate specificity studies of the brain acyltransferase activity for different acyl-CoA species (16:0, 18:0, 18:2, 20:4, 22:4, 22:6) indicate selectivity dependent on the alkylglycerophosphate concentration. At low 1-alkyl-*sn*-glycero-3-phosphate concentrations, the enzyme prefers polyunsaturated acyl-CoA species to saturated species. Based on specific activity and kinetic parameters towards a series of acyl-CoA donors, alkyl-GP acyltransferase may be different from acyl-GP (lysophosphatidic acid) acyltransferase (Fleming and Hajra, 1977).

Two human LPAAT have been cloned, LPAAT-α and LPAAT-β (West et al., 1997). Human LPAATs resemble (48% identical) yeast and bacterial enzymes in their amino acid sequences. This enzyme is encoded by a gene located on chromosome 9p34.3 (Aguado and Campbell, 1998). Overexpression of these two cDNAs in mammalian cells leads to increased LPAAT activity in cell-free extracts. This correlates with enhancement of transcription and synthesis of tumor necrosis factor-α and interleukin-6 from cells upon stimulation with interleukin-1β. LPAAT overexpression may amplify cellular signaling responses from cytokines (West et al., 1997). LPAAT has not been purified from mammalian sources.

2.2.5 Alkylacyl Glycerophosphate Phosphohydrolase

This enzyme hydrolyzes the phosphate group from alkylacyl glycerophosphate. This enzyme has not been purified and characterized from either neural or nonneural sources. It is not known whether alkylacyl glycerophosphate phosphohydrolase is different from well-characterized diacyl glycerophosphate phosphohydrolase (Lee, 1998). Diacyl glycerophosphate phosphohydrolase was partially purified from rat liver. Gel filtration of rat liver cytosol on Bio-Gel A-5m results in four peaks (Ide and Nakazawa, 1985). All show activity with either phosphatidate bound to microsomal membranes (PAmb) or phosphatidate dispersed in sonicated microsomal lipids (PAaq) as the substrate. A major part of the PAmb phosphohydrolase activity (52%) is eluted in a peak with an apparent molecular mass of 500 kDa in which the PAaq phosphohydrolase activity is very low. A major PAaq phosphohydrolase activity peak (48%) is obtained in the void volume, in which the PAaq phosphohydrolase activity is higher than the PAmb phosphohydrolase activity. The addition of 0.075% Tween 20 to the elution buffer results in one peak with molecular mass of 500 kDa. Phosphatidate phosphohydrolase is a Mg^{2+}-dependent enzyme that is inhibited by *N*-ethylmaleimide. It is involved in phospholipase D-mediated signal transduction processes.

2.2.6 CDP-Ethanolamine: Diacylglycerol Ethanolaminephosphotransferase

This enzyme catalyzes the transfer of the phosphoethanolamine head group from CDP-ethanolamine to alkylacylglycerols. It has been purified from bovine liver microsomes to homogeneity with multiple column chromatographic procedures (Mancini et al., 1999). The purification method is based on the high hydrophobicity of the protein whose charged sites appear to be masked from interaction with the chromatographic stationary phase when membranes are solubilized with an excess of nonionic detergent. The purified enzyme migrates as a single band on SDS-polyacrylamide gel electrophoresis with molecular mass of 38 kDa and has both ethanolaminephosphotransferase and cholinephosphotransferase activities. Collective evidence based upon kinetic studies suggests that both activities are Mn^{2+}-dependent and that the same catalytic site is involved in cholinephosphotransferase and ethanolaminephosphotransferase reactions (Mancini et al., 1999). Mg^{2+}-dependent CDP-choline:diacylglycerol cholinephosphotransferase (EC 2.7.8.2) activity is completely inactivated during the solubilization and purification steps (Mancini et al., 1999).

The conversion of 1-alkyl-2-acyl-*sn*-Gro*P*Etn to 1-alk-1′-enyl-2-acyl-*sn*-Gro*P*Etn (ethanolamine plasmalogen) is carried out by a cytochrome b5-dependent microsomal electron transport system. This system consists of cytochrome b5, NADH:cytochrome b5 reductase, and cyanide-sensitive Δ1-alkyl desaturase (Snyder et al., 1985). Choline plasmalogens are synthesized from PlsEtns by polar-head group modifications by a base-exchange enzyme or *N*-methyltransferases (Paltauf, 1994; Horrocks et al., 1986; Lee, 1998; Mozzi et al., 1989). Enzymes that catalyze the last step of PlsEtn synthesis are not fully characterized and so more studies are required on the isolation and characterization of these enzymes (Nagan and Zoeller, 2001; Lee, 1998).

Plasmalogens can also be synthesized from alkylglycerols, bypassing the first three steps through the action of a kinase, ATP:1-alkyl-*sn*-glycerol phosphotransferase (alkylglycerol kinase). The product 1-*O*-alkyl-2-lyso-*sn*-glycero-3-phosphate enters the synthesizing cycle after the reductase step (Nagan and Zoeller, 2001). This pathway represents a salvage pathway for plasmalogen biosynthesis from partially degraded plasmalogens and alkylacyl glycerophospholipids. In another pathway, peroxisomes utilize acetyl-CoA, a product of β-oxidation, and tetradecanoyl-CoA to form hexadecanol, which condenses with 1-acyl dihydroxyacetone phosphate to generate 1-alkyl dihydroxyacetone phosphate. This reaction is catalyzed by 1-alkyl dihydroxyacetone phosphate synthase (Whitehouse, 1997).

Chemical synthesis of 1-*O*-[9′-(1″-pyrenyl)]nonyl-*sn*-glycerol (pAG), a fluorescent ether lipid with a pyrene moiety covalently attached at the alkyl chain terminus was reported (Zheng et al., 2006) (Fig. 2.2). This ω-pyrene-labeled 1-*O*-alkyl-*sn*-glycerol is taken up by CHO-K1 and NRel-4 (a variant that is defective in dihydroxyacetone phosphate dehydrogenase) cells. Treatment of CHO-K1 and NRel-4 cells results in the incorporation of pAG into ethanolamine and choline glycerophospholipids as well as a neutral lipid fraction tentatively identified as alkyldiacylglycerols. NRel-4 cells incorporate more fluorescence in the phospholipid fraction than CHO-K1, specifically in the ethanolamine glycerophospholipids. Analysis of the fluorescent lipids demonstrates that 93% of the pAG is taken up by glycerolipids with the intact ether bond. Although the addition of 20 μM HG to the medium fully restores PlsEtn biosynthesis in NRel-4 cells, pAG only partially restores PlsEtn synthesis (Zheng et al., 2006). Incubation of cells with pAG followed by irradiation with long-wavelength (>300 nm) ultraviolet light produces cytotoxicity. NRel-4 cells exhibit an increase in sensitivity to UV light compared with CHO-K1 cells. It is proposed that this photodynamic cytotoxicity approach can be used to select for mutants that are defective in downstream steps in ether lipid biosynthesis.

2.3 Plasmalogen Synthesizing Enzymes During Brain Development

Plasmalogens are a major constituent of the myelin sheath. The levels of PlsEtn in brain tissue depend on the degree of myelination (Horrocks and Sharma, 1982). Highly unsaturated fatty acids including docosahexaenoic acid (DHA) are in brain plasmalogens. Plasmalogens are in lipid raft microdomains isolated from myelin (Rodemer et al., 2003). Initial accumulation of plasmalogen occurs in 5- and 7-days-old rat brain followed by rapid accumulation between 10 and 17 days after birth (Korey and Orchen, 1959; Wells and Dittmer, 1967). The specific activity of NADPH:alkyl dihydroxyacetone phosphate oxidoreductase is highest in microsomes from 5-days-old rat brain (El Bassiouni et al., 1975), indicating that the increased synthesis of a key intermediate used for plasmalogen synthesis occurs earlier than the main burst of galactosylcerebroside synthesis for incorporation into myelin. Collective evidence suggests that the concentration of PlsEtn increases rapidly during the intense period of myelination. Ethanolamine glycerophospholipids from the myelin sheath contain up to 70% PlsEtn. An eightfold increase in PlsEtn levels (per gram of brain tissue) occurs in human white matter during the first year of life so that PlsEtn accounts for 20% of the glycerophospholipid mass (70% of the ethanolamine glycerophospholipids). Developmental studies of human brain indicate that there is a steep rise in PlsEtn content, followed by a further rise up to 30–40 years of age. This is followed by a decline of PlsEtn levels and myelin during normal aging. At 70 years of age, the levels of PlsEtn are 18% less than that at 40 years of age (Rouser and Yamamoto, 1968; Horrocks et al., 1981). Besides brain, a decline in plasmalogen level has been observed in other tissues in normal aging and in some pathologic conditions. A negative correlation of age with serum plasmalogen-derived hexadecanal dimethylacetal (16:0 DMA) or octadecanal dimethylacetal (18:0 DMA) is observed in healthy adults (Brosche, 2001). The DMAs are formed during process of methylation of fatty acids from the aldehydes bound at the *sn*-1 position of plasmalogens. Data from 118 elderly subjects (57–94 years of age) indicate that the highest 16:0 DMA values are found in hypercholesterolemic subjects. Furthermore, there is a negative correlation between serum triacylglycerols and plasmalogen-derived 16:0 DMA ($n = 118$) suggesting a relationship between low DMA values and elevated triacylglycerol levels.

2.4 Topology and Distribution of Plasmalogens and Enzymes Synthesizing Plasmalogens

As stated earlier, dihydroxyacetone phosphate acyltransferase, alkyl dihydroxyacetone phosphate synthase, and acyl/alkyl dihydroxyacetone reductase are located in peroxisomes. Dihydroxyacetone phosphate acyltransferase and alkyl dihydroxyacetone phosphate synthase are intraperoxisomal proteins facing the peroxisomal lumen, whereas acyl-CoA reductase and acyl/alkyl dihydroxyacetone reductase are located on the side of the peroxisomal membrane facing the cytosol (Brites et al., 2004). This topology of plasmalogen synthesizing enzymes indicates that the substrates of the dihydroxyacetone phosphate (acyl-CoA and dihydroxyacetone phosphate) should either be transported from cytosol into peroxisomes or synthesized inside peroxisomes. Free fatty acid generated during the plasmalogen biosynthesis pathway should be reactivated to its CoA-ester form to be a substrate for the dihydroxyacetone phosphate acyltransferase reaction. Peroxisomes contain an acyl-CoA synthase facing the peroxisomal lumen (Brites et al., 2004). This acyl-CoA is the very long chain acyl-CoA synthase. ATP, which is required for the synthetase reaction, comes from cytosol in exchange for AMP. This transport requires PMP34, a peroxisomal adenine nucleotide transporter that belongs to a family of mitochondrial solute carrier family of transporters (Visser et al., 2002). Dihydroxyacetone phosphate, the other reactant of the reaction catalyzed by dihydroxyacetone phosphate acyltransferase, is either transported from the cytosol into peroxisomes or generated in the peroxisomal matrix by glycerol-3-phosphate dehydrogenase (Hajra et al., 2000). The peroxisomal localization of dihydroxyacetone phosphate acyltransferase and alkyl dihydroxyacetone phosphate synthase is not only important for enzymic activity but also for the stability of these enzymes (Brites et al., 2004). Dihydroxyacetone phosphate acyltransferase and alkyl dihydroxyacetone phosphate synthase are known to interact and form a heterotrimeric complex, but the active functional units of these enzymes are monomers (Biermann et al., 1998). The heterotrimeric complex may regulate plasmalogen biosynthesis by modulating and facilitating substrate channeling (Brites et al., 2004).

In myelin, PlsEtns are predominantly localized in the inner leaflet (Kirschner and Ganser, 1982). Also, in red blood cell membranes, PlsEtns are found predominantly in the inner leaflet (Marinetti and Crain, 1978). Thus at equilibrium, 79% of PlsEtns are located in the inner leaflet. In contrast, the inner leaflet has only 20% of the choline plasmalogens. Thus in the red blood cell membrane, the asymmetric distribution of plasmalogens is similar to that of diacyl glycerophospholipids.

2.5 Plasmalogens in Lipid Rafts

Membranes contain submicron-sized domains called lipid rafts. These lipid rafts are enriched in cholesterol, sphingolipids, and plasmalogens (Pike et al., 2001). They also contain proteins such as the GPI-anchored proteins. Lipid rafts do not have a characteristic morphology but their occurrence in membranes compartmentalizes cellular processes. Lipid rafts can be stabilized to form larger platforms through protein–protein and protein–lipid interactions. Lipid rafts play an important role in intracellular protein transport, membrane fusion, and platforms for signal transduction processes in which sphingolipids and plasmalogens provide second messengers (Suzuki, 2002). They also serve as platforms for cell surface antigens and adhesion molecules that are crucial for cell activation, polarization, and signaling. In neurons, lipid rafts have been found in dendrites where they sustain a variety of postsynaptic protein complexes. Changes in plasmalogen levels in pathological conditions result in alterations in the composition of lipid rafts causing abnormalities in signal transduction processes. On the basis of the importance of plasmalogens in model membranes, it is proposed that plasmalogens not only provide second messengers but also contribute to signal transduction efficiency (Farooqui and Horrocks, 2004; Wanders and Waterham, 2006).

2.6 Plasmalogens in the Nucleus

Plasmalogens are also found in the nucleus where they are associated with chromatin (Albi et al., 2004). Although the role of plasmalogens in the nucleus is not fully understood, based on the occurrence of plasmalogen-selective phospholipase A_2 (Farooqui et al., 2004) and phospholipase C (Albi et al., 2004) in the nucleus, it is proposed that plasmalogens may be associated with neural cell proliferation, differentiation, and regulation of cell cycle. Plasmalogen-derived lipid mediators mediate these processes. The molecular mechanism involved in neural cell proliferation and differentiation is not fully understood, but plasmalogen-derived lipid mediators may be involved. These lipid mediators are generated through the stimulation of plasmalogen-selective PLA_2 (Antony et al., 2003; Farooqui et al., 2004) and PLC (Albi and Magni, 2004). Stimulation of these enzymes produces arachidonic acid and alkylacyl glycerols. Both metabolites stimulate protein kinase C and enhance the generation of other lipid mediators such as eicosanoids and platelet-activating factor.

Ischemic/reperfusion injury of rat hearts results in a 50% loss of myocytic nuclear choline and ethanolamine glycerophospholipids when compared with perfused hearts of controls. The loss of nuclear choline and ethanolamine glycerophospholipids during reperfusion of ischemic myocardium is partially reversed by the PlsCho-PLA_2 inhibitor, bromoenol lactone (Williams et al., 2000). This suggests that the loss of nuclear phospholipids during ischemia/reperfusion is mediated, in part, through the activation of PlsCho-PLA_2. Western blotting studies on isolated nuclei from ischemic hearts indicate that PlsCho-PLA$_2$ is translocated to the nucleus after myocardial ischemia (Williams et al., 2000). Collective evidence suggests that the nuclear phospholipid mass decreases after myocardial ischemia by a mechanism that involves, at least in part, phospholipolysis mediated by PlsCho-PLA$_2$.

2.7 Factors Affecting Plasmalogen Biosynthesis in Brain

Plasmalogen biosynthesis is affected by endogenous and exogenous factors such as diet, age, and genetic factors (Paltauf, 1994). A deficiency of n-3 fatty acids in the diet may result in plasmalogen deficiency and abnormal signal transduction processes in neural membranes (Farooqui and Horrocks, 2001). Dietary supplements of fish oil ethyl esters reduce the arachidonate-containing species of PlsEtns, whereas molecular species having 20:5(n-3), 22:6(n-3), and/or 22:5(n-3) acyl groups are increased in the spleen, lung, and kidneys (Blank et al., 1994). In testicular tissue from rats fed with fish oil diets, the molecular species of PlsEtns containing 22:5 (n-6) acyl groups are also reduced. An increase of PlsEtns with 18:1 alk-1-enyl moieties paired with highly unsaturated *sn*-2 acyl groups are found in the tissues of rats fed with fish oil plus selachyl alcohol diacetate supplements (Blank et al., 1994). Collective evidence suggests that supplementation of n-3 fatty acid normalizes signal transduction processes and many functions of the brain, liver, heart tissues, and reproductive organs. The incorporation of n-3 fatty acids in various mammalian tissues significantly modulates arachidonic acid metabolism by inhibiting the production of eicosanoids (Horrocks and Farooqui, 2004).

Ageing modulates the activity of dihydroxyacetone phosphate acyltransferase, the first enzyme of plasmalogen biosynthesis. Although diet has no effect on dihydroxyacetone phosphate acyltransferase, the aging process influences its activity. Littermates from two generation of n-3-deficient rats were fed an equilibrated diet containing either α-linolenic acid alone or with two doses of DHA. After weaning, or 3, 9, or 21 months of diet, rat brains were used for the determination of enzymic activity and plasmalogen levels. Dihydroxyacetone phosphate acyltransferase activity and plasmalogen levels are markedly increased in rat brain at 3 months when compared with age-matched brains from weaning. The enzymic activity is significantly decreased (30% and 40%) from 3 months to 9 and 21 months, respectively. In senescence-accelerated R1 mice, the PlsEtn content reached a maximal level at 5 months and then decreased from 5 to 9 months (André et al., 2006a). No age-dependent changes were observed in brain plasmalogen contents in senescence-accelerated P8 mice. In another study, levels and acyl compositions of PlsEtn, PtdEtn, and PtdSer were determined in the frontal cortex and hippocampus from 2 and 18-month-old rats (Favrelière et al., 2000). In 18-month-old rats, the fatty acid compositions of these three glycerophospholipids show an increase of monounsaturated fatty acid (18:1 n-9 and 20:1 n-9) and a decrease in polyunsaturated fatty acid (PUFAs), essentially DHA. DHA is markedly decreased in hippocampus PtdEtn at

18 months. Both DHA and arachidonic acid are considerably lower in frontal cortex PlsEtn. Hippocampus and frontal cortex undergo specific age-induced modifications in PlsEtn and PtdEtn acyl composition. It is proposed that decreased plasmalogen levels in aged rat brain may be not only due to decreased dihydroxyacetone phosphate acyltransferase activity, but also due to increased activity of plasmalogen-selective phospholipase A_2 (André et al., 2006b; André et al., 2005a).

Administration of *myo*-inositol (*myo*-Ins) increases the levels of plasmalogens in the brain tissue (Pettegrew et al., 2001). *myo*-Ins is an important organic osmolyte that is found in brain, retina, and kidney (Nonaka et al., 1999). Its concentration increases in brain and cerebrospinal fluid with age. Its levels in cerebrospinal fluid may be an indicator of brain atrophy (Chang et al., 1996). The mechanism of action of *myo*-Ins is not fully elucidated. However, significant information is available about its biological roles. *myo*-Ins is metabolized to PtdIns, which makes up a small, but very significant, component of neural membranes. PtdIns is metabolized to PtdIns-4,5-P_2, a key intermediate in biological signaling. The possible benefit of *myo*-Ins in the management of depression, panic attacks, and obsessive-compulsive behavior may be explained by the role of *myo*-Ins as a second-messenger precursor (Patel et al., 2006; Shaldubina et al., 2007).

Acute administration of *myo*-Ins plus [2-^{13}C]ethanolamine ([2-^{13}C]Etn) significantly elevates levels of PlsEtns in whole rat brain (Hoffman-Kuczynski and Reo, 2004). This increase in PlsEtns is localized to specialized brain areas. Thus cerebellum is the brain area most affected by the *myo*-Ins plus Etn administration (Hoffman-Kuczynski and Reo, 2005). Surprisingly, earlier studies demonstrated that administration of *myo*-Ins alone minimally affects cerebellum (Patishi et al., 1996; Kofman et al., 1998). Collectively these studies suggest that the administration of *myo*-Ins plus Etn increases the ability of cerebellum to synthesize new PlsEtn molecules. The molecular mechanism of *myo*-Ins-mediated PlsEtn synthesis is still not known, but it is proposed that *myo*-inositol catabolism through the pentose phosphate cycle generates 2 mol of NADPH. This increase in NADPH level may be associated with increased PlsEtn synthesis in rat brain. An elevated PlsEtn/PtdEtn ratio can lead to tighter neural membrane packing. This may affect membrane dynamics and induce alterations in fluidity and permeability. The synthesis of new PlsEtns may be important due to its potential role as a cellular antioxidant (Hoffman-Kuczynski and Reo, 2005).

2.8 Conclusion

Plasmalogens are major constituents of neural membranes. Plasmalogen synthesizing enzymes are localized in peroxisomes, mitochondria, and endoplasmic reticulum. Although some plasmalogen synthesizing enzymes have been purified by multiple column chromatographic procedures from liver, but their purification from brain tissue has not been achieved. Two major reasons may be responsible for the lack of isolation and characterization of enzymes synthesizing brain plasmalogens. First, the activities of these enzymes in brain tissue are quite low when compared with other lipid synthesizing enzymes. Second, assays for determining activities are complex, laborious, and time consuming.

Dihydroxyacetone phosphate acyltransferase catalyzes the acylation of dihydroxyacetone phosphate with acyl-CoA with formation of 1-acyl dihydroxyacetone phosphate. The acyl group in this intermediate is then replaced by a long-chain alcohol that provides the oxygen for the ether linkage in the reaction catalyzed by alkyl dihydroxyacetone phosphate synthase. The carbonyl function in alkyl dihydroxyacetone phosphate is then reduced and acylated to give an alkyl analog of phosphatidic acid. This intermediate is dephosphorylated prior to introduction of the phosphocholine or phosphoethanolamine group (Lee, 1998; Nagan and Zoeller, 2001; Murphy, 2001). The rate-limiting step for plasmalogen biosynthesis has not been identified but dihydroxyacetone phosphate acyltransferase is essential for plasmalogen synthesis. Collective evidence suggests that the synthesis of PlsEtns is a very complex process that initially requires the participation of peroxisomal enzymes, followed by contributions from mitochondrial and endoplasmic reticulum enzymes.

References

Acar N., Gregoire S., Andre A., Juaneda P., Joffre C., Bron A. M., Creuzot-Garcher C. P., and Bretillon L. (2007). Plasmalogens in the retina: In situ hybridization of dihydroxyacetone phosphate acyltransferase (DHAP-AT) – the first enzyme involved in their biosynthesis – and comparative study of retinal and retinal pigment epithelial lipid composition. Exp. Eye Res. 84:143–151.

Aguado B. and Campbell R. D. (1998). Characterization of a human lysophosphatidic acid acyltransferase that is encoded by a gene located in the class III region of the human major histocompatibility complex. J. Biol. Chem. 273:4096–4105.

Albi E., Cataldi S., Magni M. V., and Sartori C. (2004). Plasmalogens in rat liver chromatin: New molecules involved in cell proliferation. J. Cellular Physiol. 201:439–446.

Albi E. and Magni M. P. V. (2004). The role of intranuclear lipids. Biol. Cell. 96:657–667.

André A., Chanséaume E., Dumusois C., Cabaret S., Berdeaux O., and Chardigny J. M. (2006a). Cerebral plasmalogens and aldehydes in senescence-accelerated mice P8 and R1: A comparison between weaned, adult and aged mice. Brain Res. 1085:28–32.

André A., Juanéda P., Sébédio J. L., and Chardigny J. M. (2005a). Effects of aging and dietary n-3 fatty acids on rat brain phospholipids: Focus on plasmalogens. Lipids 40:799–806.

André A., Juanéda P., Sébédio J. L., and Chardigny J. M. (2006b). Plasmalogen metabolism-related enzymes in rat brain during aging: influence of n-3 fatty acid intake. Biochimie 88:103–111.

André A., Tessier C., Brétillon L., Sébédio J. L., and Chardigny J. M. (2005b). In situ hybridization of dihydroxyacetone phosphate acyltransferase, the regulating enzyme involved in plasmalogen biosynthesis. Mol. Brain Res. 136:142–147.

Antony P., Freysz L., Horrocks L. A., and Farooqui A. A. (2003). Ca^{2+}-independent phospholipases A_2 and production of arachidonic acid in nuclei of LA-N-1 cell cultures: A specific receptor activation mediated with retinoic acid. Mol. Brain Res. 115:187–195.

Balakrishnan S., Goodman H., and Cumings J. N. (1961). The distribution of phosphorus-containing lipid compounds in the human brain. J. Neurochem. 8:276–284.

Biermann J., Just W. W., Wanders R. J., and van den Bosch H. (1999). Alkyl-dihydroxyacetone phosphate synthase and dihydroxyacetone phosphate acyltransferase form a protein complex in peroxisomes. Eur. J. Biochem. 261:492–499.

Biermann J., Schoonderwoerd K., Hom M. L., Luthjens L. H., and van den Bosch H. (1998). The native molecular size of alkyl-dihydroxyacetonephosphate synthase and dihydroxyacetonephosphate acyltransferase. Biochim. Biophys. Acta 1393:137–142.

Blank M. L., Smith Z. L., Cress E. A., and Snyder F. (1994). Molecular species of ethanolamine plasmalogens and transacylase activity in rat tissues are altered by fish oil diets. Biochim. Biophys. Acta Lipids Lipid Metab. 1214:295–302.

Brites P., Waterham H. R., and Wanders R. J. A. (2004). Functions and biosynthesis of plasmalogens in health and disease. Biochim. Biophys. Acta Mol. Cell Biol. Lipids 1636:219–231.

Brosche T. (2001). Plasmalogen levels in serum from patients with impaired carbohydrate or lipid metabolism and in elderly subjects with normal metabolic values. Arch. Gerontol. Geriatr. 32:283–294.

Cai J., Nelson K. C., Wu M., Sternberg P., Jr., and Jones D. P. (2000). Oxidative damage and protection of the RPE. Prog. Retin. Eye Res. 19:205–221.

Causeret C., Bentejac M., Albet S., Teubner B., and Bugaut M. (1997). Copurification of dihydroxyacetone-phosphate acyl-transferase and other peroxisomal proteins from liver of fenofibrate-treated rats. Biochimie 79:423–433.

Chang L., Ernst T., Poland R. E., and Jenden D. J. (1996). *In vivo* proton magnetic resonance spectroscopy of the normal aging human brain. Life Sci. 58:2049–2056.

Datta S. C., Ghosh M. K., and Hajra A. K. (1990). Purification and properties of acyl/alkyl dihydroxyacetone-phosphate reductase from guinea pig liver peroxisomes. J. Biol. Chem. 265:8268–8274.

de Vet E. C., Hilkes Y. H., Fraaije M. W., and van den Bosch H. (2000). Alkyl-dihydroxyacetonephosphate synthase. Presence and role of flavin adenine dinucleotide. J. Biol. Chem. 275:6276–6283.

de Vet E. C., IJlst L., Oostheim W., Dekker C., Moser H. W., van den Bosch H., and Wanders R. J. (1999). Ether lipid biosynthesis: Alkyl-dihydroxyacetonephosphate synthase protein deficiency leads to reduced dihydroxyacetonephosphate acyltransferase activities. J. Lipid Res. 40:1998–2003.

de Vet E. C. and van den Bosch H. (2000). Alkyl-dihydroxyacetonephosphate synthase. Cell Biochem. Biophys. 32:117–121.

El Bassiouni E. A., Piantadosi C., and Snyder F. (1975). Metabolism of alkyldihydroxyacetone phosphate in rat brain. Biochim. Biophys. Acta 388:5–11.

Engelmann B., Bräutigam C., and Thiery J. (1994). Plasmalogen phospholipids as potential protectors against lipid peroxidation of low density lipoproteins. Biochem. Biophys. Res. Commun. 204:1235–1242.

Farooqui A. A., Antony P., Ong W. Y., Horrocks L. A., and Freysz L. (2004). Retinoic acid-mediated phospholipase A_2 signaling in the nucleus. Brain Res. Rev. 45:179–195.

Farooqui A. A. and Horrocks L. A. (2001). Plasmalogens: Workhorse lipids of membranes in normal and injured neurons and glia. Neuroscientist 7:232–245.

Farooqui A. A. and Horrocks L. A. (2004). Plasmalogens, platelet-activating factor, and other ether lipids. In: Nicolaou A. and Kokotos G. (eds.), *Bioactive Lipids*. Oily Press, Bridgwater, England, pp. 107–134.

Favrelière S., Stadelmann-Ingrand S., Huguet F., De Javel D., Piriou A., Tallineau C., and Durand G. (2000). Age-related changes in ethanolamine glycerophospholipid fatty acid levels in rat frontal cortex and hippocampus. Neurobiol. Aging 21:653–660.

Fleming P. J. and Hajra A. K. (1977). 1-Alkyl-*sn*-glycero-3-phosphate:acyl-CoA acyltransferase in rat brain microsomes. J. Biol. Chem. 252:1663–1672.

Getz G. S., Bartley W., Lurie D., and Notton B. M. (1968). The phospholipids of various sheep organs, rat liver and of their subcellular fractions. Biochim. Biophys. Acta 152:325–339.

Ghosh M. K. and Hajra A. K. (1986). Subcellular distribution and properties of acyl/alkyl dihydroxyacetone phosphate reductase in rodent livers. Arch. Biochem. Biophys. 245:523–530.

Hajra A. K., Larkins L. K., Das A. K., Hemati N., Erickson R. L., and MacDougald O. A. (2000). Induction of the peroxisomal glycerolipid-synthesizing enzymes during differentiation of 3T3-L1 adipocytes – Role in triacylglycerol synthesis. J. Biol. Chem. 275:9441–9446.

Hoffman-Kuczynski B. and Reo N. V. (2004). Studies of myo-inositol and plasmalogen metabolism in rat brain. Neurochem. Res. 29:843–855.

Hoffman-Kuczynski B. and Reo N. V. (2005). Administration of myo-inositol plus ethanolamine elevates phosphatidylethanolamine plasmalogen in the rat cerebellum. Neurochem. Res. 30:47–60.

Horrocks L. A. (1972). Content, composition, and metabolism of mammalian and avian lipids that contain ether groups. In: Snyder F. (ed.), *Ether Lipids: Chemistry and Biology*. Academic Press, New York, pp. 177–272.

Horrocks L. A. and Farooqui A. A. (2004). Docosahexaenoic acid in the diet: Its importance in maintenance and restoration of neural membrane function. Prostaglandins Leukot. Essent. Fatty Acids 70:361–372.

Horrocks L. A. and Sharma M. (1982). Plasmalogens and O-alkyl glycerophospholipids. In: Hawthorne J. N. and Ansell G. B. (eds.), *Phospholipids, New Comprehensive Biochemistry, Vol. 4*. Elsevier Biomedical Press, Amsterdam, pp. 51–93.

Horrocks L. A., VanRollins M., and Yates A. J. (1981). Lipid changes in the ageing brain. In: Davison A. N. and Thompson R. H. S. (eds.), *The Molecular Basis of Neuropathology*. Edward Arnold Ltd., London, pp. 601–630.

Horrocks L. A., Yeo Y. K., Harder H. W., Mozzi R., and Goracci G. (1986). Choline plasmalogens, glycerophospholipid methylation, and receptor-mediated activation of adenylate cyclase. In: Greengard P. and Robison G. A. (eds.), *Advances in Cyclic Nucleotide and Protein Phosphorylation Research, Vol. 20*. Raven Press, New York, pp. 263–292.

Ide H. and Nakazawa Y. (1985). Phosphatidate phosphatase in rat liver: the relationship between the activities with membrane-bound phosphatidate and aqueous dispersion of phosphatidate. J Biochem. 97:45–54.

Jamal Z., Martin A., Gomez-Munoz A., and Brindley D. N. (1991). Plasma membrane fractions from rat liver contain a phosphatidate phosphohydrolase distinct from that in the endoplasmic reticulum and cytosol. J. Biol. Chem. 266:2988–2996.

Jones C. L. and Hajra A. K. (1983). Solubilization and partial purification of dihydroxyacetone-phosphate acyltransferase from guinea pig liver. Arch. Biochem. Biophys. 226:155–165.

Kirschner D. A. and Ganser A. L. (1982). Myelin labeled with mercuric chloride. Asymmetric localization of phosphatidylethanolamine plasmalogen. J. Mol. Biol. 157:635–658.

Kofman O., Agam G., Shapiro J., and Spencer A. (1998). Chronic dietary inositol enhances locomotor activity and brain inositol levels in rats. Psychopharmacology (Berl) 139:239–242.

Korey S. R. and Orchen M. (1959). Plasmalogens of the nervous system. I. Deposition in developing rat brain and incorporation of ^{14}C isotope from acetate and palmitate into the α, β-unsaturated ether chain. Arch. Biochem. Biophys. 83:381–389.

Lee T. C. (1998). Biosynthesis and possible biological functions of plasmalogens. Biochim. Biophys. Acta Lipids Lipid Metab. 1394:129–145.

Liu D. L., Nagan N., Just W. W., Rodemer C., Thai T. P., and Zoeller R. A. (2005). Role of dihydroxyacetonephosphate acyltransferase in the biosynthesis of plasmalogens and nonether glycerolipids. J. Lipid Res. 46:727–735.

Lohner K. (1996). Is the high propensity of ethanolamine plasmalogens to form non-lamellar lipid structures manifested in the properties of biomembranes? Chem. Phys. Lipids 81:167–184.

Mancini A., Del Rosso F., Roberti R., Orvietani P., Coletti L., and Binaglia L. (1999). Purification of ethanolaminephosphotransferase from bovine liver microsomes. Biochim. Biophys. Acta Lipids Lipid Metab. 1437:80–92

Marinetti G. V. and Crain R. C. (1978). Topology of amino phospholipids in the red cell membrane. J. Supramol. Struct. 8:191–213.

Mozzi R., Gramignani D., Andriamampandry C., Freysz L., and Massarelli R. (1989). Choline plasmalogen synthesis by the methylation pathway in chick neurons in culture. Neurochem. Res. 14:579–583.

Murphy R. C. (2001). Free-radical-induced oxidation of arachidonoyl plasmalogen phospholipids: Antioxidant mechanism and precursor pathway for bioactive eicosanoids. Chem. Res. Toxicol. 14:463–472.
Nagan N., Hajra A. K., Das A. K., Moser H. W., Moser A., Lazarow P., Purdue P. E., and Zoeller R. A. (1997). A fibroblast cell line defective in alkyl-dihydroxyacetone phosphate synthase: a novel defect in plasmalogen biosynthesis. Proc. Natl. Acad. Sci. USA 94:4475–4480.
Nagan N., Hajra A. K., Larkins L. K., Lazarow P., Purdue P. E., Rizzo W. B., and Zoeller R. A. (1998). Isolation of a Chinese hamster fibroblast variant defective in dihydroxyacetonephosphate acyltransferase activity and plasmalogen biosynthesis: Use of a novel two-step selection protocol. Biochem. J. 332:273–279.
Nagan N. and Zoeller R. A. (2001). Plasmalogens: Biosynthesis and functions. Prog. Lipid Res. 40:199–229.
Nonaka M., Kohmura E., Yamashita T., Yamauchi A., Fujinaka T., Yoshimine T., Tohyama M., and Hayakawa T. (1999). Kainic acid-induced seizure upregulates Na^+/*myo*-inositol cotransporter mRNA in rat brain. Brain Res. Mol. Brain Res. 70:179–186.
Ofman R., Hettema E. H., Hogenhout E. M., Caruso U., Muijsers A. O., and Wanders R. J. (1998). Acyl-CoA:dihydroxyacetonephosphate acyltransferase: Cloning of the human cDNA and resolution of the molecular basis in rhizomelic chondrodysplasia punctata type 2. Hum. Mol. Genet. 7:847–853.
Ofman R., Hogenhout E. M., and Wanders R. J. (1999). Identification and characterization of the mouse cDNA encoding acyl-CoA:dihydroxyacetone phosphate acyltransferase. Biochim. Biophys. Acta 1439:89–94.
Ofman R., Lajmir S., and Wanders R. J. A. (2001). Etherphospholipid biosynthesis and dihydroxyactetone-phosphate acyltransferase: Resolution of the genomic organization of the human GNPAT gene and its use in the identification of novel mutations. Biochem. Biophys. Res. Commun. 281:754–760.
Ofman R. and Wanders R. J. A. (1994). Purification of peroxisomal acyl-CoA:dihydroxyacetonephosphate acyltransferase from human placenta. Biochim. Biophys. Acta Protein Struct. Mol. Enzymol. 1206:27–34.
Paltauf F. (1994). Ether lipids in biomembranes. Chem. Phys. Lipids 74:101–139.
Patel N. C., DelBello M. P., Cecil K. M., Adler C. M., Bryan H. S., Stanford K. E., and Strakowski S. M. (2006). Lithium treatment effects on *myo*-inositol in adolescents with bipolar depression. Biol. Psychiat. 60:998–1004.
Patishi Y., Lubrich B., Berger M., Kofman O., Van Calker D., and Belmaker R. H. (1996). Differential uptake of *myo*-inositol in vivo into rat brain areas. Eur. Neuropsychopharmacol. 6:73–75.
Pettegrew J. W., Panchalingam K., Levine J., McClure R. J., Gershon S., and Yao J. K. (2001). Chronic myo-inositol increases rat brain phosphatidylethanolamine plasmalogen. Biol. Psychiat. 49:444–453.
Pike L. J., Han X. L., Chung K. N., and Gross R. W. (2001). Lipid rafts are enriched in plasmalogens and arachidonate-containing phospholipids and the expression of caveolin does not alter the lipid composition of these domains. FASEB J. 15:A20.
Rodemer C., Thai T. P., Brugger B., Kaercher T., Werner H., Nave K. A., Wieland F., Gorgas K., and Just W. W. (2003). Inactivation of ether lipid biosynthesis causes male infertility, defects in eye development and optic nerve hypoplasia in mice. Hum. Mol. Genet. 12:1881–1895.
Rouser G. and Yamamoto A. (1968). Curvilinear regression course of human brain lipid composition changes with age. Lipids 3:284–287.
Schlossman D. M. and Bell R. M. (1977). Microsomal *sn*-glycerol 3-phosphate and dihydroxyacetone phosphate acyltransferase activities from liver and other tissues. Evidence for a single enzyme catalizing both reactions. Arch. Biochem. Biophys. 182:732–742.
Shaldubina A., Buccafusca R., Johanson R. A., Agam G., Belmaker R. H., Berry G. T., and Bersudsky Y. (2007). Behavioural phenotyping of sodium-myo-inositol cotransporter heterozygous knockout mice with reduced brain inositol. Genes Brain Behav. 6:253–259.

Snyder F., Lee T.-C., and Wykle R. L. (1985). Ether-linked glycerolipids and their bioactive species: Enzymes and metabolic regulation. In: Martonosi A. N. (Ed.), *The Enzymes of Biological Membranes*. Plenum, New York, pp. 1–58.

Stamps A. C., Elmore M. A., Hill M. E., Kelly K., Makda A. A., and Finnen M. J. (1997). A human cDNA sequence with homology to non-mammalian lysophosphatidic acid acyltransferases. Biochem. J. 326 (Pt 2):455–461.

Stoffel W., LeKim D., and Heyn G. (1970). Metabolism of sphingosine bases. XIV. Sphinganine (dihydrosphingosine), an effective donor of the alk-1-enyl chain of plasmalogens. Hoppe Seylers. Z. Physiol. Chem. 351:875–883.

Suzuki T. (2002). Lipid rafts at postsynaptic sites: Distribution, function and linkage to postsynaptic density. Neurosci. Res. 44:1–9.

Thai T. P., Heid H., Rackwitz H. R., Hunziker A., Gorgas K., and Just W. W. (1997). Ether lipid biosynthesis: isolation and molecular characterization of human dihydroxyacetonephosphate acyltransferase. FEBS Lett. 420:205–211.

Toews A. D. and Horrocks L. A. (1976). Developmental and aging changes in protein concentration and 2′,3′-cyclic nucleosidemonophosphate phosphodiesterase activity (EC 3.1.4.16) in human cerebral white and gray matter and spinal cord. J. Neurochem. 27:545–550.

Visser W. F., van Roermund C. W., Waterham H. R., and Wanders R. J. (2002). Identification of human PMP34 as a peroxisomal ATP transporter. Biochem. Biophys. Res. Commun. 299:494–497.

Wanders R. J. A. and Waterham H. R. (2006). Peroxisomal disorders: The single peroxisomal enzyme deficiencies. Biochim. Biophys. Acta Mol. Cell Res. 1763:1707–1720.

Webber K. O. and Hajra A. K. (1993). Purification of dihydroxyacetone phosphate acyltransferase from guinea pig liver peroxisomes. Arch. Biochem. Biophys. 300:88–97.

Wells M. A. and Dittmer J. C. (1967). A comprehensive study of the postnatal changes in the concentration of the lipids of developing rat brain. Biochemistry 6:3169–3175.

West J., Tompkins C. K., Balantac N., Nudelman E., Meengs B., White T., Bursten S., Coleman J., Kumar A., Singer J. W., and Leung D. W. (1997). Cloning and expression of two human lysophosphatidic acid acyltransferase cDNAs that enhance cytokine-induced signaling responses in cells. DNA Cell Biol. 16:691–701.

Whitehouse P. J. (1997). Genesis of Alzheimer's disease. Neurology 48:2–7.

Williams S. D., Hsu F. F., and Ford D. A. (2000). Electrospray ionization mass spectrometry analyses of nuclear membrane phospholipid loss after reperfusion of ischemic myocardium. J. Lipid Res. 41:1585–1595.

Zheng H., Duclos R. I. J., Smith C. C., Farber H. W., and Zoeller R. A. (2006). Synthesis and biological properties of the fluorescent ether lipid precursor 1-*O*-[9′-(1″-pyrenyl)]nonyl-*sn*-glycerol. J. Lipid Res. 47:633–642.

Zomer A. W. M., De Weerd W. F. C., Langeveld J., and van den Bosch H. (1993). Ether lipid synthesis: Purification and identification of alkyl dihydroxyacetone phosphate synthase from guinea-pig liver. Biochim. Biophys. Acta Lipids Lipid Metab. 1170:189–196.

Chapter 3
Catabolism of Plasmalogens in Brain

3.1 Introduction

Among the glycerophospholipids, plasmalogens are a special class characterized by the presence of a long-chain enol ether (vinyl ether) bond at the *sn*-1 position of the glycerol moiety. In the plasmalogen molecule, the perpendicular orientation of arachidonyl or docosahexaenoyl chains at the *sn*-2 position of the glycerol moiety and the lack of a carbonyl group at the *sn*-1 position affect the hydrophilicity of the head groups, resulting in stronger intermolecular hydrogen bonding between the head groups (Lohner, 1996). These properties allow ethanolamine plasmalogens to adopt the inverse hexagonal phase with greater propensity, but may also be responsible for a different membrane potential compared with diacylglycerophospholipids (Lohner, 1996). Plasmalogens can be distinguished from diacylglycerolipids in two properties. First, the vinyl ether linkage of plasmalogens is acid labile and second, plasmalogens act as efficient antioxidants (Murphy, 2001). Neural membrane plasmalogens contain up to 70% of docosahexaenoic acid at the *sn*-2 position (Layden et al., 2005). Like diacylglycerophospholipids, degradation of the plasmalogens is a receptor-mediated process coupled with the stimulation of plasmalogen-selective phospholipase A_2 (PlsEtn-PLA_2). Because the enol–ether linkage is susceptible to oxidation (Weisser and Spiteller, 1996), plasmalogens undergo an epoxidation process. Their epoxides readily undergo hydrolytic cleavage to form "long chain" α-hydroxyaldehydes. These aldehydes accumulate in brain on aging and under pathological conditions (Weisser and Spiteller, 1996).

3.2 Plasmalogen-Selective Phospholipase A_2 (PlsEtn-PLA_2)

The release of arachidonic acid or docosahexaenoic acid from the *sn*-2 position of plasmalogens is catalyzed by plasmalogen-selective PLA_2 (PlsEtn-PLA_2). This PLA_2 was discovered and characterized by us (Hirashima et al., 1992; Farooqui et al., 1995a). It occurs in the cytosolic fraction of the brains from various animal species. It can be separated from cytosolic PLA_2 by Sephadex G-75 column chromatography. In mammalian brains, PlsEtn-PLA_2 accounts for the greatest proportion of total PLA_2 activity (Table 3.1) (Yang et al., 1997).

Brain PlsEtn-PLA_2 has a molecular mass of 39kDa and does not require Ca^{2+} (Hirashima et al., 1992; Farooqui et al., 1995a). It is not affected by ATP and other nucleotides in the micromolar range, but is markedly inhibited by these nucleotides at 2mM or above, the normal intracellular concentration of ATP (Fig. 3.1). Enzymic activity is also inhibited by ADP, but AMP and cAMP have no effect. This differs from the heart PlsCho-PLA_2 that is stimulated by the addition of ATP and other nucleotides and inhibited by DTNB and Triton X-100 (Hazen et al., 1991; Hazen and Gross, 1993) (Table 3.2). Canine myocardial PlsCho-PLA_2 is purified by sequential column chromatographies on anionic exchange, chromatofocusing, ATP agarose affinity chromatography, mono Q chromatography, and hydroxylapatite chromatography. It is associated with phosphofructokinase as a complex with a molecular mass of 400kDa. It is proposed that interactions between phosphofructokinase and myocardial PlsCho-PLA_2 are highly specific, and these interactions may be involved in coordinated regulation of phospholipolysis and glycolysis

Table 3.1 Proportions of PlsEtn-PLA_2 and $cPLA_2$ activities in brains of various animal species.

Animal species	PlsEtn-PLA_2 (%)	$cPLA_2$ (%)
Rat	57 ± 6	30 ± 3
Pig	60 ± 5	31 ± 4
Dog	63 ± 7	30 ± 3
Horse	60 ± 5	31 ± 5
Chicken	12 ± 4	80 ± 10

Enzymic activities were precipitated with 40% ammonium sulfate and proportions were calculated after separation with Sephadex G-75 Column chromatography (Yang et al., 1997; Hirashima et al., 1992). PlsEtn-PLA_2 activity is quantified as described in Chap. 4 (Hirashima et al., 1992). Values are means ± S.E.M, $n = 3$

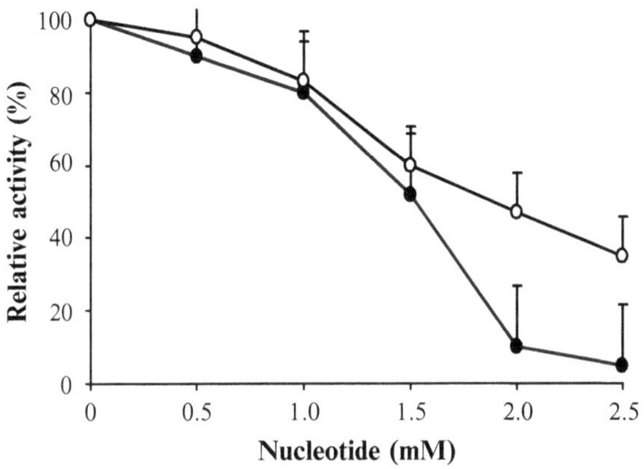

Fig. 3.1 Effect of ATP and ADP on partially purified PlsEtn-PLA_2 from bovine brain. ATP (*filled circle*) and ADP (*empty circle*). Specific activity of the control (7 $pmol^{-1}$ min^{-1} mg^{-1}) is taken as 100%. Values are means ± S.E.M., $n = 5$.

Table 3.2 Comparison of kinetic and physicochemical properties of Ca^{2+}-independent plasmalogen-selective PLA_2 from bovine brain, human myocardium, and rabbit kidney.

Property	Bovine brain PlsEtn-PLA_2	Human myocardium PlsCho-PLA_2	Rabbit kidney PlsCho-PLA_2
pH optimum	7.4	7.0	7.3
K_m value (μM)	40.0	4.0	–
V_{max} ($nmol^{-1}$ min^{-1} mg^{-1})	65.0	194,000	1,160
Substrate	PlsEtn	PlsCho	PlsCho
Localization	Cytosol	Cytosol	Cytosol
Molecular mass	39 kDa	40 kDa	28 kDa
Effect of ATP	Inhibited	Stimulated	No effect
Effect of Triton-X-100	Stimulated	Inhibited	No effect
Effect of DTNB	Inhibited	Inhibited	–

For bovine brain, human myocardium, and rabbit kidney PLA_2 data modified and summarized from (Farooqui et al., 1995a; Hazen and Gross, 1993), and (Portilla and Dai, 1996), respectively

Table 3.3 Effects of glycosaminoglycans on plasmalogen-selective PLA_2 from bovine brain.

Glycosaminoglycan	Effect
Heparan sulphate	Inhibited (IC_{50}, 15.0 μg ml^{-1})
Hyaluronic acid	Inhibited (IC_{50}, 46.0 μg ml^{-1})
Condroitin 4-sulfate	Inhibited (IC_{50}, 200.0 μg ml^{-1})
Condroitin 6-sulfate	Inhibited (IC_{50}, 250.0 μg ml^{-1})
Keratan sulfate	Inhibited (IC_{50}, 225.0 μg ml^{-1})
DNA	No effect
RNA	No effect

Data modified from (Yang et al., 1994a)

(Hazen and Gross, 1993). Myocardial PlsCho-PLA_2 also interacts with calmodulin. These interactions may also have implications in regulation of PlsCho-PLA_2 activity by the calcium transducer, calmodulin, in heart muscle (Wolf and Gross, 1996). Brain PlsEtn-PLA_2 does not interact with calmodulin.

Nonionic detergents, Triton X-100 and Tween-20, stimulate the enzymic activity of brain PlsEtn-PLA_2. These detergents inhibit the activity of heart PlsCho-PLA_2. Other detergents, such as octylglucoside, sodium deoxycholate, and sodium taurocholate, inhibit the PlsEtn-PLA_2 in a dose-dependent manner (Table 3.2). The SH-group blocking agents, dithio-*bis*-2-nitrobenzoic acid and iodoacetate, inhibit bovine brain PlsEtn-PLA_2. Other SH-group blocking agents such as iodoacetate and *N*-ethylmaleimide inhibit the enzymic activity as well but to a lesser degree. This inhibition can be reversed by dithiothreitol. PlsEtn-PLA_2 is also inhibited by polyvalent anions (citrate > sulfate > phosphate) (Farooqui et al., 1995a). The purified PlsEtn-PLA_2 is inhibited by metal ions such as Ag^+ and Hg^{2+} > Fe^{3+}. Quinacrine and nordihydroguariaretic acid inhibit the PlsEtn-PLA_2 in a dose-dependent manner (Farooqui et al., 1997).

Bovine brain PlsEtn-PLA_2 is markedly inhibited by glycosaminoglycans (Table 3.3). Heparan sulfate is the most potent inhibitor, followed by hyaluronic acid, chondroitin sulfate, and heparin (Yang et al., 1994a). The inhibition by glycosaminoglycans can be reversed by the addition of protamine sulfate. *N*-acetylneuraminic acid, gangliosides, and sialoglycoproteins also inhibit the plasmalogen-selective PLA_2 in a dose-dependent manner (Fig. 3.2). However, colominic acid (poly 2,8-*N*-acetylneuraminic acid) has no effect on its enzymic activity (Table 3.4) (Yang et al., 1994b). Interactions of plasmalogen-selective PLA_2 with glycosaminoglycans and sialoglycoconjugates may be involved in the anchoring of this enzyme to plasma membranes.

Immunocytochemical localization of plasmalogen-selective PLA_2 has been performed in neuronal and astrocytic cultures (Yang et al., 1997). The colocalization of plasmalogen-selective PLA_2 with glial fibrillary acidic protein (GFAP) suggests that this PLA_2 is predominantly associated with astrocytes (Fig. 3.3) (Farooqui and

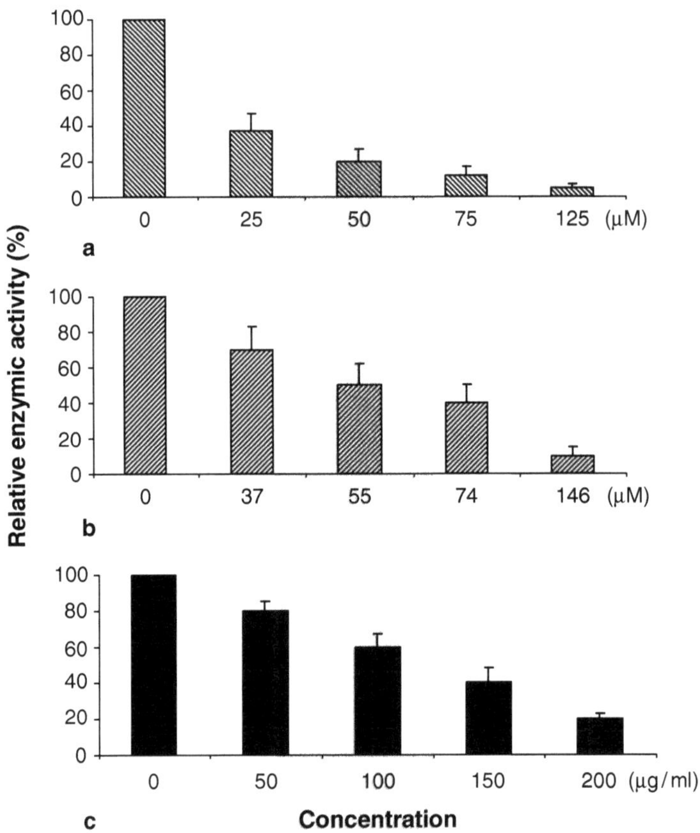

Fig. 3.2 Effect of *N*-acetyl neuraminic acid (NANA), GM3 ganglioside, and bovine submaxillary gland mucin on partially purified PlsEtn-PLA_2 from bovine brain. Specific activity of the control (7 pmol^{-1} min^{-1} mg^{-1}) is taken as 100%. Values are means ± S.E.M, $n = 5$. (**a**) NANA, (**b**) GM3 ganglioside, and (**c**) bovine submaxillary gland mucin. Data modified from Yang et al., (1994a).

Table 3.4 Effect of *N*-acetyl neuraminic acid, sialoglycoproteins, and gangliosides on brain PlsEtn-PLA_2.

Glycolipid	Effect
Ganglioside GM3	Inhibited (IC_{50}, 75.0 μg ml^{-1})
Ganglioside GM1	Inhibited (IC_{50}, 125.0 μg ml^{-1})
Cerebroside	No effect
Sulfatide	No effect
N-Acetylneuraminic acid	Inhibited (IC_{50}, 22.0 μg ml^{-1})
Colominic acid	No effect
Mucin	Inhibited (IC_{50}, 125.0 μg ml^{-1})
Cowper's gland mucin	Inhibited (IC_{50}, 250 μg ml^{-1})

Data modified from (Yang et al., 1994b)

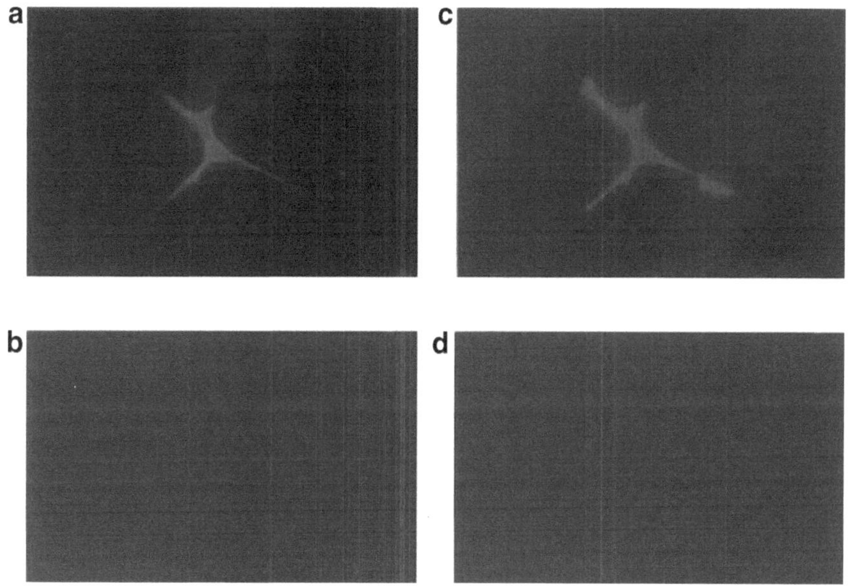

Fig. 3.3 Immunocytochemical localization of PlsEtn-PLA_2 and GFAP in astrocytic cortical cultures. Dual labeling was performed with anti-PlsEtn-PLA_2 (a) and anti-GFAP (c) and visualized with fluorescein filter for PLA_2 and a rhodamine filter for GFAP. Negative controls using preimmune sera for PlsEtn-PLA_2 (b) and GFAP (d), respectively. Note the same call can be stained with PlsEtn-PLA_2 and GFAP antibodies. Magnification: x 40.

Horrocks, 2001b). In contrast, the 85 kDa cytosolic PLA_2 is localized in neurons as well as astrocytes (Farooqui et al., 2000c).

A plasmalogen-selective PLA_2 has also been purified to homogeneity from rabbit kidney cortex through sequential column chromatography, including anion exchange, hydrophobic interaction, Mono Q, hydroxylapatite, phenyl-Sepharose, and chromatofocusing fast protein liquid chromatography (Portilla and Dai, 1996). The purified enzyme shows optimal activity at neutral pH and has specific activity of 1.2 μmol min^{-1} mg^{-1} protein. It has a molecular mass of 28 kDa and hydrolyzes PlsCho > PtdCho (Table 3.2). It selectively cleaves phospholipids containing arachidonic acid at the *sn*-2 position in comparison to oleic acid. Antibodies against the purified protein precipitate all of the soluble Ca^{2+}-independent PLA_2 activity from rabbit kidney cortex. Cloning of a full-length rat cDNA encoding PlsCho-PLA_2, using a sequence derived from the purified kidney cortex enzyme, has been performed. A cDNA from rat kidney that encodes the rat homolog of the PlsCho-PLA_2 and also a closely related isoform can also be isolated. The rat cDNA encodes a 24 kDa protein and contains the sequence G-F-S-Q-G, which fitted the active site consensus sequence G-X-S-X-G of carboxyl esterase (Portilla et al., 1998). Collective evidence suggests that PlsCho-PLA_2 enzymes are encoded by a multigene family in rats, mice, rabbits, and human. Northern blot analyses of various rat tissues indicate that the PlsCho-PLA_2 gene is ubiquitously expressed with highest mRNA in kidney and small intestine. Rat PlsCho-PLA_2 was expressed in a baculovirus expression system with expression of PlsCho-PLA_2 and lysophospholipase activities (Portilla et al., 1998).

On the basis of the effects of PLA_2 inhibitors in cell cultures of hippocampal origin, the release of arachidonic and docosahexaenoic acids in astrocytes is controlled by different isoforms of PLA_2, i.e., cPLA_2 and PlsEtn-PLA_2, respectively (Sergeeva et al., 2005). The enrichment of hippocampal tissue with DHA is neuroprotective. The incorporation of DHA in plasmalogens is required for the neuroprotective effect in hippocampal and astrocytic cultures. The protective effect is substantially higher in dentate gyrus than in CA1 and CA3 areas. Moreover, in astrocytic cultures, the release of arachidonic and docosahexaenoic acids is differently regulated through Ca^{2+}- and cAMP-dependent signal transduction pathways (Strokin et al., 2003). This supports the view that the release of arachidonic and docosahexaenoic acids is differentially modulated under normal and pathological situations (Strokin et al., 2006; Farooqui et al., 2006).

3.3 Receptor-Mediated Degradation of Plasmalogens

It is well known that interactions of agonists with receptors at the neural cell surface results in the stimulation of various isoforms of PLA_2. These isoforms include cytosolic phospholipase A_2 (cPLA_2), PlsEtn-PLA_2, and secretory phospholipase A_2 (sPLA_2). Because the acyl groups of plasmalogens turnover rapidly in brain, it is proposed that interactions of agonists with receptors at the neural cell surface results in the stimulation of PlsEtn-PLA_2 with generation of lysoplasmalogens and arachidonic acid or docosahexaenoic acid. In neural membranes, lysoplasmalogens are usually reacylated or hydrolyzed by lysoplasmalogenase. Arachidonic acid or docosahexaenoic acid are metabolized to eicosanoids or docosanoids, respectively (Farooqui and Horrocks, 2001a). The release of eicosanoids and docosanoids may constitute the first wave of second-messenger generation during the initial phase of signal transduction (Turini and Holub, 1994). The generation of lysoplasmalogens in neural membranes may induce changes in membrane fluidity and permeability and allow the influx of external Ca^{2+} via plasma membrane channels. This influx of Ca^{2+} results in the translocation and stimulation of Ca^{2+}-dependent $cPLA_2$, a PLA_2 that is specific for the release of arachidonic acid from the *sn*-2 position of PtdCho. The degradation of this glycerophospholipid produces the subsequent wave of second messenger generation in the late phase of signal transduction. Collectively, these studies suggest that plasmalogens are consumed first during second messenger generation in the initial phase of the signal transduction process. In contrast, PtdCho is utilized at later stages.

Treatment of neuron-enriched cultures with kainic acid results in a dose-dependent increase in PlsEtn-PLA_2 activity (Fig. 3.4). A kainate/AMPA antagonist, CNQX, 6-cyano-7-nitroquinoxaline-2, 3-dione, can block the increase in PlsEtn-PLA_2 activity (Fig. 3.5), suggesting that the stimulation of PlsEtn-PLA_2 is a receptor-mediated process and plasmalogen degradation is mediated by KA receptors (Farooqui et al., 2003). Bromoenol lactone, a specific PLA_2 inhibitor, also blocks the stimulation of the 39 kDa PLA_2 activity (Fig. 3.6). This suggests that BEL can be used to prevent KA-induced neurodegeneration. Collectively, our studies suggest that receptor-mediated degradation of plasmalogens involves Ca^{2+}-independent PlsEtn-PLA_2.

The stimulation of kainate receptors on the cell surface results in the generation of fatty acid (arachidonic or docosahexaenoic acid) and lysoplasmalogens. Under

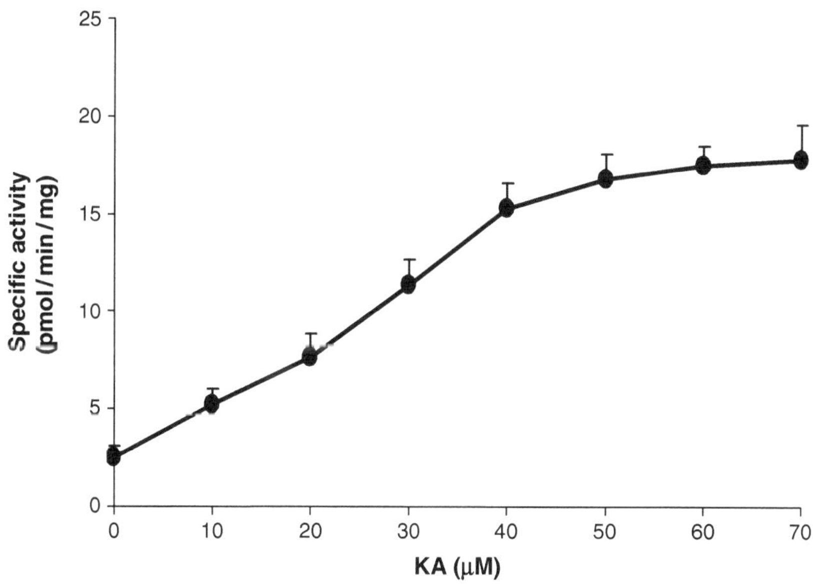

Fig. 3.4 Effect of kainic acid on PlsEtn-selective PLA_2 activity of neuron-enriched culture from rat cerebral cortex. Enzymic activity is determined by the method of Hirashima et al. (1992). Values are means ± S.E.M, $n = 5$.

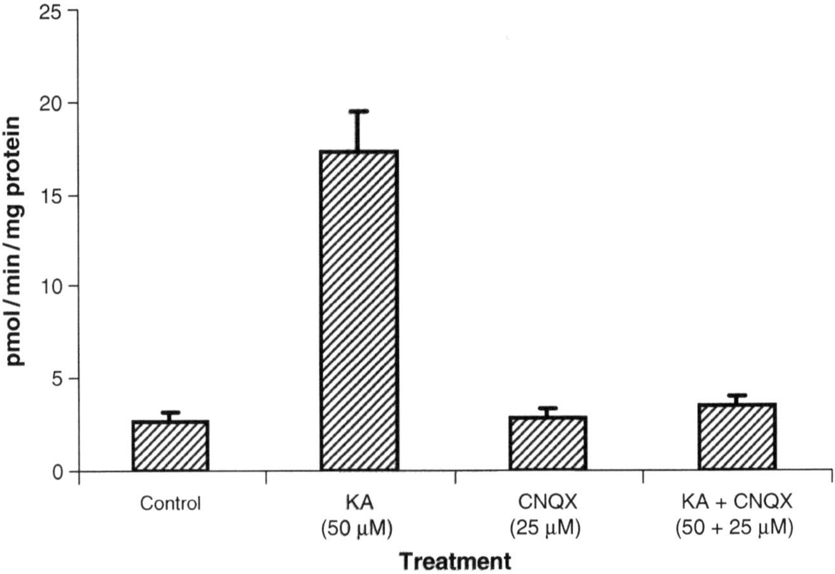

Fig. 3.5 Effect of CNQX on kainic acid-induced stimulation of PlsEtn-PLA_2 activity of neuron-enriched culture from rat cerebral cortex. Enzymic activity is determined by the method of Hirashima et al. (1992). Values are means ± S.E.M., $n = 5$.

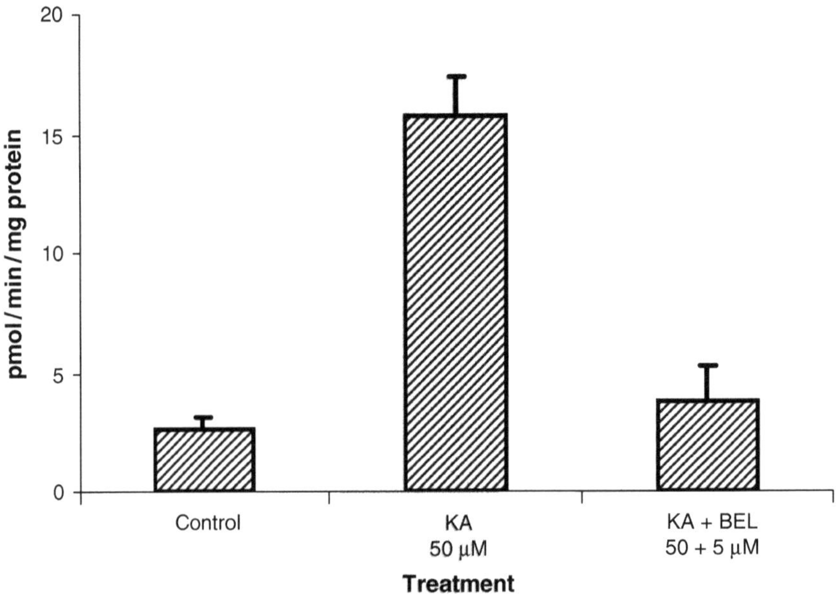

Fig. 3.6 Effect of bromoenol lactone on kainic acid-induced stimulation of PlsEtn-PLA_2 activity of neuron-enriched culture from rat cerebral cortex. Enzymic activity is determined by the method of Hirashima et al. (1992). BEL (5 µM) has no effect on basal PlsEtn-PLA_2 activity in control neuron-enriched cultures. Values are means ± S.E.M, $n = 5$.

normal conditions, lysoplasmalogens are then either reacylated or hydrolyzed by a lysoplasmalogenase and arachidonic acid is metabolized to eicosanoids (Fig. 3.7). This process may constitute the first wave of second messenger generation (the immediate phase of signal transduction) during receptor stimulation (Turini and Holub, 1994). The generation of lysoplasmalogens can induce changes in membrane permeability and fluidity and allow the influx of external Ca^{2+} via plasma membrane channels. Changes in the Ca^{2+} level result in the translocation of Ca^{2+}-dependent enzymes, including 85 kDa cytosolic PLA_2. This process induces the subsequent wave of second messenger generation (the late phase of signal transduction). Thus, plasmalogens provide second messengers that may be involved in earlier stages of signal transduction. In contrast, phosphatidylcholine is utilized at later stages of the signal transduction process.

Alterations in muscarinic cholinergic signal transduction in plasmalogen-deficient CHO cell lines are associated with a marked reduction in amyloid precursor protein (Périchon et al., 1998), suggesting that plasmalogen levels affect muscarinic cholinergic signals and amyloid precursor protein processing. This observation is of great interest because a decrease in plasmalogen levels is known to occur not

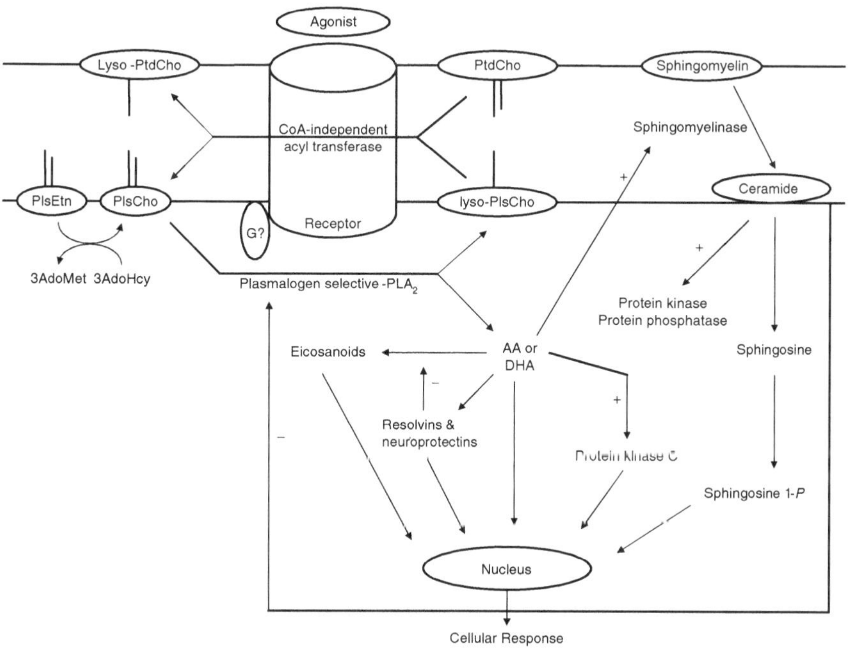

Fig. 3.7 The hypothetical receptor-mediated degradation of plasmalogens by PlsEtn-phospholipase A_2, cyclooxygenase, and lipoxygenase. Plasmenylethanolamine (PlsEtn); plasmenylcholine (PlsCho); lyso-plasmenylcholine (lyso-PlsCho); lyso-phosphatidylcholine (lyso-PtdCho); *S*-adenosyl-L-methionine (AdoMet); *S*-adenosyl-2-homocysteine (AdoHcy); plasmalogen-selective phospholipase A_2 (PlsEtn-PLA_2); and glycerophosphocholine (Gro*P*Cho); Three moles of AdoMet are consumed during the methylation reaction.

only in Alzheimer disease, but also in a variety of other neurological disorders in which muscarinic cholinergic signal transduction, amyloid precursor protein processing, and membrane integrity are abnormal (Ginsberg et al., 1998; Farooqui et al., 1997; Périchon et al., 1998; Guan et al., 1999). This supports the hypothesis that receptor-mediated increases in the hydrolysis of plasmalogens produce abnormalities in membrane integrity resulting in increased calcium influx and activation of calcium-dependent proteases, phospholipases, phosphatases, endonucleases, and nitric oxide synthases. These may cause serious brain damage (Farooqui and Horrocks, 1991).

Lysoplasmalogens activate purified cyclic AMP-dependent protein kinase (Williams and Ford, 1997) independently of cAMP, indicating the involvement of plasmalogen-selective PLA_2 in nuclear signaling. Our recent studies support this. LA-N-1 cell nuclei contain plasmalogen-selective PLA_2 activity that is stimulated by retinoic acid in a dose- and time-dependent manner (Antony et al., 2001). A pan retinoic acid receptor antagonist, BMS943 (Farooqui et al., 2004a), blocks the stimulation of plasmalogen-selective PLA_2. Thus, the stimulation of plasmalogen-selective PLA_2 is a receptor-mediated process. The activation of a nuclear transcription factor, cAMP response element-binding protein (CREB), also depends upon the activity of plasmalogen-selective PLA_2. In addition, c-fos expression is also increased in response to perfusion with 500 nM lysoplasmenylcholine (Williams and Ford, 2001). All these studies indicate that the metabolism of plasmalogens and lysoplasmalogens may be involved in signal transduction processes in the nucleus.

3.4 Regulation of PlsEtn-PLA_2

As stated earlier that bovine brain PlsEtn-PLA_2 is inhibited by glycosaminoglycans and sialoglycoconjugates. The interactions among PlsEtn-PLA_2, glycosaminoglycans, and sialoglycoconjugates may be involved in anchoring this enzyme to neural membranes where its plasmalogens are localized. Role of glycosphingolipids as regulators of signal transduction processes has been recently appreciated through the identification of sphingolipid-derived lipid mediators such as ceramide, ceramide 1-phosphate, sphingosine, and sphingosine 1-phosphate (Pettus et al., 2004a; Pyne, 2004). Sphingomyelinase, an enzyme that generates ceramide, decreases the levels of plasmalogens in rat brain slices (Latorre et al., 1999). This effect can be mimicked by C_2-ceramide (Latorre et al., 2003). The decrease in plasmalogens by sphingomyelinase or C_2-ceramide is prevented by quinacrine, ganglioside, and bromoenol lactone, which are inhibitors of plasmalogen-selective PLA_2 activity (Farooqui and Horrocks, 2001b). It is interesting to note that addition of the caspase-3 inhibitor, acetyl-L-aspartyl-L-glutamyl-L-valyl-L-aspartyl-cholomethylketone (Ac-DEVD-CMK), partially blocks the ceramide-induced stimulation of plasmalogen-selective PLA_2 without altering sphingomyelinase-elicited ceramide accumulation (Latorre et al., 1999). This suggests the involvement at the nuclear level of plasmalogen hydrolysis in signal transduction related to apoptotic cell death (Farooqui et al., 2004a). Arachidonic acid is known to stimulate sphingomyelinase and ceramide activates plasmalogen-selective PLA_2 activity (Farooqui et al., 2000b). Thus, a close interaction between a plasmalogen-selective generated second messenger, arachidonic acid, and the sphingomyelinase generated second messenger, ceramide (Farooqui et al., 2004b; Farooqui et al., 2007) occurs during receptor-mediated signal transduction processes. Ceramide and ceramide 1-phosphate stimulate protein kinase C, while sphingosine and sphingosine 1-phosphate inhibit protein kinase C, Ca^{2+}/calmodulin-dependent kinase, Na^+/K^+-ATPase, CTP/phosphocholine cytidylyltransferase, and PLC (Farooqui et al., 2007). Both ceramide 1-phosphate and sphingosine 1-phosphate promote the induction of COX-2 (Pettus et al., 2004b). Ceramide 1-phosphate also induces translocation of PLA_2 to the nuclear membrane. Thus, sphingolipid-derived metabolites may act in concert to regulate the production of eicosanoids from arachidonic acid, a fatty acid that modulates sphingomyelinase (Farooqui et al., 2004b; Farooqui et al., 2007). Collectively, these studies suggest that interactions between plasmalogen and sphingolipid-derived lipid mediators regulate cellular functions such as proliferation, migration, cytoskeletal organization, inflammation, and differentiation (Pyne, 2004; Pettus et al., 2004b; Farooqui et al., 2007). The regulation of PlsEtn-PLA_2 may be a complex process. In brain tissue, it may vary from one cell type to another. Regulation of PlsEtn-PLA_2 may have superimposed mechanisms. Thus, it is important not only to understand the molecular mechanism of receptor-mediated plasmalogen degradation, but also interactions of plasmalogen-derived lipid mediators with other neural membrane constitutents such as glycosaminoglycans and sialoglycoconjugates.

3.5 Turnover of Plasmalogen in Brain

The turnover of arachidonic acid in ethanolamine and choline plasmalogens in brain is much greater than indicated by most labeling experiments because plasmalogens are not pulse-labeled. The fractional turnover rate for arachidonic acid in choline plasmalogens is about 2% per hour. The corresponding rate for docosahexaenoic acid is more than 7%. The synthesis rates in adult rat brain microsomes give half lives of 15 min for PlsCho and 2.9 h for PlsEtn (Rintala et al., 1999). The specific radioactivity of the arachidonic acid in choline plasmalogen molecular species at 24 h is greater than that in any corresponding molecular species of PtdCho or PtdIns, the glycerophospholipids labeled initially with arachidonic acid (Horrocks, 1989). Although its fatty acids turn over very rapidly, the PlsCho and PlsEtn are not pulse-labeled in the glycerol or alkenyl moieties, so those portions of the molecules turn over slower.

Ether glycerophospholipids in rat brain are synthesized and turned over rapidly (Rosenberger et al., 2002). Because the ether side-chains originate from fatty alcohols, hexadecanol labeled with tritium at the *sn*-1-position has used. Part of the hexadecanol is oxidized to palmitic acid, but the tritium is lost during this process.

Theoretically, in ether glycerophospholipids, all side chains are labeled with tritium. Plasmalogen synthesis steps include desaturation of half of the hydrogen atoms at the *sn*-1 position of the alkyl group. The turnover rates and half-lives can then be calculated from transfer coefficients and the rates of synthesis. PakH is the precursor for PakCho and PakEtn. PakEtn is the precursor for PlsEtn, which is the precursor for PlsCho (Rosenberger et al., 2002).

The incorporation of radioactivity into myelin glycerophospholipids is very slow and always less than 10% of the total at times up to 2 h when it is 3% (Rosenberger et al., 2002). The incorporation is predominantly into gray matter as shown with radioautographic studies. Much more radioactivity is localized in the synaptosomal fraction than in the microsomal fraction. The synaptosomal fraction contains plasma membranes and nerve-endings with their mitochondria and synaptic vesicles. The microsomal fraction contains membranes from plasma membranes and endoplasmic reticulum. These results (Rosenberger et al., 2002) suggest the dynamic role of ether glycerophospholipids in plasticity and signaling processes in the brain.

With tritium-labeled glycerol, young rats (18-days old) synthesize PlsEtn at the rate of 32 nmol g^{-1} min^{-1} (Masuzawa et al., 1984). This rate is in excellent agreement with the rate found for older rats with hexadecanol. Turnover rates measured with the loss of radioactivity (Miller et al., 1977; Miller and Morell, 1978) are much longer than rates estimated from synthesis (Horrocks, 1969). The longer rates are due to recycling of radioactivity and to sequestration into relatively stable pools of glycerophospholipids.

3.6 Remodeling of Plasmalogens (Reacylation/Deacylation Reactions)

Action of PlsEtn-PLA_2 on plasmenylcholine or plasmenylethanolamine and degradation of plasmanylcholine or plasmanylethanolamine by $cPLA_2$ generates lysoplasmenycholine or lysoplamenylethanolamine and lysoplasmanylcholine or lysoplasmanyethanolamine, respectively. These lysophospholipids are reacylated by acyltransferasess and transacylases (Farooqui et al., 2000a). 1-Alkenyl-GPC/acyl-CoA acyltransferase activity has been reported to occur in brain and heart tissues. This enzyme differs from the corresponding 1-alkyl-GPC/acyl-CoA acyltransferase in kinetic properties, molecular weight, and substrate specificity (Arthur and Choy, 1986; Arthur et al., 1987). During transacylation reaction, an acyl group is transferred from donor phospholipids to the acceptor phospholipids. This type of transacylation reaction does not involve PLA_2 for the release of fatty acid. Both CoA-dependent and CoA-independent transacylation reactions have been in mammalian tissues, and there is strong evidence that the acyl groups transfer usually occurs at the *sn*-2 position of glycerol moiety. Thus, remodeling of arachidonate and docosahexaenoate in plasmalogens involves CoA-independent transacylase. This results in the accumulation of arachidonate and docosahexaenoate into plasmalogens (Fonteh and Chilton, 1992).

3.7 Degradation of Plasmalogens by Phospholipase C

A neutral active phospholipase C (PLC) activity has been partially purified by anion exchange, hydroxylapatite, chromatofocusing, and gel filtration chromatographies from heart myocardium (Wolf and Gross, 1985). The partially purified enzyme hydrolyzes PlsCho as well PtdCho. Myocardial PLC shows optimal activity between pH optimum 7 and 8, requires divalent cations for maximal activity, and does not hydrolyze PtdIns or sphingomyelin. The K_m and V_{max} values for PlsCho are 20 μM and 237 μmol h^{-1} mg^{-1} protein. Myocardial cytosol contains a potent inhibitor that masks PLC activity until it is removed during the purification procedure. It is proposed that plasmenylcholine biosynthesis in myocardium is initiated by PLC-mediated hydrolysis of PlsEtn. This results in the generation of 1-O-alk-1′-enyl-2-acyl-*sn*-glycerol. This metabolite is utilized by choline phosphotransferase for the generation of PlsCho (Ford and Gross, 1988). Thus, myocardial plasmenylcholine biosynthesis occurs by polar head group remodeling that utilizes endogenous 1-O-alk-1′-enyl-2-acyl-*sn*-glycerol as a synthetic intermediate.

3.8 Nonenzymic Oxidation of Plasmalogens in Brain

Plasmalogens generate two classes of aldehydes in the presence of reactive oxygen species (ROS) (Stadelmann-Ingrand et al., 2001). This suggests the existence of two pathways for the nonenzymic oxidation of plasmalogens. One involving singlet oxygen-mediated oxidation of plasmalogens results in generation of aldehyde with carbon length of 15:0, 17:0, and 17:1 and the other is ·OH-mediated oxidation essentially leading to the formation of α-hydroxyaldehyde (16:0-OH and 18:0-OH) (Felde and Spiteller, 1995).

During lipid peroxidation, the vinylether bond is attacked by lipid peroxides (lipidperoxy radicals). This process results in the formation of epoxides. These epoxides undergo rapid hydrolytic cleavage to form long chain α-hydroxyaldehydes (Felde and Spiteller, 1995). These aldehydes have higher chemical reactivity than other lipid peroxidation-mediated products such as malondialdehyde and 4-hydroxynonenal. Under physiological conditions, they react with amines to produce unstable imines and can be trapped by hydrogenation with $NaBH_4$ (Mlakar and Spiteller, 1996). On the basis of the chemical analysis, it is speculated that plasmalogen oxidation is an intramolecular reaction promoted by polyunsaturated fatty acid at the *sn*-2 position of glycerol moiety. This process results in the generation of hydroperoxyl radicals, and these radicals are able to cause epoxidation of enolether double bond in plasmalogens (Fig. 3.8) (Weisser and Spiteller, 1996). Plasmalogen-derived aldehydes may have several effects in the brain tissue. They may react with neural membrane ethanolamine glycerophosphoplipids and then form adducts that alter neurotransmitter release in the brain tissue (Stadelmann-Ingrand et al., 2004; Farooqui and Horrocks, 2007). In neutrophils, plasmalogen-derived α-hydroxyaldehydes cause oxidative burst that may result in more damage

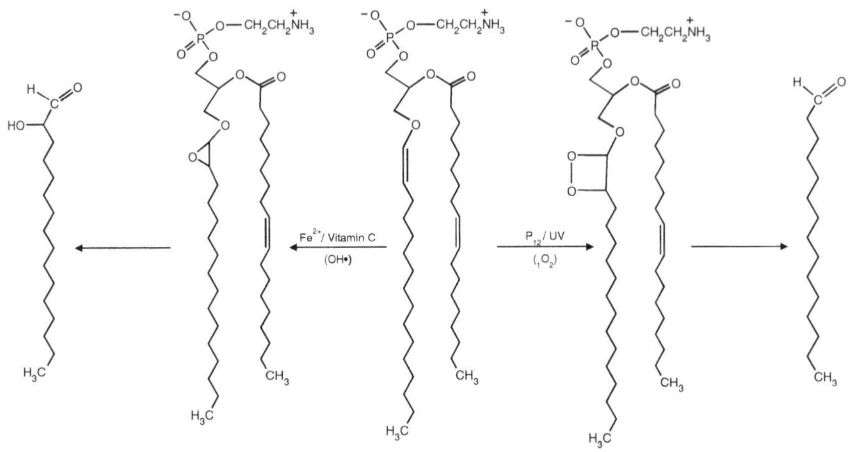

Fig. 3.8 Generation of long chain fatty aldehyde through non-enzymic oxidation of plasmalogens. Generation of long-chain fatty aldehydes from plasmalogens. Singlet oxygen-mediated oxidation of plasmalogen results in generation of an aldehyde via dioxetane intermediate (Stadelmann-Ingrand et al., 2001) and hydroxylradical attack on plasmalogen generates alpha-hydroxyaldehyde via formation of epoxide (Weisser et al., 1997).

to surrounding tissue. Collective evidence suggests that high levels of plasmalogen-derived aldehydes during aging and disease processes may promote neural cell injury (Farooqui and Horrocks, 2007).

Oxygen radical also attacks vinyl ether bond in plasmalogens. It abstracts a hydrogen atom from the vinyl ether bond and through the formation of allylic hydroperoxides, epoxides, and hemiacetal intermediates, plasmalogen molecule undergoes decomposition process generating aldehydes and lysoplasmalogens (Morand et al., 1988). Choline plasmalogen-oxidized products involving the *sn*-1 position alone include 1-formyl-2-arachidonyl or docosahexaenoyl lipid derivatives, whereas oxidation products involving the *sn*-2 position alone include chain-shortened omega-aldehyde radyl substituents at *sn*-2. Collectively, these studies suggest that oxidation products of esterified arachidonic or docosahexaenoic acid groups in plasmalogens are different from the specific free radical oxidation at the 1′-alkenyl position (Khaselev and Murphy, 2000b). It is stated that the mechanism of formation of these oxidized products may involve cooperation between the *sn*-1 vinyl ether substituent and the arachidonoyl or docosahexaenoyl substituent at *sn*-2 of the glycerophospholipid. As arachidonic or docosahexaenoic acid is found in high amounts in most plasmalogen glycerophospholipids, the susceptibility of plasmalogens to free radical oxidation likely involves concomitant oxidation of the arachidonyl or docosahexaenoyl radyl groups esterified at the *sn*-2 position (Khaselev and Murphy, 2000a).

Studies on age- and strain-related levels of aldehydes and α-hydroxyaldehydes in the brain homogenates of SAM P8 and R1mice at weaning, 5 months and 9 months have indicated the presence of 15:0 and 17:0 aldehydes. These results are similar to those obtained during in vitro oxidation through UV irradiation or Fe^{2+}/ascorbate induced-oxidation (Stadelmann-Ingrand et al., 2001). Both strains have high levels of aldehydes and α-hydroxyaldehydes, and these levels remain constant between adult and aged mice (André et al., 2006). This is in contrast with studies on aged human and bovine brains where levels of plasmalogen oxidative products are markedly increased (Weisser et al., 1997; Weisser and Spiteller, 1996). Collectively, these studies suggest that levels cerebral plasmalogen aldehydes are not related with aging process in the senescence-accelerated mice P8 and R1 (André et al., 2006).

Plasmalogens decrease copper-induced oxidation of polyunsaturated fatty acids in low-density lipoproteins (Hahnel et al., 1999; Zommara et al., 1995). This is due to the formation of copper-plasmalogen complex. Although the exact mechanism associated with inhibitory process remains unknown, but it is proposed that the enol ether double bond in plasmalogen complexes with Cu^{2+} in 1:1 stoichiometry (Hahnel et al., 1999). In this way, plasmalogens not only provide binding sites for copper within the lipid phase of the low-density lipoproteins, but may play a major role in protection of lipoproteins and cellular membranes from oxidative stress-mediated injury (Hahnel et al., 1999).

Myeloperoxidase is a heme enzyme that catalyzes the generation of reactive chlorinating species (RCS) such as HOCl, OCl^-, and Cl_2 from hydrogen peroxide in leukocytes of the innate immune system (Kim et al., 2004).

$$H_2O_2 + Cl^- \rightarrow H_2O + HOCl$$

This enzyme is present in atherosclerotic lesions and provides a source for the generation of proinflammatory chlorinated reactants contributing to the endothelial dysfunction. The modification of high-density lipoprotein (HDL) by hypochlorous acid/hypochlorite (HOCl/OCl)-generated in vivo by the myeloperoxidase-hydrogen peroxide-chloride system of activated phagocytes forms a proatherogenic lipoprotein particle that binds to and is internalized by endothelial cells (Marsche et al., 2004). RCS target vinylether bond of plasmalogens. RCS attack on plasmalogens results in the generation of α-chloro fatty aldehydes (α-ClFALD) and an *sn*-1 lyso-PlsCho (Albert et al., 2002; Thukkani et al., 2003a; Thukkani et al., 2003b) (Fig. 3.9). Plasmalogen-derived α-ClFALD not only decreases the expression nitric oxide synthase and inhibits its activity in endothelial cells, but it also acts as a chemoattractant in neutrophils (Marsche et al., 2004). The levels of α-ClFALD and lyso-PtdCho molecular species are elevated in human atherosclerotic lesions. Collectively, these studies suggest that vinyl ether bond of PlsCho is preferentially targeted by RCS compared with that of aliphatic *sn*-2 alkenes of diacylglycerophospholipids (Wildsmith et al., 2006). This attack results in the production of lyso-PtdCho molecular species (Lessig et al., 2007). Plasmalogens containing an oleic acid residue in *sn*-2 position are converted by moderate amounts of HOCl primarily into 1-lyso-2-oleoyl-*sn*-glycero-3-phosphocholine and at elevated HOCl concentrations into the corresponding chlorohydrin species. In contrast, plasmalogens containing docosahexaenoic acid generate 1-lyso-2-docosahexaenoyl-glycerophosphocholine upon HOCl treatment (Wildsmith et al., 2006; Lessig et al., 2007). On the basis of these results, it can be speculated that like nonneural cells (Wildsmith et al., 2006),

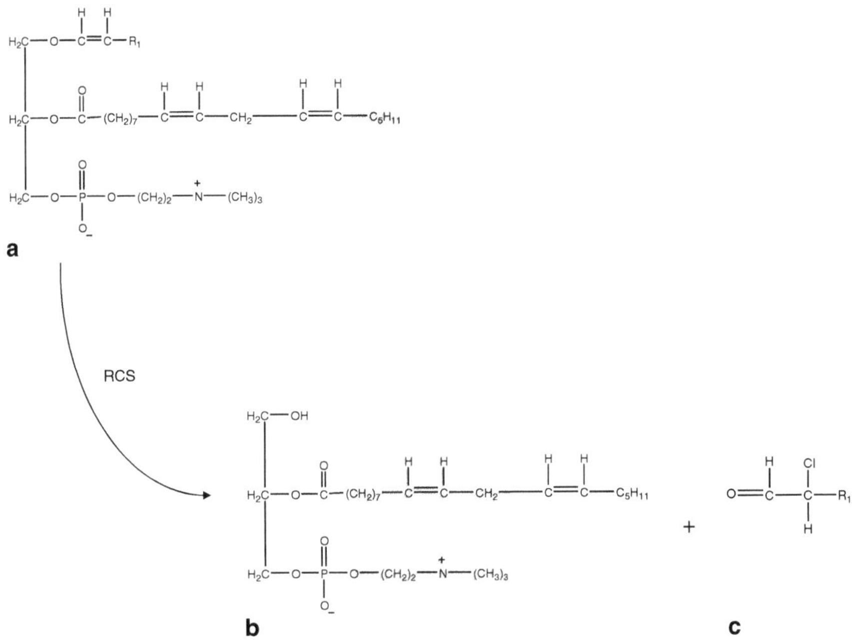

Fig. 3.9 Generation of α-chloro-fatty aldehyde and lyso-PtdCho from choline plasmalogen. Choline plasmalogen containing 18:2 at the *sn*-2 position (**a**); unsaturated lysophosphatidylcholine containing 18:2 at the *sn*-2 position (**b**); α-chlorofatty aldehyde (2-ClHDA and 2-cholooctadecanal, ClODA) (**c**), and reactive chlorinating species (RCS).

the action of HOCl on neural membrane plasmalogens may cause cytotoxicity through the generation of lyso-PtdCho and chlorinated fatty aldehyde. This aldehyde forms Schiff bases with primary amines of ethanolamine glycerophospholipids and lysine inducing alterations in the neural membrane signal transduction processes associated with phospholipase D (PLD) (Wildsmith et al., 2006). Collective evidence suggests that the modification of high-density lipoprotein (HDL) by hypochlorous acid/hypochlorite (HOCl/OCl)-generated in vivo by the myeloperoxidase-hydrogen peroxide-chloride system of activated phagocytes forms a proatherogenic lipoprotein particle that binds to and is internalized by endothelial cells.

3.9 Plasmalogen-Derived Lipid Mediators and Their Importance in Brain

PlsEtn-PLA$_2$ hydrolyzes plasmalogen molecule into arachidonic or docosahexaenic acid and lysoplasmalogen. Very little is known about the interactions between upstream lipid acyltransferases and downstream cyclooxygenase (COX), lipoxygenase (LOX), and epoxygenase (EPOX) (Farooqui et al., 1995b; Farooqui and Horrocks, 2001a). These interactions may modulate the intensity of signal transduction processes at cellular and subcellular levels (O'Banion, 1999; Consilvio et al., 2004; Bosetti et al., 2004). The complexity of this problem becomes obvious when one considers the coupling mechanisms of PlsEtn-PLA$_2$ with different receptors at cellular and subcellular levels and tries to relate them to neuronal and glial cell functions (Farooqui and Horrocks, 2005; Bosetti et al., 2004). In the brain tissue, the released arachidonic acid is oxidized to prostaglandins, leukotrienes, and thromboxanes (Fig. 3.10) (Phillis et al., 2006). Prostaglandins act on EP1, EP2, EP3, and EP4 receptors coupled through G proteins. During oxidative stress, isoprostanes, isoketals, and 4-hydroxynonenal (4-HNE), the nonenzymic oxidative products of arachidonic acid, are also generated in the brain tissue (Farooqui and Horrocks, 2007). Isoprostanes are structurally similar to prostaglandins (Fig. 3.11). They exert their effects through the thromboxane receptors, TPα and TPβ (Farooqui and Horrocks, 2007). Isoprostanes are potent prooxidants, vasoconstrictors, and mitogens (Fam and Morrow, 2003). 4-HNE, a nine-carbon α, β-unsaturated aldehyde, is a signaling molecule at low concentration but at high concentration has ability to react with lysine, cysteine, and histidine residue in proteins. It also undergoes a Michael addition reaction with glutathione and reacts with deoxyguanosine and aminoglycerophospholipids to form complex products (Farooqui and Horrocks, 2007).

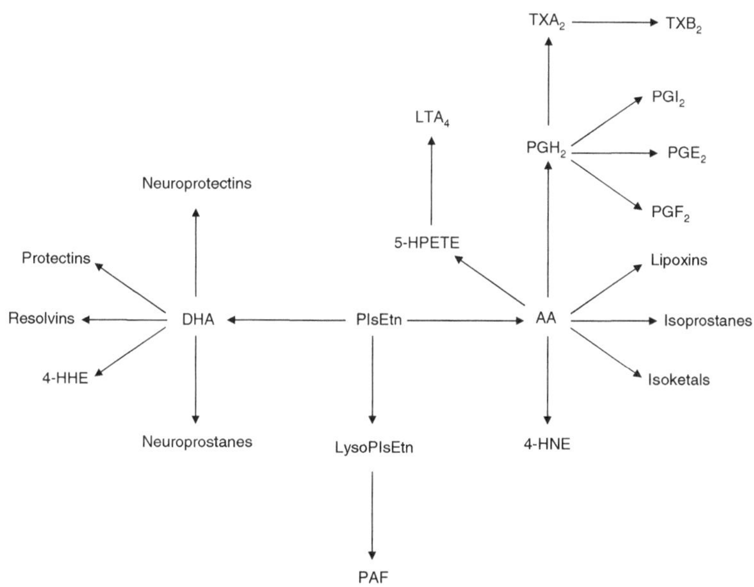

Fig. 3.10 Lipid mediators generated from plasmalogens, arachidonate, and docosahexenoate through the action of PLA$_2$, cyclooxygenase, lipoxygenase, and other down stream enzymes. *PlsEtn* Ethanolamine plasmalogen; *AA* Arachidonate; *DHA* docosahexaenoate; *4-HHE* 4-hydroxyhexenal; *lyso-PlsEtn* lysoplasmalogen; *PAF* platelet-activating factor; *TXA2* thromboxanes; *PGE$_2$* and *PGF$_2$* prostaglandins; *PGI$_2$* prostacyclin; *LTA$_4$* leukotriene; and *4-HNE* 4-hydroxynonenal.

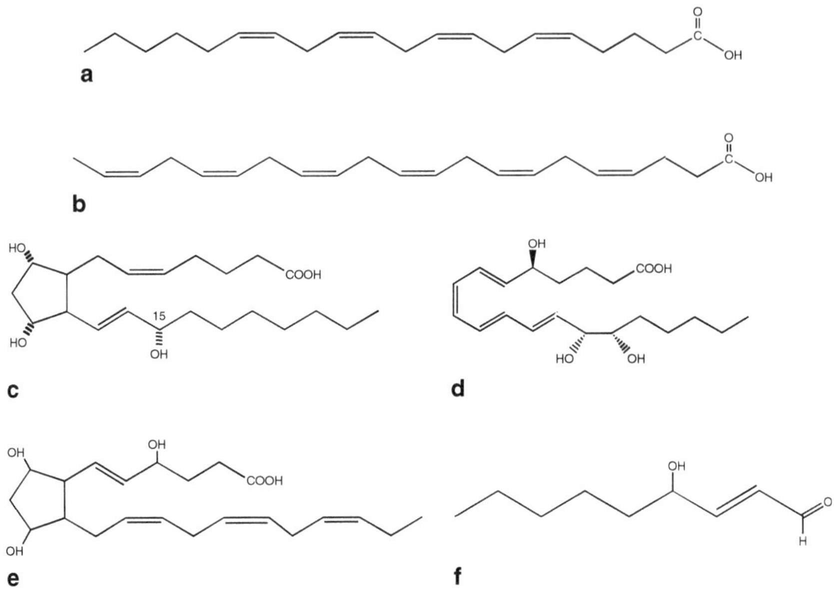

Fig. 3.11 Chemical structures of plasmalogen derived lipid mediators: Arachidonate (**a**), docosahexaenoate (**b**), 15-F_{2c} isoprostane (**c**), lipoxin B4 (**d**), neuroprostane (**e**), 4-HNE (**f**).

The action of lipoxygenases on arachidonic acid-derived HPETE and HETE leads to the formation of lipoxins (LXA_4 and LXB_4), a group of trihydroxytetraene eicosanoids involved in the resolution of acute inflammation by modulating key steps in leukocyte trafficking and preventing neutrophil-mediated acute tissue injury (Serhan, 1994; Serhan and Levy, 2003; Kantarci and Van Dyke, 2003). These lipoxins are potent antiinflammatory and proresolving molecules that act through specific G protein-coupled receptors, ALX and LXA receptors (Norel and Brink, 2004; Chiang et al., 2005). Lipoxin A receptors are expressed on spinal astrocytes both in vivo and in vitro, and lipoxins regulate spinal nociceptive processing through their action upon astrocytic activation (Svensson et al., 2007).

Actions of a 15-lipoxygenase-like enzyme on DHA produce 17S-resolvins, 10-, 17S-docosatrienes, and protectins (Fig. 3.12) (Hong et al., 2003; Marcheselli et al., 2003; Serhan et al., 2004; Serhan and Savill, 2005). These second messengers are collectively called as docosanoids. They are potent endogenous antiinflammatory and proresolving chemical lipid mediators (Serhan, 2006).They antagonize the effects of eicosanoids, modulate leukocyte trafficking, and downregulate the expression of cytokines in glial cells (Hong et al., 2003; Marcheselli et al., 2003; Serhan et al., 2004). The specific receptors for these bioactive lipid metabolites occur in neural and nonneural tissues. These receptors include resolvin D receptors (resoDR1), resolvin E receptors (resoER1), and neuroprotectin D receptors (NPDR). Characterization of these receptors in the brain tissue is in progress (Marcheselli

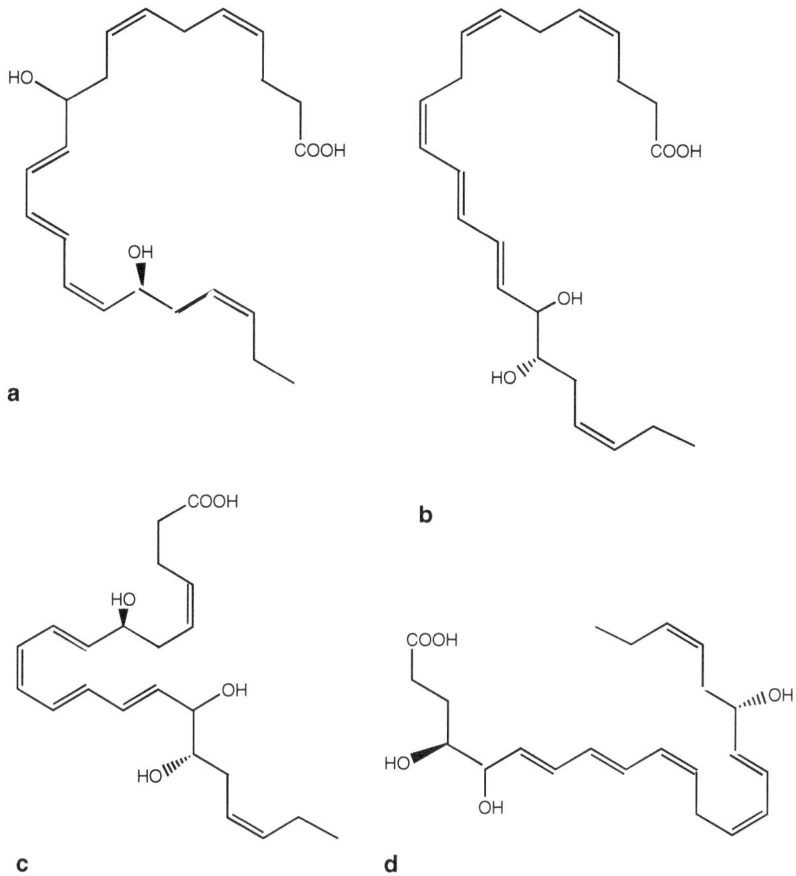

Fig. 3.12 Chemical structures of resolvin and docosatrienes. These lipid mediators are generated through enzymic oxidation of DHA: 10,17 *S*-Docosatriene (**a**), 16,17 *S*-docosatrienes (**b**), 7S,16,17 S-resolvin (**c**), and 4S, 5,17 S-resolvin.

et al., 2003; Serhan et al., 2004; Mukherjee et al., 2004). Nonenzymic oxidation of DHA produces 4-hydroxyhexenal (4-HHE), neuroprostanes, and neuroketals.

PAF (1-O-alkyl-2-acetyl-*sn*-glycerophosphocholine) is a potent proinflammatory agent in infectious and inflammatory diseases (Snyder, 1995). PAF can be synthesized from plasmalogens in neural cells (Ishii et al., 2002; Tokuoka et al., 2003). PAF exerts its biological effects by activating the PAF receptors that consequently activate leukocytes, stimulate platelet aggregation, and induce the release of cytokines and expression of cell adhesion molecules (Snyder, 1995; Maclennan et al., 1996; Ishii et al., 2002; Honda et al., 2002). The binding of PAF to intracellular sites elicits gene expression in neuronal and glial cell lines (Bazan et al., 1994; Tokuoka et al., 2003). PAF also stimulates the inducible isoform of PLA_2 and cyclooxygenase-2 (COX-2). COX-2 is encoded by an immediate early gene and is responsible for prostaglandin synthesis in neuropathological processes. PAF receptors are also involved in the release of PGE_2 from astrocytes. This release of prostaglandin E_2 is closely associated with pathophysiology of inflammatory pain (Watkins et al., 2001). PAF is also an essential component of the intricate mechanisms by which immune cells such as leukocytes are recruited to their targets (Zimmerman et al., 1996). Collective evidence suggests that PAF-mediated neuroinflammation is closely associated with short and long-term responses during cell stimulation and neural trauma (Bazan et al., 1997).

3.10 Lysoplasmalogens in Brain

The hydrolysis of plasmalogens by plasmalogen-selective phospholipase A_2 produces arachidonic or docosahexaenoic acid and ethanolamine or choline lysoplasmalogen (Farooqui and Horrocks, 2001b). In the brain tissue, lysoplasmalogens are either rapidly reacylated to maintain normal levels of plasmalogens in neural membranes or hydrolyzed by a lysoplasmalogenase (Jurkowitz-Alexander et al., 1989). This enzyme converts lysoplasmalogen into fatty aldehyde and *sn*-glycero-3-phosphobase. It is located in the microsomal fraction of rat brain and liver. It has been purified from rat liver microsomes, using multiple column chromatographic procedures (Jurkowitz-Alexander et al., 1989). Lysoplasmalogens can also be converted to an acetylated platelet-activating factor analog either by a CoA-independent transacetylase activity or via acetyl transferase activity (Lee, 1998). Like lyso-PtdCho, lyso-PlsEtn and lyso-PlsCho are amphiphilic molecules. They act as detergents and affect the integrity of neural membranes by interacting with individual proteins or by affecting the biophysical properties of neural membranes. Lyso-PlsEtn and lyso-PlsCho increase membrane fluidity (Han and Gross, 1991) and modulate activities of various enzymes. Thus lyso-PlsCho activates cAMP-dependent protein kinase (PKA) (Williams and Ford, 1997), suggesting that lysoplasmalogen may be involved in the signal transduction processes. The activation of PKA by lysoplasmalogens suggests that these lyso-glycerophospholipids may be crucial in the modulation of PKA activation. This process is similar to the modulation of PKC by diacylglycerol lipase. Other lyso-glycerophospholipids, such as lyso-PlsEtn and lyso-PtdSer, have no effect on cAMP-dependent protein kinase activity (Williams and Ford, 1997).

Studies on the effect of lyspphospholipids on contraction in intact isolated rabbit ventricular myocytes have indicated that lyso-PlsCho mediates spontaneous contractions of intact isolated rabbit ventricular myocytes significantly faster than lyso-PtdCho (Caldwell and Baumgarten, 1998). The median time for the development of lyso-PlsCho-induced spontaneous activity is less than half of that required for lyso-PtdCho to induce depolarization of myocytes. This effect of lyso-PlsCho on myocytes is retarded by lanthanides. Thus Gd^{3+} and La^{3+} are equally effective inhibitors of the lyso-PlsEtn-induced current and equally delay the onset of spontaneous contractions. However, the characteristics of lanthanide block indicate that Gd^{3+}-sensitive, poorly selective, stretch-activated channels do not participate during this process. Instead, results are consistent with the view that lanthanides increase phospholipid ordering and may thereby oppose membrane perturbations induced by lyso-PlsCho (Caldwell and Baumgarten, 1998). Collectively, these studies suggest that by altering the action potential and inducing depolarization, high levels of lyso-PlsCho in the heart tissue produce spontaneous contraction, which may cause arrhythmia (Caldwell and Baumgarten, 1998). Lysoplasmalogens also inhibit the plasma membrane Na^+, K^+-ATPase activity in myocytes and proximal tubule in kidney (McHowat et al., 1998; Schonefeld et al., 1996).

3.11 Conclusion

Plasmalogens are major constituents of neural membranes. Their degradation is catalyzed by Ca^{2+}-independent cytosolic PLA_2. This enzyme is found in brains of many animal species. Plasmalogen-selective PLA_2 has been purified to homogeneity from bovine brain cytosol. The purified enzyme is markedly inhibited by glycosaminoglycans. Plasmalogen-selective-PLA_2 is also inhibited by sialic acid, gangliosides, sialoglycoproteins. Interactions with glycosaminoglycocans, gangliosides, and sialoglycoproteins may be involved in regulation of enzymic activity. In neural membranes, enrichment of arachidonic acid and docosahexaenoic acid at the *sn*-2 position of glycerol moiety occurs through reacylation/deacylation reaction. Under physiological conditions, acylation/deacylation reactions not only maintain levels of plasmalogens in neural membranes, but also stabilize them. Under pathological conditions, metabolism of arachidonic acid and docosahexaenoic acid produces a variety of lipid mediators. Thus, arachidonic acid generates prostaglandins, leukotrienes, thromboxanes, and lipoxins. Prostaglandins, leukotrienes, and thromboxanes are vasoconstrictive and proinflammatory, and their accumulation cause neural cell injury (Phillis et al., 2006). Lipoxins are antiinflammatory and their accumulation may protect cells from neural cell injury. In contrast, the breakdown of docosahexaenoic acid produces docosatrienes, resolvins, protectins, and neuroprotectins (Bazan, 2005; Serhan, 2005). These lipid mediators are collectively called as docosanoids and their generation protects cells from the neural injury.

References

Albert C. J., Crowley J. R., Hsu F. F., Thukkani A. K., and Ford D. A. (2002). Reactive brominating species produced by myeloperoxidase target the vinyl ether bond of plasmalogens – Disparate utilization of sodium halides in the production of alpha-halo fatty aldehydes. J. Biol. Chem. 277:4694–4703.

André A., Chanséaume E., Dumusois C., Cabaret S., Berdeaux O., and Chardigny J. M. (2006). Cerebral plasmalogens and aldehydes in senescence-accelerated mice P8 and R1: A comparison between weaned, adult and aged mice. Brain Res. 1085:28–32.

Antony P., Freysz L., Horrocks L. A., and Farooqui A. A. (2001). Effect of retinoic acid on the Ca^{2+}-independent phospholipase A_2 in nuclei of LA-N-1 neuroblastoma cells. Neurochem. Res. 26:83–88.

Arthur G. and Choy P. C. (1986). Acylation of 1-alkenyl-glycerophosphocholine and 1-acyl-glycerophosphocholine in guinea pig heart. Biochem. J. 236:481–487.

Arthur G., Zaborniak C. L., and Choy P. C. (1987). The presence of 1-radyl-glycerophosphoethanolamine acyltransferase activity in guinea pig heart mitochondria. Biochem. Cell Biol. 65:1016–1021.

Bazan N. G. (2005). Neuroprotectin D1 (NPD1): A DHA-derived mediator that protects brain and retina against cell injury-induced oxidative stress. Brain Pathol. 15:159–166.

Bazan N. G., Fletcher B. S., Herschman H. R., and Mukherjee P. K. (1994). Platelet-activating factor and retinoic acid synergistically activate the inducible prostaglandin synthase gene. Proc. Natl. Acad. Sci. USA 91:5252–5256.

Bazan N. G., Packard M. G., Teather L., and Allan G. (1997). Bioactive lipids in excitatory neurotransmission and neuronal plasticity. Neurochem. Int. 30:225–231.

Bosetti F., Langenbach R., and Weerasinghe G. R. (2004). Prostaglandin E_2 and microsomal prostaglandin E synthase-2 expression are decreased in the cyclooxygenase-2-deficient mouse brain despite compensatory induction of cyclooxygenase-1 and Ca^{2+}-dependent phospholipase A_2. J. Neurochem. 91:1389–1397.

Caldwell R. A. and Baumgarten C. M. (1998). Plasmalogen-derived lysolipid induces a depolarizing cation current in rabbit ventricular myocytes. Circ. Res. 83:533–540.

Chiang N., Arita M., and Serhan C. N. (2005). Anti-inflammatory circuitry: Lipoxin, aspirin-triggered lipoxins and their receptor ALX. Prostaglandins Leukot. Essent. Fatty Acids 73:163–177.

Consilvio C., Vincent A. M., and Feldman E. L. (2004). Neuroinflammation, COX-2, and ALS – a dual role? Exp. Neurol. 187:1–10.

Fam S. S. and Morrow J. D. (2003). The isoprostanes: Unique products of arachidonic acid oxidation – A review. Curr. Medicinal Chem. 10:1723–1740.

Farooqui A. A., Antony P., Ong W. Y., Horrocks L. A., and Freysz L. (2004a). Retinoic acid-mediated phospholipase A_2 signaling in the nucleus. Brain Res. Rev. 45:179–195.

Farooqui A. A. and Horrocks L. A. (1991). Excitatory amino acid receptors, neural membrane phospholipid metabolism and neurological disorders. Brain Res. Rev. 16:171–191.

Farooqui A. A. and Horrocks L. A. (2001a). Plasmalogens, phospholipase A_2, and docosahexaenoic acid turnover in brain tissue. J. Mol. Neurosci. 16:263–272.

Farooqui A. A. and Horrocks L. A. (2001b). Plasmalogens: Workhorse lipids of membranes in normal and injured neurons and glia. Neuroscientist 7:232–245.

Farooqui A. A. and Horrocks L. A. (2005). Signaling and interplay mediated by phospholipases A_2, C, and D in LA-N-1 cell nuclei. Reprod. Nutr. Develop. 45:613–631.

Farooqui A. A. and Horrocks L. A. (2007). *Glycerophospholipids in the Brain: Phospholipases A_2 in Neurological Disorders*. Springer, New York, pp. 1–394.

Farooqui A. A., Horrocks L. A., and Farooqui T. (2000a). Deacylation and reacylation of neural membrane glycerophospholipids. J. Mol. Neurosci. 14:123–135.

Farooqui A. A., Horrocks L. A., and Farooqui T. (2000b). Glycerophospholipids in brain: Their metabolism, incorporation into membranes, functions, and involvement in neurological disorders. Chem. Phys. Lipids 106:1–29.

Farooqui A. A., Horrocks L. A., and Farooqui T. (2007). Interactions between neural membrane glycerophospholipid and sphingolipid mediators: A recipe for neural cell survival or suicide. J. Neurosci. Res. 85:1834–1850.

Farooqui A. A., Ong W. Y., and Horrocks L. A. (2003). Plasmalogens, docosahexaenoic acid, and neurological disorders. In: Roels F., Baes M., and de Bies S. (eds.), *Peroxisomal Disorders and Regulation of Genes*. Kluwer Academic, London, pp. 335–354.

Farooqui A. A., Ong W. Y., and Horrocks L. A. (2004b). Biochemical aspects of neurodegeneration in human brain: involvement of neural membrane phospholipids and phospholipases A_2. Neurochem. Res. 29:1961–1977.

Farooqui A. A., Ong W. Y., and Horrocks L. A. (2006). Inhibitors of brain phospholipase A_2 activity: Their neuropharmacological effects and therapeutic importance for the treatment of neurologic disorders. Pharmacol. Rev. 58:591–620.

Farooqui A. A., Ong W. Y., Horrocks L. A., and Farooqui T. (2000c). Brain cytosolic phospholipase A_2: Localization, role, and involvement in neurological diseases. Neuroscientist 6:169–180.

Farooqui A. A., Rapoport S. I., and Horrocks L. A. (1997). Membrane phospholipid alterations in Alzheimer disease: Deficiency of ethanolamine plasmalogens. Neurochem. Res. 22:523–527.

Farooqui A. A., Wells K., and Horrocks L. A. (1995a). Breakdown of membrane phospholipids in Alzheimer disease: Involvement of excitatory amino acid receptors. Molec. Chem. Neuropathol. 25:155–173.

Farooqui A. A., Yang H.-C., and Horrocks L. A. (1995b). Plasmalogens, phospholipases A_2, and signal transduction. Brain Res. Rev. 21:152–161.

Felde R. and Spiteller G. (1995). Plasmalogen oxidation in human serum lipoproteins. Chem. Phys. Lipids 76:259–267.

Fonteh A. N. and Chilton F. H. (1992). Rapid remodeling of arachidonate from phosphatidylcholine to phosphatidylethanolamine pools during mast cell activation. J. Immunol. 148:1784–1791.

Ford D. A. and Gross R. W. (1988). Identification of endogenous 1-O-alk-1′-enyl-2-acyl-*sn*-glycerol in myocardium and its effective utilization by choline phosphotransferase. J. Biol. Chem. 263:2644–2650.

Ginsberg L., Xuereb J. H., and Gershfeld N. L. (1998). Membrane instability, plasmalogen content, and Alzheimer's disease. J. Neurochem. 70:2533–2538.

Guan Z. Z., Wang Y. A., Cairns N. J., Lantos P. L., Dallner G., and Sindelar P. J. (1999). Decrease and structural modifications of phosphatidylethanolamine plasmalogen in the brain with Alzheimer disease. J. Neuropathol. Exp. Neurol. 58:740–747.

Hahnel D., Huber T., Kurze V., Beyer K., and Engelmann B. (1999). Contribution of copper binding to the inhibition of lipid oxidation by plasmalogen phospholipids. Biochem. J. 340:377–383.

Han X. and Gross R. W. (1991). Proton nuclear magnetic resonance studies on the molecular dynamics of plasmenylcholine/cholesterol and phosphatidylcholine/cholestrol bilayers. Biochim. Biophys. Acta Biomembr. 1063:129–136.

Hazen S. L., Ford D. A., and Gross R. W. (1991). Activation of a membrane-associated phospholipase A_2 during rabbit myocardial ischemia which is highly selective for plasmalogen substrate. J. Biol. Chem. 266:5629–5633.

Hazen S. L. and Gross R. W. (1993). The specific association of a phosphofructokinase isoform with myocardial calcium-independent phospholipase A_2. Implications for the coordinated regulation of phospholipolysis and glycolysis. J. Biol. Chem. 268:9892–9900.

Hirashima Y., Farooqui A. A., Mills J. S., and Horrocks L. A. (1992). Identification and purification of calcium-independent phospholipase A_2 from bovine brain cytosol. J. Neurochem. 59:708–714.

Honda Z., Ishii S., and Shimizu T. (2002). Platelet-activating factor receptor. J. Biochem. 131:773–779.

Hong S., Gronert K., Devchand P. R., Moussignac R. L., and Serhan C. N. (2003). Novel docosatrienes and 17S-resolvins generated from docosahexaenoic acid in murine brain, human blood, and glial cells – Autacoids in anti-inflammation. J. Biol. Chem. 278:14677–14687.

Horrocks L. A. (1969). Metabolism of the ethanolamine phosphoglycerides of mouse brain myelin and microsomes. J. Neurochem. 16:13–18.

Horrocks L. A. (1989). Sources of brain arachidonic acid uptake and turnover in glycerophospholipids. Ann. N Y Acad. Sci. 559:17–24.

Ishii S., Nagase T., and Shimizu T. (2002). Platelet-activating factor receptor. Prostaglandins Other Lipid Mediat. 68–69:599–609.

Jurkowitz-Alexander M., Ebata H., Mills J. S., Murphy E. J., and Horrocks L. A. (1989). Solubilization, purification, and characterization of lysoplasmalogen alkenylhydrolase (lysoplasmalogenase) from rat liver microsomes. Biochim. Biophys. Acta 1002:203–212.

Kantarci A. and Van Dyke T. E. (2003). Lipoxins in chronic inflammation. Crit. Rev. Oral Biol. Med. 14:4–12.

Khaselev N. and Murphy R. C. (2000a). Peroxidation of arachidonate containing plasmenyl glycerophosphocholine: facile oxidation of esterified arachidonate at carbon-5. Free Radic. Biol. Med. 29:620–632.

Khaselev N. and Murphy R. C. (2000b). Structural characterization of oxidized phospholipid products derived from arachidonate-containing plasmenyl glycerophosphocholine. J. Lipid Res. 41:564–572.

Kim S. Y., Min D. S., Choi J. S., Choi Y. S., Park H. J., Sung K. W., Kim J., and Lee M. Y. (2004). Differential expression of phospholipase D isozymes in the hippocampus following kainic acid-induced seizures. J. Neuropathol. Exp. Neurol. 63:812–820.

Latorre E., Aragonés M. D., Fernández I., and Catalán R. E. (1999). Platelet-activating factor modulates brain sphingomyelin metabolism. Eur. J. Biochem. 262:308–314.

Latorre E., Collado M. P., Fernández I., Aragonés M. D., and Catalán R. E. (2003). Signaling events mediating activation of brain ethanolamine plasmalogen hydrolysis by ceramide. Eur. J. Biochem. 270:36–46.

Layden B. T., Abukhdeir A. M., Malarkey C., Oriti L. A., Salah W., Stigler C., Geraldes C. F. G. C., and de Freitas D. M. (2005). Identification of Li^+ binding sites and the effect of Li^+ treatment on phospholipid composition in human neuroblastoma cells: A Li-7 and P-31 NMR study. Biochim. Biophys. Acta Mol. Basis Dis. 1741:339–349.

Lee T. C. (1998). Biosynthesis and possible biological functions of plasmalogens. Biochim. Biophys. Acta Lipids Lipid Metab. 1394:129–145.

Lessig J., Schiller J., Arnhold J., and Fuchs B. (2007). Hypochlorous acid-mediated generation of glycerophosphocholine from unsaturated plasmalogen glycerophosphocholine lipids. J. Lipid Res. 48:1316–1324.

Lohner K. (1996). Is the high propensity of ethanolamine plasmalogens to form non-lamellar lipid structures manifested in the properties of biomembranes?. Chem. Phys. Lipids 81:167–184.

Maclennan K. M., Smith P. F., and Darlington C. L. (1996). Platelet-activating factor in the CNS. Prog. Neurobiol. 50:585–596.

Marcheselli V. L., Hong S., Lukiw W. J., Tian X. H., Gronert K., Musto A., Hardy M., Gimenez J. M., Chiang N., Serhan C. N., and Bazan N. G. (2003). Novel docosanoids inhibit brain ischemia-reperfusion-mediated leukocyte infiltration and pro-inflammatory gene expression. J. Biol. Chem. 278:43807–43817.

Marsche G., Heller R., Fauler G., Kovacevic A., Nuszkowski A., Graier W., Sattler W., and Malle E. (2004). 2-Chlorohexadecanal derived from hypochlorite-modified high-density lipoprotein-associated plasmalogen is a natural inhibitor of endothelial nitric oxide biosynthesis. Arterioscler. Thromb. Vasc. Biol. 24:2302–2306.

Masuzawa Y., Sugiura T., Ishima Y., and Waku K. (1984). Turnover rates of the molecular species of ethanolamine plasmalogen of rat brain. J. Neurochem. 42:961–968.

McHowat J., Liu S., and Creer M. H. (1998). Selective hydrolysis of plasmalogen phospholipids by Ca^{2+}-independent PLA_2 in hypoxic ventricular myocytes. Am. J. Physiol. Cell Physiol. 274: C1727–C1737.

Miller S. L., Benjamin J. A., and Morell P. (1977). Metabolism of glycerophospholipids of myelin and microsomes in rat brain. Reutilization of precursors. J. Biol. Chem. 252:4025–4037.

Miller S. L. and Morell P. (1978). Turnover of phosphatidylcholine in microsomes and myelin in brains of young and adult rats. J. Neurochem. 31:771–777.

Mlakar A. and Spiteller G. (1996). Distinction between enzymic and nonenzymic lipid peroxidation. J. Chromatogr. A 743:293–300.

Morand O. H., Zoeller R. A., and Raetz C. R. H. (1988). Disappearance of plasmalogens from membranes of animal cells subjected to photosensitized oxidation. J. Biol. Chem. 263:11597–11606.

Mukherjee P. K., Marcheselli V. L., Serhan C. N., and Bazan N. G. (2004). Neuroprotectin D1: A docosahexaenoic acid-derived docosatriene protects human retinal pigment epithelial cells from oxidative stress. Proc. Natl. Acad. Sci. USA 101:8491–8496.

Murphy R. C. (2001). Free-radical-induced oxidation of arachidonoyl plasmalogen phospholipids: Antioxidant mechanism and precursor pathway for bioactive eicosanoids. Chem. Res. Toxicol. 14:463–472.

Norel X. and Brink C. (2004). The quest for new cysteinyl-leukotriene and lipoxin receptors: Recent clues. Pharmacol. Ther. 103:81–94.

O'Banion M. K. (1999). Cyclooxygenase-2: Molecular biology, pharmacology, and neurobiology. Crit. Rev. Neurobiol. 13:45–82.

Périchon R., Moser A. B., Wallace W. C., Cunningham S. C., Roth G. S., and Moser H. W. (1998). Peroxisomal disease cell lines with cellular plasmalogen deficiency have impaired muscarinic cholinergic signal transduction activity and amyloid precursor protein secretion. Biochem. Biophys. Res. Commun. 248:57–61.

Pettus B. J., Bielawska A., Subramanian P., Wijesinghe D. S., Maceyka M., Leslie C. C., Evans J. H., Freiberg J., Roddy P., Hannun Y. A., and Chalfant C. E. (2004a). Ceramide 1-phosphate is a direct activator of cytosolic phospholipase A_2. J. Biol. Chem. 279:11320–11326.

Pettus B. J., Chalfant C. E., and Hannun Y. A. (2004b). Sphingolipids in inflammation: Roles and implications. Curr. Mol. Med. 4:405–418.

Phillis J. W., Horrocks L. A., and Farooqui A. A. (2006). Cyclooxygenases, lipoxygenases, and epoxygenases in CNS: Their role and involvement in neurological disorders. Brain Res. Rev. 52:201–243.

Portilla D., Crew M. D., Grant D., Serrero G., Bates L. M., Dai G. H., Sasner M., Cheng J., and Buonanno A. (1998). cDNA cloning and expression of a novel family of enzymes with calcium-independent phospholipase A2 and lysophospholipase activities. J. Am. Soc. Nephrol. 9:1178–1186.

Portilla D. and Dai G. (1996). Purification of a novel calcium-independent phospholipase A_2 from rabbit kidney. J. Biol. Chem. 271:15451–15457.

Pyne S. (2004). Lysolipids: Sphingosine 1-phosphate and lysophosphatidic acid. In: Nicolaou A. and Kokotos G. (eds.), *Bioactive Lipids*. The Oily Press, Bridgwater, England, pp. 85–106.

Rintala J., Seemann R., Chandrasekaran K., Rosenberger T. A., Chang L., Contreras M. A., Rapoport S. I., and Chang M. C. J. (1999). 85 kDa cytosolic phospholipase A_2 is a target for chronic lithium in rat brain. NeuroReport 10:3887–3890.

Rosenberger T. A., Oki J., Purdon A. D., Rapoport S. I., and Murphy E. J. (2002). Rapid synthesis and turnover of brain microsomal ether phospholipids in the adult rat. J. Lipid Res. 43:59–68.

Schonefeld M., Noble S., Bertorello A. M., Mandel L. J., Creer M. H., and Portilla D. (1996). Hypoxia-induced amphiphiles inhibit renal Na^+,K^+-ATPase. Kidney Int. 49:1289–1296.

Sergeeva M., Strokin M., and Reiser G. (2005). Regulation of intracellular calcium levels by polyunsaturated fatty acids, arachidonic acid and docosahexaenoic acid, in astrocytes: Possible involvement of phospholipase A_2. Reprod. Nutr. Develop. 45:633–646.

Serhan C. N. (1994). Lipoxin biosynthesis and its impact in inflammatory and vascular events. Biochim. Biophys. Acta 1212:1–25.

Serhan C. N. (2005). Novel ω-3-derived local mediators in anti-inflammation and resolution. Pharmacol. Ther. 105:7–21.

Serhan C. N. (2006). Novel chemical mediators in the resolution of inflammation: Resolvins and protectins. Anesthesiol. Clinics N. Am. 24:341–364.

Serhan C. N., Arita M., Hong S., and Gotlinger K. (2004). Resolvins, docosatrienes, and neuroprotectins, novel omega-3-derived mediators, and their endogenous aspirin-triggered epimers. Lipids 39:1125–1132.

Serhan C. N. and Levy B. (2003). Novel pathways and endogenous mediators in anti-inflammation and resolution. Chem. Immunol. Allergy 83:115–145.

Serhan C. N. and Savill J. (2005). Resolution of inflammation: The beginning programs the end. Nature Immunol. 6:1191–1197.

Snyder F. (1995). Platelet-activating factor: the biosynthetic and catabolic enzymes. Biochem. J. 305:689–705.

Stadelmann-Ingrand S., Favrelière S., Fauconneau B., Mauco G., and Tallineau C. (2001). Plasmalogen degradation by oxidative stress: Production and disappearance of specific fatty aldehydes and fatty α-hydroxyaldehydes. Free Radical Biol. Med. 31:1263–1271.

Stadelmann-Ingrand S., Pontcharraud R., and Fauconneau B. (2004). Evidence for the reactivity of fatty aldehydes released from oxidized plasmalogens with phosphatidylethanolamine to form Schiff base adducts in rat brain homogenates. Chem. Phys. Lipids 131:93–105.

Strokin M., Chechneva O., Reymann K. G., and Reiser G. (2006). Neuroprotection of rat hippocampal slices exposed to oxygen-glucose deprivation by enrichment with docosahexaenoic acid and by inhibition of hydrolysis of docosahexaenoic acid-containing phospholipids by calcium independent phospholipase A_2. Neuroscience 140:547–553.

Strokin M., Sergeeva M., and Reiser G. (2003). Docosahexaenoic acid and arachidonic acid release in rat brain astrocytes is mediated by two separate isoforms of phospholipase A_2 and is differently regulated by cyclic AMP and Ca^{2+}. Brit. J. Pharmacol. 139:1014–1022.

Svensson C. I., Zattoni M., and Serhan C. N. (2007). Lipoxins and aspirin-triggered lipoxin inhibit inflammatory pain processing. J. Exp. Med. 204:245–252.

Thukkani A. K., Albert C. J., Wildsmith K. R., Messner M. C., Martinson B. D., Hsu F. F., and Ford D. A. (2003a). Myeloperoxidase-derived reactive chlorinating species from human monocytes target plasmalogens in low density lipoprotein. J. Biol. Chem. 278:36365–36372.

Thukkani A. K., McHowat J., Hsu F. F., Brennan M. L., Hazen S. L., and Ford D. A. (2003b). Identification of α-chloro fatty aldehydes and unsaturated lysophosphatidylcholine molecular species in human atherosclerotic lesions. Circulation 108:3128–3133.

Tokuoka S. M., Ishii S., Kawamura N., Satoh M., Shimada A., Sasaki S., Hirotsune S., Wynshaw-Boris A., and Shimizu T. (2003). Involvement of platelet-activating factor and LIS1 in neuronal migration. Eur. J. Neurosci. 18:563–570.

Turini M. E. and Holub B. J. (1994). The cleavage of plasmenylethanolamine by phospholipase A_2 appears to be mediated by the low affinity binding site of the TxA_2/PGH_2 receptor in U46619- stimulated human platelets. Biochim. Biophys. Acta Lipids Lipid Metab. 1213:21–26.

Watkins L. R., Milligan E. D., and Maier S. F. (2001). Glial activation: A driving force for pathological pain. Trends Neurosci. 24:450–455.

Weisser M. and Spiteller G. (1996). Increase of aldehydic compounds derived from plasmalogens in the brain of aged cattle. Chem. Phys. Lipids 82:173–178.

Weisser M., Vieth M., Stolte M., Riederer P., Pfeuffer R., Leblhuber F., and Spiteller G. (1997). Dramatic increase of alpha-hydroxyaldehydes derived from plasmalogens in the aged human brain. Chem. Phys. Lipids 90:135–142.

Wildsmith K. R., Albert C. J., Hsu F. F., Kao J. L. F., and Ford D. A. (2006). Myeloperoxidase-derived 2-chlorohexadecanal forms Schiff bases with primary amines of ethanolamine glycerophospholipids and lysine. Chem. Phys. Lipids 139:157–170.

Williams S. D. and Ford D. A. (1997). Activation of myocardial cAMP-dependent protein kinase by lysoplasmenylcholine. FEBS Lett. 420:33–38.

Williams S. D. and Ford D. A. (2001). Calcium-independent phospholipase A_2 mediates CREB phosphorylation and c-fos expression during ischemia. Am. J. Physiol. Heart Circ. Physiol. 281:H168–H176.

Wolf M. J. and Gross R. W. (1996). The calcium-dependent association and functional coupling of calmodulin with myocardial phospholipase A2 – Implications for cardiac cycle-dependent alterations in phospholipolysis. J. Biol. Chem. 271:20989–20992.

Wolf R. A. and Gross R. W. (1985). Identification of neutral active phospholipase C which hydrolyzes choline glycerophospholipids and plasmalogen selective phospholipase A2 in canine myocardium. J. Biol. Chem. 260:7295–7303.

Yang H.-C., Farooqui A. A., and Horrocks L. A. (1994a). Effects of glycosaminoglycans and glycosphingolipids on cytosolic phospholipases A_2 from bovine brain. Biochem. J. 299:91–95.

Yang H.-C., Farooqui A. A., and Horrocks L. A. (1994b). Effects of sialic acid and sialoglycoconjugates on cytosolic phospholipases A_2 from bovine brain. Biochem. Biophys. Res. Commun. 199:1158–1166.

Yang H.-C., Farooqui A. A., Rammohan K. W., Haun S. E., and Horrocks L. A. (1997). Occurrence and characterization of plasmalogen-selective phospholipase A_2 in brain of various animal species. J. Neurochem. 69:S205C.

Zimmerman G. A., Elstad M. R., Lorant D. E., McIntyre T. M., Prescott S. M., Topham M. K., Weyrich A. S., and Whatley R. E. (1996). Platelet-activating factor (PAF): Signalling and adhesion in cell-cell interactions. Adv. Exp. Med. Biol. 416:297–304.

Zommara M., Tachibana N., Mitsui K., Nakatani N., Sakono M., Ikeda I., and Imaizumi K. (1995). Inhibitory effect of ethanolamine plasmalogen on iron- and copper-dependent lipid peroxidation. Free Radical Biol. Med. 18:599–602.

Chapter 4
Assay and Purification of Plasmalogen-Selective Phospholipase A_2 and Lysoplasmalogenase Activities

4.1 Introduction

Plasmalogens represent a special group of glycerophospholipids characterized by a vinyl ether bond at the *sn*-1 position and arachidonic acid or docosahexaenoic acid at the *sn*-2 position, and the *sn*-3 carbon usually has a phospholipid head group like choline or ethanolamine at the glycerol backbone. In all mammalian cells, these glycerophospholipids are especially rich in brain, heart, and red blood cells. While choline plasmalogen is rich in heart, ethanolamine plasmalogen is rich in brain white matter. On the basis of various studies, it is suggested that ethanolamine plasmalogens are abundant in cholesterol-rich biomembranes having long life spans, such as nervous system myelin and red blood cells (Farooqui and Horrocks, 2004). Although the role of plasmalogens is not fully understood, collective evidence suggests that besides being structural component and reservoir for arachidonic and docosahexaenoic acids in neural membranes, plasmalogens play an important role in signal transduction processes, membrane dynamics, membrane fusion, and protection against oxidative stress (Farooqui and Horrocks, 2001). On the basis of two-dimensional NMR studies, it is proposed that choline and ethanolamine plasmalogens have a different glycerol backbone conformation with respect to the membrane interface than diacylglycerophospholipids (Han and Gross, 1990). This unique conformation motif is selectively recognized by enzymes responsible for receptor-mediated breakdown of plasmalogen (Farooqui et al., 2003). The stimulation of kainate type of glutamate receptors on neuronal cell surface results in the stimulation of the Ca^{2+}-independent plasmalogen-selective PLA_2 (PlsCho-PLA_2 and PlsEtn-PLA_2) and generation of arachidonic or docosahexaenoic acids and lysoplasmalogen (Farooqui et al., 2003). Arachidonic and docosahexaenoic acids are metabolized to eicosanoids and docosanoids, respectively. Lysoplasmalogen is either reacylated to plasmalogen or hydrolyzed by lysoplasmalogenase (Farooqui et al., 2003; Farooqui and Horrocks, 2007). Plasmalogen-selective-PLA_2 has been purified and characterized from various sources including heart, brain, and kidney (Hazen and Gross, 1993; Hirashima et al., 1992; Portilla and Dai, 1996). The activity of this enzyme can be determined by radiochemical and fluorometric procedures (Farooqui and Horrocks, 1988).

4.2 Determination of PlsEtn and PlsCho-PLA_2 by Radiochemical Procedures

Radiochemical procedures are discontinuous, laborious, and time consuming. These procedures require separation of the radioactive substrate from the radioactive products. Briefly, labeled plasmalogens and their hydrolyzed products, lysoplasmalogens and labeled fatty acids, are extracted (Bligh and Dyer, 1959) using chloroform, methanol, and acetic acid followed by TLC. Spots are visualized, identified using standards, scraped into scintillation fluid, and counted. Radiolabeled substrates and fatty acids can also be separated by the Dole extraction procedure (Dole, 1956) using 2-propanol, heptane, and sulfuric acid. This separation procedure is not as efficient as the TLC procedure. Although the majority of glycerophospholipids and lyso-glycerophospholipids generated by PLA_2 remain in the aqueous phase, small portion is always extracted into the heptane phase and interfere with the measurement of radioactivity in fatty acids.

4.2.1 Preparation of Radiolabled [^{3}H] Plasmenylcholine (Choline Plasmalogen)

Although the preparation of the radiolabeled plasmalogen is laborious and gives low yield, radiochemical procedures are the most sensitive methods for the determination of activities of PlsEtn-PLA_2 activity in the brain tissue, cell culture preparations, and subcellular fractions. The sensitivity of radiochemical procedures depends on the specific radioactivity of the labeled glycerophospholipids. As stated above, radiochemical procedures are discontinuous, laborious, and time consuming, and require the separation of radioactive substrate from the radioactive products.

Lysoplasmalogen (Fig. 4.1) is prepared by the method of Gross and his collaborators (Han et al., 1992) with slight modification. Bovine heart phosphatidylcholine (2.5 g) is hydrolyzed with 100 ml of 0.05 M KOH in 1/1 methanol/benzene (v/v). The solution is stirred for 40 min at 40°C under a stream of nitrogen gas. The reaction is allowed to proceed until 98% of PtdCho has been hydrolyzed. The reaction is stopped by adding 25 ml of ethyl formate, and the reaction mixture is evaporated to almost dryness by rotary evaporation (Han et al., 1992). The residue is suspended in chloroform and dried again. The residue is extracted with 100 ml of chloroform, 100 ml of methanol, and 90 ml water and transferred to a separatory funnel (Bligh and Dyer, 1959). After shaking, the choloroform layer is collected. The aqueous phase is reextracted with 100 ml of chloroform. The combined chloroform extracts are evaporated to dryness, resuspended in 5 ml chloroform, and applied to tandem Dynamax Macro-HPLC silica columns (each 21.4 mm × 25 cm). The column is eluted with a linear gradient from 100% chloroform/0% methanol to 0% chloroform/100% methanol over 2 l at 15 ml/min. The choline lysoplasmalogens and other

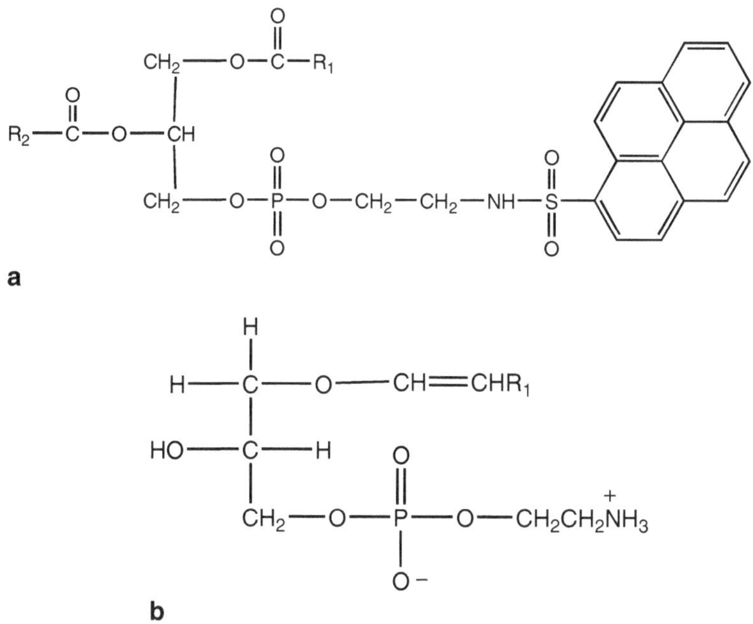

Fig. 4.1 Structures of pyrene-labeled ethanolamine glycerophospholipid (**a**) and ethanolamine lysoplasmalogen (**b**).

ether-linked lysophospholipids are eluted at 60–70% of methanol. The lysoplasmenylcholine is then separated from other ether-linked choline lyso-phospholipids by reverse phase HPLC utilizing an Econosil octadecyl silica column (10 mm × 25 cm) with isocratic elution using methanol/water/acetonitrile (57/23/20, v/v). Fractions containing lysoplasmenylcholine are concentrated and stored at −80°C under nitrogen. The recovery of lysoplasmalogen from bovine heart is low. Bovine (2.5 g) gives 150 mg choline lysoplasmalogen.

Lysoplasmenylethanolamine can also be prepared from phosphatidylethanolamine (PtdEtn) preparations that contain a mixture of the 1,2 diacyl- and 1-*O*-alkenyl-2-acylglycerophospholipids. PtdEtns are subjected to mild alkaline methanolysis for 20 min at room temperature. The addition of chloroform and water with vigorous mixing, but without acidification, results in a preferential retention of the lysoplasmenylethanolamine in the alkaline aqueous phase and complete separation of the methyl esters into the chloroform phase (Hanahan et al., 1990). Neutralization of the alkaline phase with dilute acetic acid, followed by addition of chloroform, allows recovery of the lysoplasmenylethanolamine in the chloroform phase in very high yields (75–80%) (Hanahan et al., 1990). In contrast, a preparation of choline glycerophospholipids rich in plasmenylcholine, treated in exactly the same manner, yields lysoplasmenylcholine molecular species that is not retained in the alkaline phase, but partitioned primarily into the chloroform-rich phase together with the methyl esters. Lysoplasmalogen can be characterized by thin-layer chromatography and compositional analysis. In addition, fast atom bombardment mass spectral analysis of the intact lysoplasmenylethanolamine together with gas chromatography–mass spectrometry of the dimethyl acetals derived from the 1-*O*-alkenyl chains allows further proof of the structure and an assessment of the purity of this compound (Hanahan et al., 1990).

4.2.2 Labeling of Lysoplasmenylcholine at the Sn-2 Position

Lysoplasmenylcholine (71 mmol) is mixed with 70 mmol of free arachidonic acid and 50 mCi of [^{3}H]arachidonic acid in a 2 ml conical reaction vial and evaporated to dryness under a stream of nitrogen. Lipid residue is placed in a vacuum desiccator in the presence of recrystallized *N,N′*-dimethyl-4-aminopyridine (DMAP) and phosphorus pentaoxide overnight. Lipid residue is dissolved in 1.7 ml chloroform and 80 mmol of dried recrystallized DMAP is added to the lipid mixture. The reaction is initiated by adding 20 mmol *N,N′*-dicyclohexylcarbodiimide (DCC) dissolved in freshly distilled chloroform. Multiple increments (10 additions) of 0.1–0.15 mol equivalents of DCC are added every 4 h with continuous stirring in dark. The reaction is terminated by the addition of methanol, and the lipid phase is dried under nitrogen. The residue is extracted with 200 ml chloroform/methanol (1/1, v/v). After shaking for 1 min, 90 ml water is added and mixture is shaken and kept at room temperature. The upper chloroform layer is collected and dried under nitrogen. The radiolabeled choline plasmalogen is purified from the reaction mixture by HPLC employing a linear gradient of hexane/2-propanol/water from 48.5/48.5/3 to 46/46/8 (v/v) at flow rate of 10 ml/min and 4 h total duration. Purified radiolabeled plasmalogen is stored under nitrogen at −80°C.

Alternatively, radioactive plasmalogen can be prepared by a biochemical procedure. In this procedure, rat liver microsomes are used to transfer labeled arachidonic or oleic acid in the presence of ATP and Coenzyme A. Assay mixture consists of 2 mmol of lysoplasmalogen, 20 mCi of 1–[^{14}C] arachidonic or oleic acid (56 mCi/mmol), 1.6 of unlabeled fatty acid in 0.1 M sodium phosphate buffer, pH 7.4 containing 600 mmol of $MgCl_2$, 6 mmol of CoA, 600 mmol of ATP, and 60 mg rat liver microsomal protein in a total volume of 4 ml. After incubating for 2 h at 37°C, the reaction is stopped by extracting the lipid with chloroform/methanol (2/1, v/v), and labeled plasmalogen is isolated by silicic acid column chromatography.

4.2.3 Determination of PlsCho-PLA$_2$ Activity

Assay mixture consist of 50 µl of 200 mM Tris-HCl buffer pH 7.0 containing 4 mM EGTA, 5% glycerol, 2 mM [^{3}H]plasmenylcholine, and 50 µl enzyme solution in a total volume of 200 µl. Reaction is incubated for 30 min at 37°C and stopped by the addition of 100 µl of butanol, vortexed, and centrifuged at 1,000 × *g* for 10 min to separate the phases. The released [^{3}H]arachidonic acid is separated by Silica gel G plates developed in petroleum ether/diethylether/acetic acid (70/30/1, v/v). Spots are visualized and identified using standards. The [^{3}H]arachidonic acid spots is scraped, and quantified by liquid scintillation counting. In radiochemical assay, interference problems can be encountered in studies requiring the use of PLA$_2$ inhibitors. Many PLA$_2$ inhibitors are amphiphilic and therefore can interfere with the extraction procedure and give false values.

4.2.4 Determination of PlsEtn-PLA$_2$ by Fluorometric Assay

The fluorometric assay for PlsEtn-PLA$_2$ is reasonably sensitive and inexpensive. Ethanolamine glycerophospholipids are labeled with pyrene sulfonylchloride (Fig. 4.2) (see Sect. 4.2.6). Yield of fluorescent-labeled plasmalogen is better than radiolabeled substrate. Despite the major drawback of the bulky pyrene group at the ethanolamine head group of pyrene-labeled plasmalogen, this assay procedure still provides several advantages for the intial investigation of PlsEtn-PLA$_2$ (Farooqui and Horrocks, 2007). Pyrene is the most frequently used fluorophore for PLA$_2$ assays. Its most characteristic features are long excised state lifetime and concentration-dependent formation of excimers (Somerharju, 2002). Pyrene is hydrophobic and its attachment does not significantly distort the conformation of the labeled glycerophospholipid molecule. Its monomer emits at 382 and 400 nm, but as an excited dimer its emission shifts from 400 to 480 nm. There is very little interference by detergents when fluorescent substrate is used and fluorescence of GroPEtn(Pyr) can be measured in the presence of detergents.

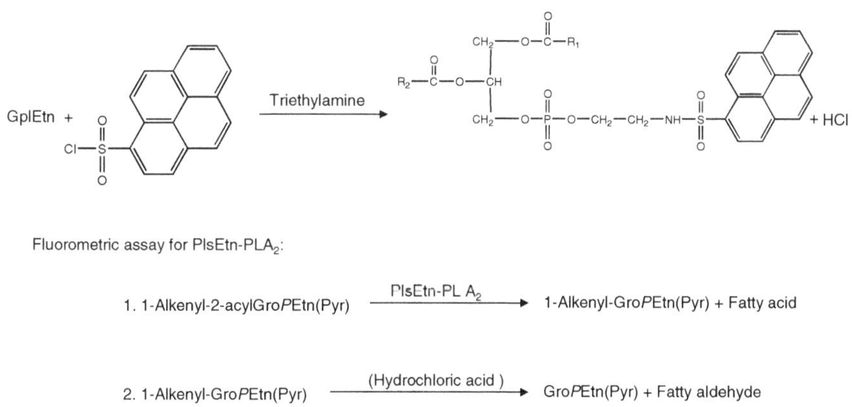

Fig. 4.2 Labeling of ethanolamine glycerophospholipids with pyrenesulfonyl chloride and principle behind fluorometric assay for plasmalogen-selective PLA$_2$. The scheme shows the reactions involved in the fluorometric assay of the action of plasmalogen-selective-PLA$_2$ on pyrene labeled 1-alkenyl-2-acyl-*sn*-Gro*P*Etn and subsequent generation of fatty aldehyde by hydrochloric acid vapors.

4.2.5 Purification of Ethanolamine Plasmalogen

Bovine brain ethanolamine glycerophospholipids (100 mg) are dried under a stream of nitrogen (Hirashima et al., 1990a). The residue is vortexed vigorously for 10 min, and the probe sonicated on ice for 3 min in 4.0 ml of 0.5 M NaCl in 40 mM Tris-HCl, pH 7.6; 4.0 ml of sodium deoxycholate (15 mg/ml) in Tris-HCl, pH 7. 6; 6.0 ml of bovine serum albumin (15 mg/ml) in 40 mM Tris-HCl, pH 7.6; and 2.0 ml of 50 mM $CaCl_2$ in 40 mM HCl, pH 7.6. Bring the final volume to 20 ml with 40 mM Tris-HCl, pH 7.6. Reaction is started with the addition of 6,000 U of *Rhizopus delemar* lipase. The reaction tube is filled with nitrogen and gently stirred for 4 h at room temperature. The reaction is stopped with the addition of 4× total volume of chloroform/methanol (2/1, v/v). The mixture is kept at room temperature for 20 min. The lower phase is transferred to a flask and evaporated to dryness and then dissolve in 1 ml of chloroform. This solution is loaded on a 25 g Unisil silicic acid column (2.5 × 30 cm^2), which has been previously equilibrated in 100% chloroform. The column is washed with 300 ml of chloroform. Free fatty acids and other neural lipids are eluted during this step. Ethanolamine plasmalogen is eluted with 500 ml of chloroform/methanol (4/1, v/v). This fraction is dried and then dissolved in 10 ml of chloroform containing a few drops of methanol. It is stored in a glass vial under nitrogen. The purity of ethanolamine plasmalogen is determined by 2D TLC using a Silica G plate (20 × 20 cm^2) under the solvent system of chloroform/methanol/ammonium hydroxide (65/25/4, v/v). Lipids are hydrolyzed with concentrated HCl fumes before developing the second dimension. Ethanolamine plasmalogen is scraped and quantified by phosphorus assay (Hirashima et al., 1990a).

4.2.6 Labeling of Ethanolamine Plasmalogen with Pyrenesulfonyl Chloride

Ethanolamine plasmalogen (3.6 mmol) dried under nitrogen is mixed with 0.75 ml of pyrene sulfonylchloride in chloroform (1.72 mg/ml) in a test tube. About 0.75 ml of a 0.32% solution of triethanolamine in methanol is added to the above mixture (Fig. 4.2). The reaction tube is filled with nitrogen, capped tight, and then wrapped with aluminum foil. This mixture is kept away from light for 4 h at room temperature. Reaction is stopped by drying under nitrogen. The dried mixture is dissolved in 0.3 ml of chloroform and applied on to a Silica G plate (20 × 20 cm^2). Plate is developed in solvent system of chloroform/methanol/ammonium hydroxide (85/22/4, v/v). Area with pyrene-labeled ethanolamine plasmalogen band is visualized under UV light. It is scrapped, and extracted three times with 12 ml methanol. Extracts are combined, filtered, and concentrated under reduced pressure. The concentration and purity of labeled plamenylethanolamine are determined by the assay of phosphorus in fluorescent spots after two-dimension TLC using the solvent system of choloroform/methanol/ammonium hydroxide (65/25/4, v/v) for both first- and second-dimensional development. Pyrenesulfonyl-labeled plasmenylethanolamine is stored in methanol at –20°C in the dark.

4.2.7 Determination of PlsEtn-PLA_2 Activity with Pyrene-Labeled Plasmalogen

The enzymic activity is determined by the measurement of Gro*P*Etn(Pyr) fluorescence. Pyrene-labeled plasmalogen (74 nmol) is dried under a stream of nitrogen and suspended in 100 mM MOPS buffer (pH 7.4) containing 5 mg fatty acid-free bovine serum albumin, 0.1% Triton X-100, and 4 mM EGTA in a final volume of 0.9 ml. The suspension is sonicated for 1 min on ice with 5-s pulses at each cycle. The reaction is started by the addition of 100 µl of plasmalogen-selective PLA_2. The mixture is incubated for 2 h at 37°C with gentle agitation. The reaction is stopped by the addition of 4 ml of chloroform/methanol (92/1, v/v), vortexed followed by the addition of 159 µl of water, and then vortexed again. The reaction mixture is centrifuged at 1,000 × *g* for 10 min. The upper phase is discarded, and the lower phase (1.2 ml) is collected and dried under nitrogen and is redissolved in 30 µl of choloroform. Samples are applied to Silica G plates ($10 \times 10\,cm^2$). One plate is subjected to acid hydrolysis by placing it in a tank containing fumes of concentrated HCl for 5 min, and other plates are processed without acid exposure. Plates are developed under the solvent system of chloroform/methanol/ammonium hydroxide (65/25/4, v/v) (Hirashima et al., 1990b; Hirashima et al., 1992; Farooqui et al., 1999). The spots of Gro*P*Etn(Pyr) are scraped and extracted with 2 ml of methanol. The fluorescence of Gro*P*Etn(Pyr) is measured using excitation and emission wavelengths 340 and 378 nm, respectively. PlsEtn-PLA_2 activity can be calculated from the fluorescence of Gro*P*Etn(Pyr) after subtracting appropriate control values (Hirashima et al., 1990b; Farooqui et al., 1999). This method is simple, convenient, and sensitive. It is absolutely specific for plasmalogen. It can be used for the purification and characterization of PlsEtn-PLA_2 (Hirashima et al., 1992).

Several precautions should be taken into consideration during radiochemical and fluorometric assay procedures. Plasmalogen are labile and unstable. Prolonged exposure of plasmalogens to atmosphere may result in their breakdown through vinyl ether linkage cleavage and radical-trapping reactions (Thompson et al., 2004). Plasmalogens should always be stored under N_2 or Ar at –80°C. All solvent should be pretreated with a powerful basic dehydrating agent to remove moisture and acidity from containers and test tubes (Thompson et al., 2004). The purity of plasmalogen should be regularly checked. If the purity of plasmalogen is less than 95%, then it should be purified again using *Rhizopus delemar* lipase digestion procedure (Hirashima et al., 1990a) as described earlier.

4.2.8 Continuous Spectrophotometric Determination of PlsEtn-PLA_2

Continuous spectrophotometric assay for PlsEtn-PLA_2 has also been described (Hirashima et al., 1989a; Jurkowitz-Alexander et al., 1991). This procedure requires the use of lysoplasmalogenase (Fig. 4.3). Lysoplasmalogenase hydrolyzes lysoplasmalogen into glycerophosphocholine and free aldehyde. Alcohol dehydrogenase quantitatively converts the free aldehyde to an alcohol with the concomitant oxidation of NADH. The disappearance of NADH can be followed spectrophotometrically at 340 nm (Fig. 4.3). The assay mixture consists of 60 mM glycylglycine buffer (pH 7.2), 1 mM dithiothreitol (DTT), 9.2 mM NADH, 11 μg (0.3 units) horse liver alcohol dehydrogenase, 50 μM lyso-PlsCho, and 0.35 mg fatty acid-free bovine serum albumin. The cuvettes are placed in the Beckman DU-65 spectrophotometer and the absorbance at 340 nm is monitored over a 5–10 min period after the addition of 20 ng to 20 μg lysoplasmalogenase. Although this procedure is less sensitive than the radiochemical and fluorometric procedures, it provides rapid, reliable, less expensive, and continuous determination of PlsEtn-PLA_2 using NADH and partially purified lysoplasmalogenase.

4.2.9 Determination of Lysoplasmalogenase

Lysoplasmalogenase, the enzyme that liberates free aldehyde and glycerophosphocholine/glycerophosphoethanolamine (Gro*P*Cho/ Gro*P*Etn) from lyso-choline/ethanolamine plasmalogen, is found in brain and liver microsomes. Spectrophotometric and spectrofluorometeric procedures for lysoplasmalogenase determination have been described.

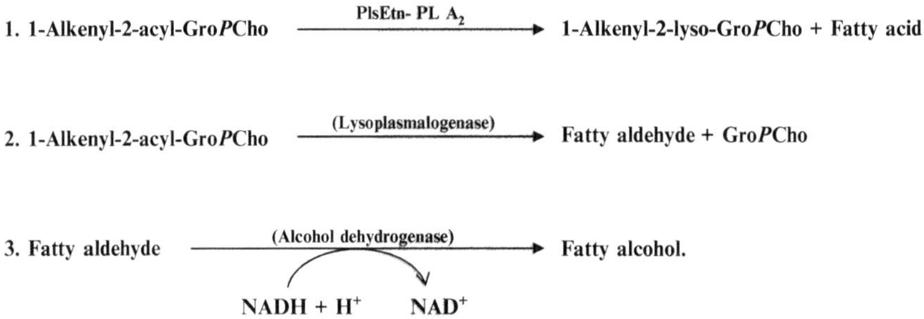

Fig. 4.3 The principle behind continuous spectrophotometric determination of PlsEtn-PLA_2. In this assay lysoplasmalogen generated by plasmalogen-selective PLA_2 is utilized by lysoplasmalogenase, which hydrolyzes it into fatty aldehyde. The fatty aldehyde is estimated through alcohol dehydrogenase reaction.

4.2.9.1 Continuous Spectrophotometric Procedure for Lysoplasmalogenase

The coupling assay for lysoplasmalogenase requires choline lysoplasmalogen as substrate and alcohol dehydrogenase as a coupling enzyme (Fig. 4.4). Rat liver microsomes are used as a source of lysoplasmalogenase. Lysoplasmalogenase hydrolyzes the alkenyl ether bond of lysoplasmalogen at the *sn*-1 position, forming free fatty aldehyde and glycerophosphocholine. The generation of free aldehyde is measured directly by coupling with alcohol dehydrogenase. The rate of fatty aldehyde formation is directly proportional to the loss of NADH that is monitored by the decrease in absorbance at 340 nm. In this assay, the amount of alcohol dehydrogenase is chosen to ensure that it is not rate-limiting. This procedure is simple, rapid, convenient, and reproducible. All reagents are available commercially except the lysoplasmalogenase that is assayed (Jurkowitz-Alexander and Horrocks, 1990).

$$\text{Lysoplasmalogen} + H_2O \rightarrow \text{glycerophosphocholine} + \text{aldehyde}$$

The reaction is carried out at 37°C in a total volume of 0.5 ml using quartz cuvettes. Reaction mixture consist of 80 mM glycylglycine or MOPS buffer (pH 7.0) containing 1 mM DTT, 0.25 mM NADH, 0.4 units of horse liver alcohol dehydrogenase per ml, 60 μM choline lysoplasmalogen or 40 μM ethanolamine lysoplasmalogen substrate, and 0.35 mg fatty acid-free bovine serum albumin per ml. The mixture is incubated for 2 min at 37°C and the reaction is started by the addition of lysoplasmalogenase. The cuvettes are mixed rapidly and the absorbance is monitored at 340 nm in Beckman DU-65 spectrophotometer equipped with a kinetic software package programmed to determine the rate of absorbance change per min and to calculate linear regression curves for each cuvette (Jurkowitz-Alexander et al., 1989, 1991).

Several precautions should be taken during this continuous spectrophotometric procedure. The precautions include minimizing light scattering changes due to the turbidity and presence of contaminating enzymes that may lead to side reactions that interfere with coupling assay, and avoiding the use of Triton X-100 because this detergent strongly inhibits lysoplasmalogenase activity (Jurkowitz-Alexander et al., 1989; Hirashima et al., 1989a, b).

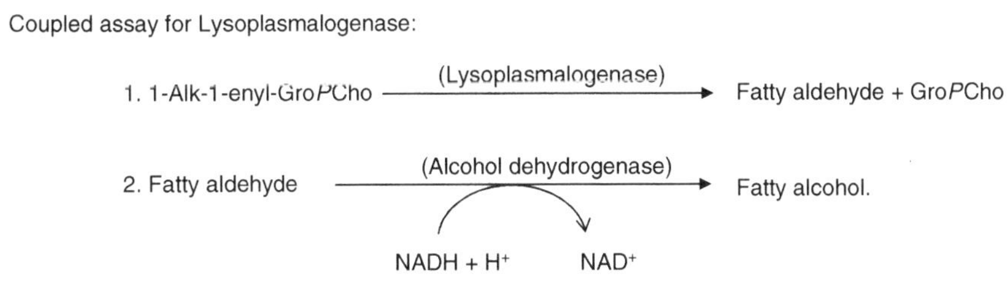

Fig. 4.4 The principle behind spectrofluorometric assay for lysoplasmalogenase. Lysoplasmalogenase generates fatty aldehyde, which is estimated through the utilization of NADH by alcohol dehydrogenase.

Using the above assay procedure, lysoplasmalogenase has been partially purified (200-fold) by multiple column chromatographic procedures (Jurkowitz-Alexander et al., 1989). The purified enzyme shows optimal activity between pH 6.8 and 7.4. It is not affected by divalent metal ions (Ca^{2+}, Mg^{2+}, and Mn^{2+}). Lysoplasmalogenase is inhibited by detergents such as Triton X-100 and sodium deoxycholate, and sulfhydryl group blocking reagents such as *p*-chloromercuribenzoate and iodoacetate. The purified enzyme has no activity toward plasmalogens. For lyso-choline plasmalogen, K_m and V_{max} values were 5.5 μM and 11.7 μmol/min/mg protein, respectively. For ethanolamine lysoplasmalogen, the K_m and V_{max} values were 40 μM and 14 μmol/min/mg, respectively (Jurkowitz-Alexander et al., 1989).

4.2.9.2 Continuous Spectrofluorometric Procedure for Lysoplasmalogenase

A continuous spectrofluorometric assay for the determination of lysoplasmalogenase using horse liver alcohol dehydrogenase as a coupling enzyme has also developed (Hirashima et al., 1989b, c). During this assay, lysoPlsCho is hydrolyzed by

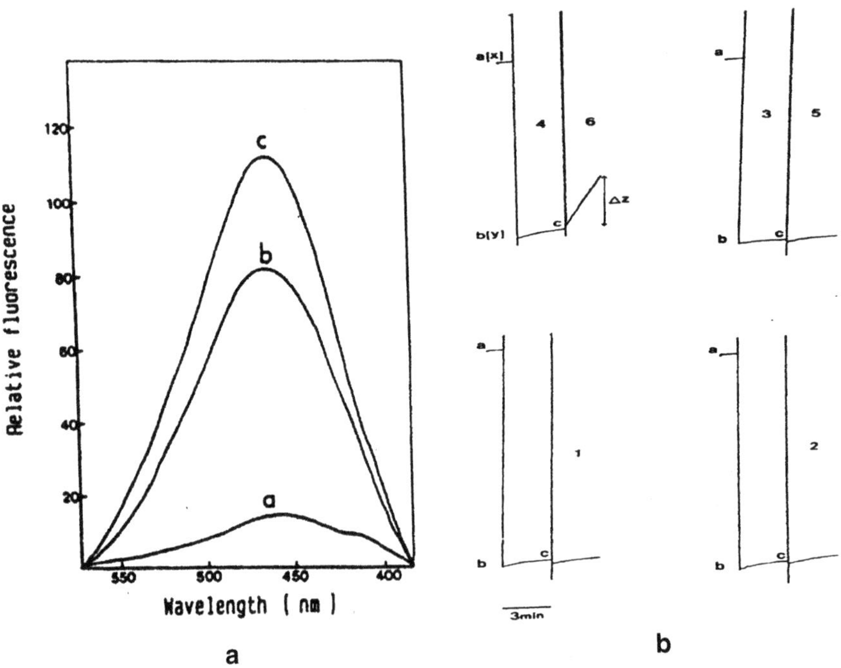

Fig. 4.5 Spectrofluorometric procedure for the determination of lysoplasmalogenase. **A**: (a) Before the addition of NADH, (b) 30 min after the beginning of reaction, (c) immediately after the addition of 10 nmol of NADH. **B**: Representation of the decrease in the fluorescence in the spectrofluorimetric lysoplasmalogenase assay. Traces 1, 2, 3, 4, 5, and 6 correspond to the tube numbers in the text. Reproduced with kind permission from Hirashima et al. (1989c) Biochem. J. 260:605–608, Portland Press, London, UK.

lysoplasmalogenase to Gro*P*Cho and free aldehyde. Alcohol dehydrogenase quantitatively converts the free aldehyde to an alcohol with oxidation of the NADH. NADH is fluorescent, whereas NAD^+ is not, so the disappearance of NADH is measured spectrofluorometrically at 340 nm (excitation) and 460 nm (emission), respectively (Fig. 4.5A).

Assay mixture consists of 50 μM lyso-PlsCho, 1 mM DTT, 10 μM NADH, 0.35 mg fatty acid-free albumin, 38.5 μg alcohol dehydrogenase, an appropriate amount of lysoplasmalogenase, and 0.2 M MOPS buffer (pH 7.4) in a total volume of 1 ml (Hirashima et al., 1989b, c). After 3 min preincubation, the difference of relative fluorescence before and after the addition of NADH to the reaction mixture is monitored. The reaction is initiated by the addition of lysoplasmalogenase and the decrease of fluorescence is recorded in Perkin-Elmer L 55 spectrofluorimeter at 37°C. The excitation and emission monochromators are set at 340 and 460 nm, respectively. The difference of relative fluorescence (y–x) before and after the addition of 10 nmol NADH to the reaction mixture is measured. After incubation, the decrease of fluorescence (Δz) is recorded (Fig. 4.5B). The NADH change is calculated from the following formula.

$$\Delta\text{NADH}\,(\text{nmol}) = \Delta z / y - x \times 10\,(\text{nmol})$$

Control incubations are critical for this assay. The standard experiments include the following incubations (Fig. 4.6).

Tube number	1	2	3	4	5	6
Alcohol dehydrogenase	-	+	-	+	-	+
Substrate	-	-	+	+	+	+
Lysoplasmalogenase	+	+	-	-	+	+

Fig. 4.6 Experimental design for the determination of lysoplasmalogenase activity by fluorimetric procedure.

Fluorescence changes in incubation tubes 6 and 5 are measured after monitoring changes in tube 4 and 3, respectively. The decreases in NADH in incubation tubes 2 and 1 are measured 3 min after the addition of NADH. The rate of lysoplasmalogenase activity can be calculated from the following equation.

$$\text{Rate of lysoplasmalogenase} = \text{Rate}\,(6-5) - \text{Rate}\,(4-3+2-1).$$

Fluorescence changes in control cuvettes are very small and stable when compared to absorbance changes monitored spectrophotometrically at 340 nm. This procedure requires an excess of alcohol dehydrogenase so that the rate of NAD^+ formation becomes proportional to the lysoplasmalogenase activity (Hirashima et al., 1989b). This assay procedure is tenfold more sensitive than a spectrophotometric procedure. The lower limit of detection is 20 pmol/min/ml (Hirashima et al., 1989b).

4.3 Activities of Plasmalogen-Selective PLA_2 in Brains of Various Animal Species and Cultured Cells of Neuronal and Glial Origin

Pig, dog, rat, bovine, and human brain cytosols contain PlsEtn-PLA_2 activity (Table 4.1). This enzyme can be precipitated by 40% ammonium sulfate and separated by Sephadex G-75 column chromatography in the presence of 1 M KCl (Hirashima et al., 1992; Yang et al., 1997). Fresh bovine brain shows the highest and the fresh rat brain displays the lowest specific activity. The specific activity of PlsEtn-PLA_2 in frozen human brain (kept in freezer for 2–3 years) is very low but higher than fresh chicken brain (Table 4.1). Antibodies raised against purified bovine brain PLA_2 cross react with PlsEtn-PLA_2 activity in homogenate. Western blotting indicates the presence of PlsEtn-PLA_2 in all animal species (Yang et al., 1997) and cultured cells of neuronal and glial origin.

4.4 Determination of Lysoplasmalogenase Activity in Rat Liver and Brain Microsomes

Lysoplasmalogenase is present in rat liver and brain microsomes (Table 4.2). It can be solubilized using octyl glucoside. Liver and brain microsomal lysoplasmalogenases have similar properties (Gunawan and Debuch, 1981; Gunawan and Debuch, 1982; Jurkowitz-Alexander et al., 1989). Under physiological conditions, the brain tissue contains low activities of PlsEtn-PLA_2 and lysoplasmalogenase and high levels of plasmalogens, while liver tissue has high activities of plasmalogen-degrading enzymes (Gunawan and Debuch, 1982; Jurkowitz-Alexander et al., 1989) and low

Table 4.1 Specific activities of PlsEtn-PLA_2 in brains of various animal species and cultured cells of neuronal glial origin.

Animal species	Specific activity (pmol/min/mg protein)
Rat[a]	70.0 ± 5.73 (n = 10)
Bovine[a]	85.8 ± 6.72 (n = 10)
Human[b]	47.5 ± 5.73 (n = 7)
Dog[a]	75.7 ± 6.95 (n = 3)
Pig[a]	83.9 ± 7.65 (n = 3)
Chicken[a]	10.9 ± 2.75 (n = 8)
PC12 cell homogenate	34.4 ± 6.83 (n = 5)[c]
Neuro2A homogenate	7.05 ± 0.97 (n = 4)[c]
SKNSH homogenate	5.25 ± 1.69 (n = 5)[c]
U1242MG homogenate	9.68 ± 1.35 (n = 4)[c]

[a]Data adopted from (Yang et al., 1997)

[b]Frozen Human brains were obtained at the autopsy, stored in freezer for 2–3 years and used for PlsEtn-PLA_2 activity determination. For other animal species fresh brains tissue were used

[c]Data adopted from (Hirashima et al., 1990b)

Table 4.2 Specific activity of lysoplasmalogenase in brain and liver microsomes.

Enzyme source	Specific activity (nmol/min/mg protein)	Reference
Rat liver microsomes	40.0	Jurkowitz-Alexander et al., 1989; Hirashima et al., 1989b
Rat brain microsomes	0.05	Gunawan and Debuch, 1982; Hirashima et al., 1989b
Rat intestinal mucosal microsomes	60.0	Jurkowitz et al., 1999
Rat intestinal homogenate	10.0	Jurkowitz et al., 1999

Adopted from Jurkowitz-Alexander et al., 1989; Hirashima et al., 1989b; Gunawan and Debuch, 1982; Jurkowitz et al., 1999

Table 4.3 Kinetic properties of brain PlsEtn-PLA$_2$ and lysoplasmalogenase.

Enzyme	Property	Value	Reference
PlsEtn-PLA$_2$	pH optimum	7.4	Farooqui et al., 1995
	K_m value (μM)	40.0	Farooqui et al., 1995
	V_{max} (pmol/min/mg)	65.0	Farooqui et al., 1995
Lysoplasmalogenase	pH optimum	6.6	Jurkowitz-Alexander et al., 1989
	K_m value (μM)	5.5	Jurkowitz-Alexander et al., 1989
	V_{max} (nmol/min/mg)	11.7	Jurkowitz-Alexander et al., 1989
	pH optimum	6.6	Jurkowitz-Alexander et al., 1989
	K_m value (μM)	42	Jurkowitz-Alexander et al., 1989
	V_{max} (nmol/min/mg)	13.6	Jurkowitz-Alexander et al., 1989

Adopted from (Farooqui et al., 1995; Jurkowitz-Alexander et al., 1989)

plasmalogen levels. This suggests that PlsEtn-PLA$_2$ and lysoplasmalogenase may participate in modulating levels of plasmalogens and lysoplasmalogen in brain and liver. Kinetic properties of these enzymes are shown in Table 4.3. Treatment of microsomal preparation with proteases has no effect on the lysoplasmalogenase activity. However, exogenous phospholipase A$_2$ inactivates this enzyme. Addition of PtdCho with varying chain length of fatty acids results in partial restoration of enzymic activity. Other glycerophospholipids such as PtdEtn and PtdIns do not restore the enzymic activity (Hirashima et al., 1989a). Like Triton X-100, sodium deoxycholate also inhibits lysoplasmalogenase activity. Using spectrofluorometric procedure, lysoplasmalogenase activity is determined in rat brain and liver microsomes, and it is found that activity in brain microsomal fraction is approximately 500–700 times lower than liver microsomes. Brain and liver lysoplasmalogenase activity is strongly inhibited by SH-blocking agents such as *p*-choloromercuribenzoate and iodoacetamide and this inhibition can be prevented by DTT. Lipids like mono- or doradylglycerophospholipids, sphingomyelin, and saturated fatty acid have no effect on lysoplasmalogenase activity (Hirashima et al., 1989a).

Plasmalogens are major glycerophospholipids of neural membranes. They provide neural membranes with suitable fluidity and permeability. They are precursors for proinflammatory and anti-inflammatory lipid mediators that are generated through the action of $cPLA_2$, plasmalogen-selective PLA_2, and lysoplasmalogenase. Assay procedures described above are simple, rapid, sensitive, and continuous. Determination of plasmalogen-selective PLA_2 and lysoplasmalogenase in the normal human brain and brain from patients with neurological disorders may lead to better understanding of the role of plasmalogens and lysoplasmalogens in normal and diseased brain tissues.

4.5 Purification of Plasmalogen-Selective PLA_2 from Brain

Plasmalogen-selective PLA_2 can be purified from bovine brain using multiple column chromatographic procedures (Hirashima et al., 1992). Briefly, bovine brain (100 g) is homogenized in 50 mM Tris-HCl buffer (pH 7.4) containing 5 mM DTT, 100 μM phenylmethyl-sulfonylfluoride (PMSF), and 25 μM EGTA (buffer A) and centrifuged at 100,000 × *g* for 60 min. Cytosolic fraction (300 ml) is brought to 40% saturation with solid ammonium sulfate. After centrifugation at 4°C at 12,500 × *g*, the supernatant is discarded and the precipitated proteins are dissolved in a minimal amount of the above buffer (1–1.5 ml). The solution is dialyzed overnight against 2.5 l of 50 mM Tris-HCl buffer(pH 7.4) containing 5 mM DTT, 100 μM PMSF, and 25 μM EGTA. The dialysate is loaded onto a column of Ultrogel AcA54 (2.5 × 75 cm^2) previously equilibrated with 10% glycerol in 50 mM Tris-HCl buffer (pH 7.4) containing 5 mM DTT, 100 μM PMSF, 1 M KCl, and 25 μM EGTA (buffer B). The column is developed at a flow rate of 24 ml/h. Two peaks having PLA_2 activity are obtained. Peak I hydrolyzes with PtdCho, whereas Peak II is active with PlsEtn. Proteins in Peak II are concentrated by the addition of 90% ammonium sulfate, dissolved in buffer A containing 10% glycerol and dialyzed against 2 l of the same buffer. The dialyzate is loaded to a hydrophobic column of PlsEtn-AffiGel 10 (0.6 × 5 cm^2). After washing with buffer A containing 10% glycerol, the column is eluted with a gradient formed from 30 ml of buffer A containing 10% glycerol plus 30 ml 1% Triton X-100 in the same buffer. To remove Triton X-100, the active fractions are loaded onto a hydroxylapatite column (0.6 × 2 cm^2) and eluted with 1 ml of 10% glycerol, 100 μM PMSF, 2 mM DTT, and 50 mM potassium phosphate buffer, pH 7.4. The eluate is subsequently loaded onto an HRLC MA7Q column previously equilibrated with 50 mM HEPES buffer, pH 7.4. Active plasmalogen-selective PLA_2 is eluted using a nonlinear salt gradient (0–500 mM NaCl) (Hirashima et al., 1992).

4.6 Purification of Lysoplasmalogenase from Liver

Lysoplasmalogenase has been partially purified from rat liver microsomes (Jurkowitz-Alexander et al., 1989). Briefly, rat liver microsomes are solubilized with 55 mM octylglucoside (at the detergent-to-protein ratio of 1.8 by weight). The solubilized microsomal protein (150 mg) is loaded onto a DEAE-cellulose column (1 × 24 cm^2), which is previously equilibrated with 20 mM Tris HCl buffer (pH 7.4)/0.5 mM DTT/17 mM octylglucoside. After isocratic elution with 55 ml of the above buffer, lysoplasmalogenase is eluted from DEAE-cellulose column using a linear KCl gradient formed from 75 ml of 20 mM Tris HCl buffer (pH 7.4)/0.5 mM DTT/17 mM octylglucoside plus 75 ml of 1 M KCl in the same buffer. Active fractions are pooled and subjected to HPHT column connected to Pharmacia FPLC system equilibrated with 20 mM potassium phosphate buffer (pH 6.9)/0.2 mM $CaCl_2$/17 mM octylglucoside/0.25 mM DTT/0.25 mM phenylmethylsulfonic acid. Active fractions are filtered (0.2 μm pore size) and injected onto the column. Isocratic elution with 5 ml is followed by a linear gradient formed from 15 ml of 20 mM potassium phosphate buffer (pH 6.9)/0.2 mM $CaCl_2$/17 mM octylglucoside/0.25 mM DTT/0.25 mM phenylmethylsulfonic acid plus 15 ml of the above buffer containing 1.0 M KCl (Jurkowitz-Alexander et al., 1989).

4.7 Conclusion

Some progress has been made on the development of rapid and sensitive assay procedures for plasmalogen-hydrolyzing enzymes, but this progress has been quite slow. The development of rapid and sensitive PLA_2 assay procedures requires the selection of a glycerophospholipid substrate with appropriate interfacial properties (physical form) in aqueous solution, an appropriate procedure for the separation of substrate from products, and finally the quantification of the product generated by PLA_2 (Reynolds et al., 1991). Radiochemical procedures are sensitive, but suffer from the disadvantage of being discontinuous, requiring separation of the radiochemical substrate from the labeled products. They are also time-consuming and expensive. Furthermore, handling radioactive compounds is undesirable not only due to safety issues and the high cost of radioactive materials, but also because of high disposal costs for long-lived radioactive material. At present, the most sensitive assay procedures for plasmalogen-hydrolyzing enzymes are continuous and discontinuous spectrophotometric and fluorometric procedures. These assay procedures are less sensitive than radiochemical assay procedures. They are preferred because of their rapidity, reliability, reproducibility, and availability of their substrate in large quantity at a low cost (Hirashima et al., 1989b, 1990b; Farooqui et al., 1999, Jurkowitz-Alexander et al., 1989). Plasmalogen-selective PLA_2 and lysoplasmalogenase can be partially purified using multiple column chromatographic procedures.

References

Bligh E. G. and Dyer W. J. (1959). A rapid method of total lipid extraction and purification. Can. J. Biochem. Physiol. 37:911–917.

Dole V. P. (1956). A relation between non-esterified fatty acids in plasma and the metabolism of glucose. J. Clin. Invest. 35:150–154.

Farooqui A. A. and Horrocks L. A. (1988). Methods for the determination of phospholipases, lipases and lysophospholipases. In: Boulton A. A., Baker G. B., and Horrocks L. A. (eds.), *Neuromethods, Vol. 7: Lipids and Related Compounds*. Humana Press, New Jersey, pp. 179–209.

Farooqui A. A. and Horrocks L. A. (2001). Plasmalogens: Workhorse lipids of membranes in normal and injured neurons and glia. Neuroscientist 7:232–245.

Farooqui A. A. and Horrocks L. A. (2004). Plasmalogens, platelet-activating factor, and other ether lipids. In: Nicolaou A. and Kokotos G. (eds.), *Bioactive Lipids*. Oily Press, Bridgwater, England, pp. 107–134.

Farooqui A. A. and Horrocks L. A. (2007). *Glycerophospholipids in the Brain: Phospholipases A_2 in Neurological Disorders*, pp. 1–394. Springer, New York.

Farooqui A. A., Ong W. Y., and Horrocks L. A. (2003). Plasmalogens, docosahexaenoic acid, and neurological disorders. In: Roels F., Baes M., and de Bies S. (eds.), *Peroxisomal Disorders and Regulation of Genes*. Kluwer Academic/Plenum Publishers, London, pp. 335–354.

Farooqui A. A., Yang H.-C., Hirashima Y., and Horrocks L. A. (1999). Determination of plasmalogen-selective phospholipase A_2 activity by radiochemical and fluorometric assay procedures. In: Doolittle M. H. and Reue K. (eds.), *Mammalian Lipases and Phospholipases. Methods in Molecular Biology*. Humana Press, Totowa, NJ, pp. 39–47.

Farooqui A. A., Yang H.C., and Horrocks L. A. (1995). Plasmalogens, phospholipases A_2, and signal transduction. Brain Res. Rev. 21:152–161.

Gunawan J. and Debuch H. (1981). Liberation of free aldehyde from 1-(1-alkenyl)-*sn*-glycero-3-phosphoethanolamine (lysoplasmalogen) by rat liver microsomes. Hoppe-Seyler's Z. Physiol. Chem. 362:445–452.

Gunawan J. and Debuch H. (1982). Lysoplasmalogenase – A microsomal enzyme from rat brain. J. Neurochem. 39:693–699.

Han X. L. and Gross R. W. (1990). Plasmenylcholine and phosphatidylcholine membrane bilayers possess distinct conformational motifs. Biochemistry 29:4992–4996.

Han X. L., Zupan L. A., Hazen S. L., and Gross R. W. (1992). Semisynthesis and purification of homogeneous plasmenylcholine molecular species. Anal. Biochem. 200:119–124.

Hanahan D. J., Nouchi T., Weintraub S. T., and Olson M. S. (1990). Novel route to preparation of high purity lysoplasmenylethanolamine. J. Lipid Res. 31:2113–2117.

Hazen S. L. and Gross R. W. (1993). The specific association of a phosphofructokinase isoform with myocardial calcium-independent phospholipase A_2. Implications for the coordinated regulation of phospholipolysis and glycolysis. J. Biol. Chem. 268:9892–9900.

Hirashima Y., Jurkowitz-Alexander M. S., Farooqui A. A., and Horrocks L. A. (1989a). Continuous spectrophotometric assay of phospholipase A_2 activity hydrolyzing plasmalogens using coupling enzymes. Anal. Biochem. 176:180–184.

Hirashima Y., Farooqui A. A., and Horrocks L. A. (1989b). Assay procedures and properties of plasmalogenase, lysoplasmalogenase and plasmalogen specific phospholipase A_2. In: Stobaugh R. E. (ed.), *Frontiers of Chemistry: Biotechnology*. American Chemical Society, Washington, DC, pp. 91–102.

Hirashima Y., Farooqui A. A., and Horrocks L. A. (1989c). Fluorimetric coupled enzyme assay for lysoplasmalogenase activity in liver. Biochem. J. 260:605–608.

Hirashima Y., Farooqui A. A., Murphy E. J., and Horrocks L. A. (1990a). Purification of plasmalogens using *Rhizopus delemar* lipase and *Naja naja naja* phospholipase A_2. Lipids 25:344–348.

Hirashima Y., Mills J. S., Yates A. J., and Horrocks L. A. (1990b). Phospholipase A_2 activities with a plasmalogen substrate in brain and in neural tumor cells: A sensitive and specific assay using pyrenesulfonyl-labeled plasmenylethanolamine. Biochim. Biophys. Acta. 1074:35–40.

Hirashima Y., Farooqui A. A., Mills J. S., and Horrocks L. A. (1992). Identification and purification of calcium-independent phospholipase A_2 from bovine brain cytosol. J. Neurochem. 59:708–714.

Jurkowitz M. S., Horrocks L. A., and Litsky M. L. (1999). Identification and characterization of alkenyl hydrolase (lysoplasmalogenase) in microsomes and identification of a plasmalogen-

active phospholipase A_2 in cytosol of small intestinal epithelium. Biochim. Biophys. Acta Lipids Lipid Metab. 1437:142–156.
Jurkowitz-Alexander M., Ebata H., Mills J. S., Murphy E. J., and Horrocks L. A. (1989). Solubilization, purification, and characterization of lysoplasmalogen alkenylhydrolase (lysoplasmalogenase) from rat liver microsomes. Biochim. Biophys. Acta. 1002:203–212.
Jurkowitz-Alexander M. S., Hirashima Y., and Horrocks L. A. (1991). Coupled enzyme assays for phospholipase activities with plasmalogen substrates. Methods Enzymol. 197:79–89.
Jurkowitz-Alexander M. S. and Horrocks L. A. (1990). Lysoplasmalogenase: Solubilization and partial purification from liver microsomes. Meth. Enzymol. 197:483–490.
Portilla D. and Dai G. (1996). Purification of a novel calcium-independent phospholipase A_2 from rabbit kidney. J. Biol. Chem. 271:15451–15457.
Reynolds L. J., Washburn W. N., Deems R. A., and Dennis E. A. (1991). Assay strategies and methods for phospholipases. Methods Enzymol. 197:3–23.
Somerharju P. (2002). Pyrene-labeled lipids as tools in membrane biophysics and cell biology. Chem. Phys. Lipids 116:57–74.
Thompson D. H., Shin J. W., Boomer J., and Kim J. M. (2004). Preparation of plasmenylcholine lipids and plasmenyl-type liposome dispersions. In: Duzgunes N. (ed.), *Liposomes, Part D. Methods in Enzymology*. Academic Press, San Diego, pp. 153–168.
Yang H.C., Farooqui A. A., Rammohan K. W., Haun S. E., and Horrocks L. A. (1997). Occurrence and characterization of plasmalogen-selective phospholipase A_2 in brain of various animal species. J. Neurochem. 69:205.

Chapter 5
Roles of Plasmalogens in Brain

5.1 Introduction

Plasmalogens play important roles in mammalian brain (Lee, 1998; Nagan and Zoeller, 2001; Farooqui and Horrocks, 2001). Beside being a structural component of cellular membranes and a major reservoir for arachidonic and docosahexaenoic acids (AA and DHA), plasmalogens are also involved in transport of ions across plasma membranes (Gross, 1985), membrane fusion (Lohner, 1996), protection of cellular membranes against oxidative stress (Zoeller et al., 1988; Engelmann et al., 1994), and the efflux of cholesterol from cells mediated by high-density lipoprotein (HDL) (Fig. 5.1) (Mandel et al., 1998). Plasmalogens are also found in the nucleus, where they may be involved in cellular differentiation (Bichenkov and Ellingson, 1999; Albi et al., 2004). The occurrence of plasmalogens in the synaptic cleft suggests that these phospholipids not only play an important role in synaptogenesis, but may also be involved in vesicle formation during neurotransmitter release (Farooqui and Horrocks, 2001). Plasmalogens may also be important in membrane dynamics, allowing the formation of inverted hexagonal structures, a property that may not only contribute to membrane fusion property, but also be important in modulating the membrane fluidity and permeability (Lohner, 1996).

5.2 Roles of Plasmalogens in Brain

5.2.1 Plasmalogens as Neural Membrane Components

Neural membranes are enriched in plasmalogens. Plasmalogens impart membranes with unique biophysical and biochemical characteristics. The perpendicular orientation of the *sn*-2 acyl chain at the neural membrane surface and the lack of a carbonyl group at the *sn*-1 position in plasmalogens affect the hydrophilicity of the head group, resulting in stronger intermolecular hydrogen bonding between the head groups (Lohner, 1996). These properties allow ethanolamine plasmalogens to adopt the inverse hexagonal phase and may be responsible for different

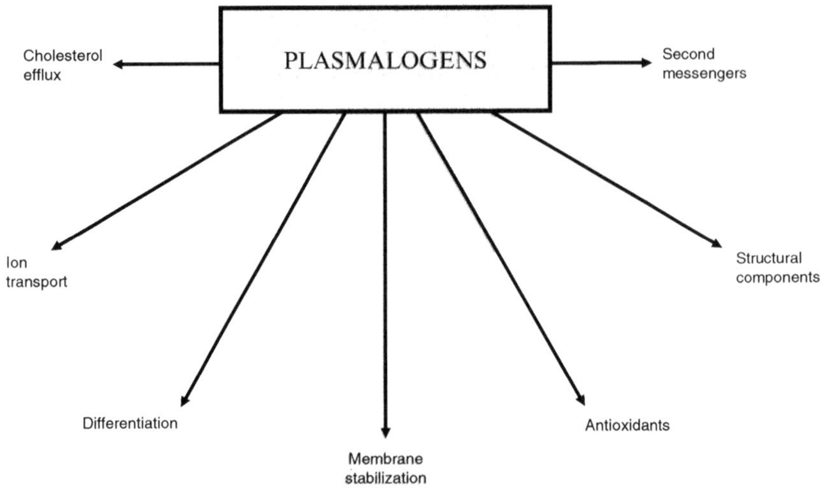

Fig. 5.1 Proposed roles of plasmalogens in brain.

membrane properties such as lipid packing, fluidity, permeability, and interactions with neural membrane receptors and ion channels, compared to other glycerophospholipids (Horrocks and Sharma, 1982).

5.2.2 *Plasmalogens as a Storage Depot for Second Messengers and Lipid Mediators*

Plasmalogens are enriched in lipid rafts (Pike et al., 2002). Lipid rafts serve as suitable platforms for signal transduction processes where plasmalogens may play an important role. The hydrolysis of plasmalogens by PlsEtn-PLA_2 results in the release of arachidonic and docosahexaenoic acids. These fatty acids are implicated in both physiological (synaptic plasticity) and pathophysiological (neurodegenerative) processes (Katsuki and Okuda, 1995). For example, at low concentration, arachidonic acid modulates ion channels, and regulates the activity of many enzyme proteins (Table 5.1) (Farooqui et al., 1997). High concentrations of arachidonic acid have profound adverse effects on the ATP-producing capacity of mitochondria by uncoupling oxidative phosphorylation, and inducing the efflux of Ca^{2+} and K^{+} from mitochondria. In addition, arachidonic acid is also associated with the regulation of gene expression (Jump et al., 1994). Enzymically, arachidonic acid is metabolized to prostaglandins, leukotrienes, and thromboxanes (Fig 5.2). These metabolites are collectively called eicosanoids (Phillis et al., 2006). At low concentration, these molecules act as second messengers. At high concentrations, eicosanoids produce oxidative stress, inflammation, and neurodegeneration. Eicosanoids act through specific superficial or intracellular receptors modulating signal transduction

Table 5.1 Effect of arachidonic acid on enzymic activities in neural and nonneural tissues.

Enzyme	Effect	Reference
Protein kinase C	Stimulation	Farooqui et al., 1997
Diacylglycerol kinase	Stimulation	Rao et al., 1994
Choline acetyltransferase	Stimulation	Chalimoniuk et al., 2004
Protein kinase A	Stimulation	Doolan and Keenan, 1994
NADPH oxidase	Stimulation	Sakata et al., 1987
Nitric oxide synthase	Stimulation	Toborek et al., 2000
Caspase-3	Stimulation	Garrido et al., 2001; Liu et al., 2001

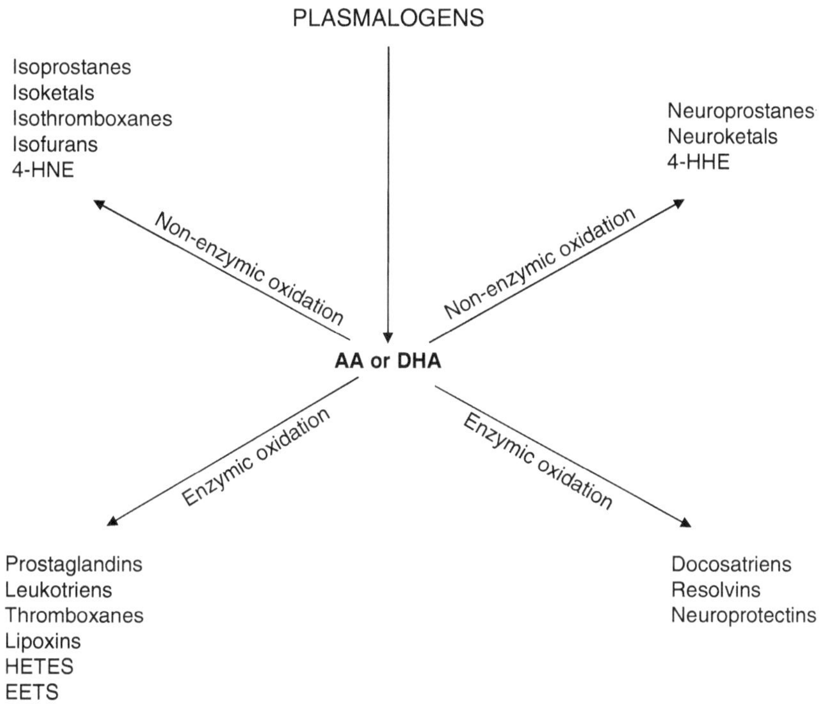

Fig. 5.2 Enzymic and nonenzymic oxidation products of arachidonic acid and docosahexaenoic acid. *4-HNE* 4-hydroxy-2-nonenal, *4-HHE* 4-hydroxyhexenal, *HETEs* hydroxyeicosatetraenoic acids, and *EETs* 5,6; 8,9; 11,12; 14,15 *cis*-epoxyeicosatrienoic acids.

pathways and gene transcription. Thus, PGD_2 activates the DP receptors, PGE_2 activates the EP receptors, and $PGF_{2\alpha}$, PGI_2, and TXA_2, respectively, stimulate the FP, IP, and TP receptors (Coleman et al., 1994). Eicosanoid receptors are typically G-protein-coupled receptors with seven transmembrane segments that have an extracellular amino terminus and an intracellular carboxyl terminus. These receptors are involved in the generation of cyclic AMP, diacylglycerol, and phosphatidyl 1,4,5-trisphosphate, and in the modulation of Ca^{2+} ion influx. Besides eicosanoids, lipoxins and HETEs are also derived from enzymic oxidation of arachidonic acid (Phillis et al., 2006) (Fig. 5.2).

At high concentration, arachidonic acid acts directly on the plasma membrane through its detergent-like action (Gamberucci et al., 1997). This involves the formation of micelles, which are aggregates of fatty acid molecules formed because of their poor solubility in aqueous solutions. These micelles disrupt cell membranes and create pores permeable to Na^+ and Ca^{2+} ions (Sawyer and Andersen, 1989). The presence of cations, especially Ca^{2+}, and higher ionic strength enhance the formation of micelles (Tanford, 1980). Nonenzymic oxidation products of arachidonic acid include isoprostanes (IsoP), 4-hydroxynonenal (4-HNE), and isoketals (IsoKs) (Fig. 5.3) (Fam and Morrow, 2003). These metabolites cause cellular injury through their oxidative-stress-mediated effects. F_2-IsoPs and 4-HNE have emerged as the most reliable markers of oxidative stress in vivo.

The minimum requirement for the generation of an IsoP is a polyunsaturated fatty acid with three contiguous, methylene-interrupted double bonds (Basu, 2004). The mechanism by which IsoPs are formed is analogous to the formation of prostaglandins by COX enzymes (Morrow et al., 1999). Unlike prostaglandins, the formation of IsoP in situ initially takes place at the esterified arachidonic acid on the glycerophospholipid molecule (Fam and Morrow, 2003). The structural difference between IsoPs and prostaglandins is that in IsoPs the side chains are cis to the cyclopentane ring, whereas in prostaglandins they have the trans orientation. IsoPs are

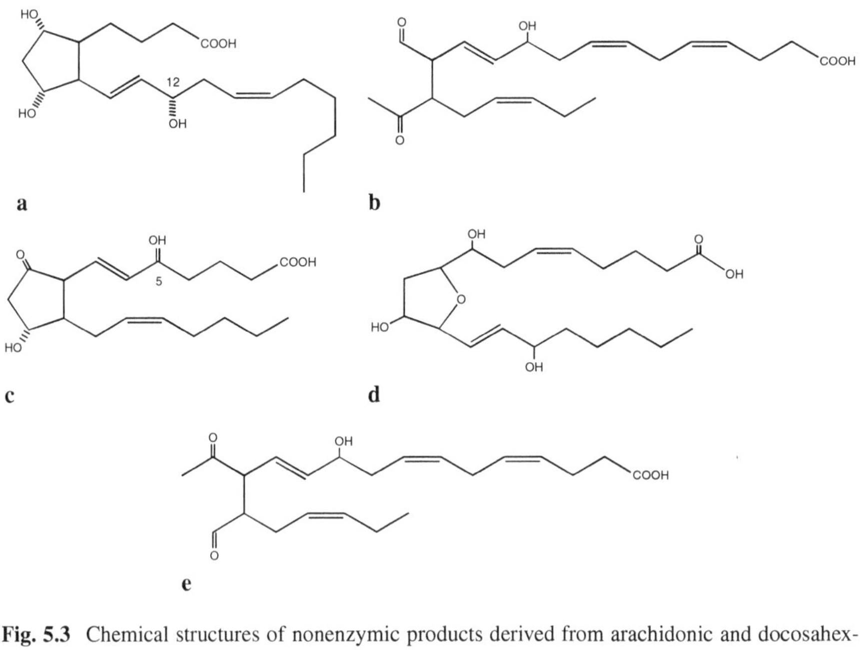

Fig. 5.3 Chemical structures of nonenzymic products derived from arachidonic and docosahexaenoic acids. 12-F_{2t}-IsoP (**a**), D4-neuroketal (**b**), 5-E_{2t}-IsoP (**c**), isofuran (**d**), and E4-neuroketal (**e**).

attractive indices of lipid peroxidation because of the specificity of their formation, their chemical stability, and the development of sensitive and specific methods for their measurement using mass spectrometry and radioimmunoassay (Fam and Morrow, 2003). IsoPs act through prostaglandin and thromboxane-like receptors and their effects can be blocked by thromboxane receptor antagonists (Takahashi et al., 1992; Morrow et al., 1996; Opere et al., 2005). Thromboxane receptors, like prostaglandin receptors, are linked to different sets of G proteins, resulting in distinct biological effects on brain and other body tissues (Lahaie et al., 1998). The presence of IsoP receptors has not been reported in the brain tissue, but their occurrence has been proposed in smooth muscle cells (Habib and Badr, 2004).

The formation of isoketal (IsoK), another arachidonic-acid-derived oxidation product, has also been reported. Unlike the F_2-IsoP, the IsoKs result in modification of biologically important proteins rather than activation of specific receptors (Davies et al., 2004). IsoKs are highly reactive γ-ketoaldehydes that form pyrrole adducts with the ε-amino group of lysine residues on protein (Davies et al., 2004). IsoKs inhibit the activity of proteasomes in glial cells, with an IC_{50} of 330 nM, and induce cell death with an IC_{50} of 670 nM. Intrahemispheric injections of 15-E_2-IsoK disrupt the blood–brain barrier. IsoKs have been detected in body tissues as well as biological fluids.

Like arachidonic acid, docosahexaenoic acid also undergoes enzymic and nonenzymic oxidation. Enzymic oxidation of docosahexaenoic acid is catalyzed by 15-lipoxygenase-like enzyme. Multistep enzymic oxidation of docosahexaenoic acid results in the generation of docosatrienes, resolvins, and neuroprotectins (Serhan, 2004, 2005; Bazan, 2005). Nonenzymic oxidation of docosahexaenoic acid results in formation of neuroprostanes and neuroketals (Roberts II et al., 1998; Nourooz-Zadeh et al., 1999; Roberts II and Fessel, 2004; Yin et al., 2005) (Fig. 5.3). Generation of neuroprostanes and neuroketals may cause alterations in neural membrane fluidity and permeability, resulting in oxidative stress and impairment in normal neuronal function (Fam and Morrow, 2003; Yin et al., 2005).

Nonenzymic peroxidation of arachidonic acid and docosahexaenoic acid also generates 4-hydroxynonenal (4-HNE), and 4-hydroxyhexenal (4-HHE), respectively (Fig. 5.4). These reactive aldehydes are important mediators of neural cell damage because of their ability to covalently modify biomolecules with disruption of important cellular function (Esterbauer et al., 1991; Lin et al., 2005; Farooqui and Horrocks, 2006). 4-HNE is a nine-carbon α,β-unsaturated aldehyde. At low concentration, it modulates cellular signaling in brain tissue (Keller and Mattson, 1998), but at high concentration it produces alterations in intracellular redox status, produces cytotoxicity, and promotes numerous cellular stress signaling responses that ultimately alter gene expression and cell viability (West and Marnett, 2005). 4-HNE affects signaling pathways through increased basal and GTP-stimulated phospholipase C and adenylate cyclase activities and decreased ornithine decarboxylase activities (Rossi et al., 1993). In 3T3-L1 adipose cells, 4-HNE not only upregulates COX-2 mRNA and its protein expression, but also upregulates p38 MAP-kinase (p38 MARK) phosphorylation in a dose-dependent manner (Zarrouki et al., 2007).

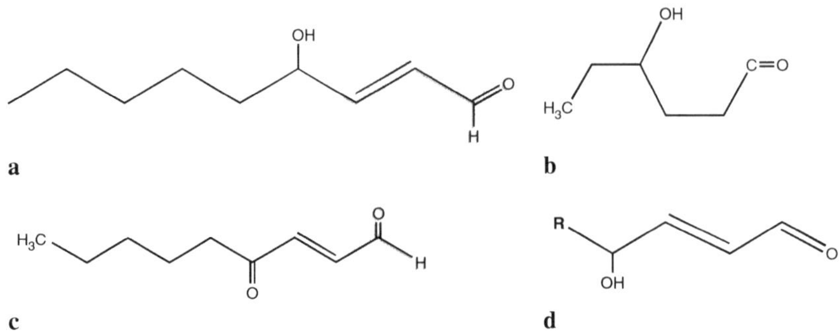

Fig. 5.4 Chemical structures of 4-hydroxy-2-nonenal (4-HNE), 4-hydroxyhexenal (4-HHE), 4-oxo-2-nonenal, and 4-hydroxy-2-alkenal (4-HAE). These aldehydes have been derived from arachidonic, docosahexaenoic, and linoleic acids respectively. 4-HNE (**a**), 4-HHE (**b**), 4-oxo-2-nonenal (**c**), and 4-HAE (**d**).

Pretreatment of 3T3-L1 cells with a selective inhibitor of p38MAPK (PD 169316) prevents 4-HNE- and glucose-oxidase-induced COX-2 expression. It is shown that oxidative stress induces COX-2 expression through the production of 4-HNE, which activates p38MAPKinase, suggesting that 4-HNE links oxidative stress and chronic inflammation through the activation of cyclooxygenase (Zarrouki et al., 2007). In addition, 4-HNE has been reported to activate signaling via c-jun N-terminal kinase and to inhibit other regulatory mechanisms such as NF-κB and the proteasomal degradation pathway (Page et al., 1999).

Plasmalogen-derived arachidonic and docosahexaenoic acids modulate dopaminergic, noradrenergic, glutamatergic, and serotonergic neurotransmission (Chalon et al., 1998; Zimmer et al., 2000; Högyes et al., 2003), ion channels, receptors (Nishikawa et al., 1994; Xiao and Li, 1999; Yehuda et al., 2002), memory processes (Nishikawa et al., 1994; Fujita et al., 2001; Fujimoto et al., 1989), and gene expression (Fernstrom, 1999; Farkas et al., 2000; Nakanishi et al., 1994). The release of docosahexaenoic acid may be involved in vesicle formation during neurotransmitter release. It markedly affects activities of many enzymes (Table 5.2). This fatty acid is metabolized to docosanoids. Both docosahexaenoic acid and docosanoids have been reported to antagonize the effects of some eicosanoids (Serhan, 2004, 2005). As stated earlier, the nonenzymic oxidation of arachidonic and docosahexaenoic acids result in the generation of 4-HNE and eicosanoids and 4-hydroxyhexenal (4-HHE), neuroprostals, and neuroketals respectively (Roberts II et al., 1998). These nonenzymic lipid mediators cause oxidative stress and neural injury.

Collective evidence suggests that under normal conditions, nearly all of the released arachidonic acid and docosahexaenoic acid is recycled back into neural membrane glycerophospholipids, particularly in plasmalogens (Rapoport, 1999; Farooqui et al., 2000). However, under pathological conditions accumulation of arachidonic and docosahexaenoic acid not only triggers an uncontrolled ROS

Table 5.2 Effect of docosahexaenoic acid (DHA) on enzymic activities associated with signal transduction.

Enzymic activity	Effect	Reference
Phospholipase A_2	Inhibition	Farooqui et al., 2006
Protein kinase C	Inhibited	Hirafuji et al., 2003
Cyclooxygenase	Inhibited	Calder, 2004
Glutathione peroxidase	Increased	Songur et al., 2004
MAP kinase	Inhibited	Hirafuji et al., 2003
Cyclin-dependent kinase	Inhibited	Hirafuji et al., 2003
Caspase-3	Inhibited	Akbar et al., 2005
Neutral sphingomyelinase	Inhibited	Wu et al., 2005
HMG-CoA reductase	Inhibited	Duncan et al., 2005; Choi et al., 1989
Endothelial nitric oxide synthase	Increased	Li et al., 2007

production and arachidonic acid cascade, but also induces the NF-κB-mediated cytokine expression. These processes are closely associated with oxidative stress and inflammation in brain tissue (Farooqui et al., 2007a, b; Sun et al., 2007).

5.2.3 Plasmalogens in Regulation of Enzymic Activities

Alkyl-linked diacylglycerols, including 1-oleoyl-2-acetyl-*sn*-glycerol, have been shown to inhibit protein kinase C activity (Daniel et al., 1988). Thus incubation of SF3271 cell with 1-*O*-hexadecylglycerol (1-O-HDG) increases the mass of ether-linked glycerophospholipid. Bradykinin initiates a transient increase in cytosolic Ca^{2+} concentration in both control and 1-O-HDG-supplemented cells, indicating that the initial receptor-linked events are not affected by 1-O-HDG supplementation (Clark and Murray, 1995). 1-O-HDG supplementation retards the bradykinin-induced activation of phospholipase D, but has no effect on the stimulation of mitogen-activated protein kinase activity. These results suggest that modulation of the ether lipid composition of membranes can alter PKC isozyme translocation and indicate that a PKC isozyme other than PKCα, most likely PKCζ, is involved in mitogen-activated protein kinase activation. Lysoplasmalogen, a PLA_2-derived product of plasmalogens, also modulates Na^+, K^+-ATPase activity (Portilla and Dai, 1996).

5.2.4 Plasmalogens in Membrane Fusion

During membrane fusion process, two adjacent membranes approach each other, establish a molecular contact, and eventually merge into one continuous membrane. This process demands flexibility of the membrane, which is largely governed by (a) the thermotropic state of the hydrocarbon interior, (b) the lateral diffusion coefficient of the lipid molecules, and (c) the spontaneous curvature of the membrane leaflets (Wilschut et al., 1985). Membrane fusion is the cornerstone of major transmembrane transport processes such as exocytosis and endocytosis and synaptic neurotransmission. Understanding the molecular events in membrane fusion poses an outstanding challenge to neuroscientists, because of its fundamental relevance to neurotransmitter release, drug delivery systems, and gene transfection (Hughson, 1995).

Plasmalogens have a high propensity to form an inverse hexagonal phase, which is the essential transitory phase for membrane fusion events (Wilschut et al., 1985). Vesicular fusion process is very sensitive to the amount of ethanolamine plasmalogen. Small reduction either in vinyl ether content and/or the arachidonic acid or docosahexaenoic acid content of vesicles markedly decreases a number of successful membrane fusion events (Lohner et al., 1991). It is suggested that the intrinsic nature of vinyl ether chain in plasmalogens facilitates the lowering of transition temperature that facilitates the initial stages of membrane fusion process. In vitro, vesicles containing PlsEtn (PtdCho/PlsEtn/PtdSer, 45:45:10) undergo fusion sixfold faster than do vesicles containing PtdEtn (Lohner et al., 1991). The fusion rate also depends on the fatty acid composition. Thus, PlsEtn containing 20:4 at the *sn*-2 position fuses fivefold faster than does the corresponding PlsEtn with an 18:1 acyl chain. The fusion behavior of plasmalogens also depends on highly selective interaction and affinity of a fusion protein with vesicles containing PlsEtn (Glaser and Gross, 1995). The purified preparations of fusion protein have glyceraldehyde-3-phosphate dehydrogenase activity and are not affected by Ca^{2+}, but have an obligatory requirement for PlsEtn and cholesterol. Protein-mediated fusion of biomembranes also requires soluble *N*-ethylmaleimide-sensitive factor attachment protein receptors (SNARE) proteins. SNARE-mediated close contract is an essential determinant of fusion specificity (Parlati et al., 2000). It is suggested that the pairing of vesicle v-SNAREs with target membrane t-SNAREs plays a central role in intracellular membrane fusion. In vitro on average, the fusion protein catalyzes one fusion event between two vesicles every millisecond. Whether this process occurs in vivo and plays a role in fusion of neurotransmitter vesicles is not established. However, the occurrence of high levels of plasmalogens in the synaptic plasma membrane and their interaction with a fusion protein suggest that plasmalogens may be involved in membrane fusion events such as endocytosis and exocytosis during neurotransmission, hormone release, and membrane vesicle trafficking (Glaser and Gross, 1995).

5.2.5 Plasmalogens in Ion Transport

On the basis of their interaction with Ca^{2+}-ATPase and sodium–calcium exchanger, plasmalogens have been proposed to play an important role in calcium transport (Bick et al., 1991; Gross, 1985). The sodium–calcium exchanger is predominantly found in the plasma membrane of excitatory cells, and is an important component of excitation–secretion as well as excitation–contraction machinery (Ford and Hale,

1996). Based on reconstitution studies, it is proposed that plasmalogens provide a critical lipid environment in which anionic glycerophospholipids serve as boundary lipids for the regulation of the trans-sarcolemmal sodium–calcium exchanger.

The levels of PlsEtn in red blood cells are associated with maximal activity of the Na^+/K^+ pump (Duhm et al., 1993). The putative preferential lipid–protein interaction of PlsEtn with the membrane-embedded portion of the pump may induce a conformational change of the protein, thereby hindering the access of intracellular Na^+ to its binding site. Similarly, the presence of PlsCho in the vesicles may modulate the function of gramicidin ion channels. Modulation of ion channels by DHA, a component of plasmalogens, suggests that these glycerophospholipids may be involved in the maintenance of ion pumps in neural membranes (Young et al., 2000).

5.2.6 Plasmalogens in High-Density Lipoprotein

Plasmalogens are synthesized in liver and secreted as component of circulating blood lipoproteins. The levels of plasmalogens in human serum are ~0.1–0.3 mM, with PlsCho and PlsEtn ratio ranging from 0.5 to 1.5 (Maeba et al., 2007). These levels are positively correlated with HDLs. Although levels of plasmalogens are significantly decreased (40%) in elderly subjects, compared to healthy young subjects, but the PlsCho/PlsEtn ratio remains the same in young and elderly subjects. The PlsCho/PlsEtn ratio correlates positively with low-density lipoprotein (LDL) particle size and negatively with apo A-II and fasting triacylglycerol levels. Collectively, these studies suggest that plasmalogens act as a marker for atherogenic small dense LDL (Maeba et al., 2007).

5.2.7 Plasmalogens, Cholesterol Oxidation, Efflux and Atherosclerosis

Murine macrophage-like cell lines (RAW 264.7, RAW.108, and RAW.12) lack the ability to synthesize plasmalogens. RAW.108 is deficient in dihydroxyacetone phosphate acyltransferase, whereas RAW.12 lacks dihydroxyacetone phosphate acyltransferase and $\Delta1'$-desaturase. In these cell lines, the cellular plasmalogen content is related to HDL-mediated cholesterol efflux. This suggestion is supported by two observations. The cellular HDL-mediated cholesterol efflux is decreased in plasmalogen-deficient fibroblasts and RAW 264.7 and RAW 108 macrophage-like cell lines. The HDL-mediated cholesterol efflux is enhanced when cells are treated with 1-*O*-hexadecyl-*sn*-glycerol, a compound that restores the level of plasmalogens (Mandel et al., 1998).

HDL-mediated reverse cholesterol transport may be one of several important mechanisms by which this antiatherogenic lipoprotein modulates the development of atherosclerosis. A reduction in the plasmalogen content of normal aortas with increasing donor age was more pronounced in arteriosclerotic aortas (Buddecke and Andresen, 1959). Based on the effect of plasmalogens on cholesterol efflux, it is proposed that these glycerophospholipids may play a crucial role in the pathogenesis of arteriosclerosis. Because plasmalogens are more susceptible to oxidation than are phosphatidylcholine and sphingomyelin, HDL and plasmalogens may be the preferred targets of lipid peroxidation before the bulk of polyunsaturated glycerophospholipids in LDL are attacked (Hofer et al., 1996). The importance of oxidized LDL in arteriosclerosis is supported by the finding of a higher oxidative susceptibility of LDL in patients with a higher degree of coronary arteriosclerosis. It is interesting to note that the PlsEtn content is lower by 20% in red cell membrane lipids in hyperlipidemic patients than in normolipidemic donors, suggesting that plasmalogens play an important role in cholesterol efflux (Engelmann et al., 1992).

NRel-4 and NZel-1 cells are CHO cell mutants, characterized by defects in dihydroxyacetone phosphate acyltransferase and alkyl dihydroxyacetone phosphate synthase respectively. Defects in these enzymes result in plasmalogen deficiency in these mutants. Defective ethanolamine plasmalogens synthesis alters cholesterol transport in the CHO cell mutants (Munn et al., 2003). Defective cholesterol transport can be restored when intermediates of ethanolamine plasmalogen biosynthesis are introduced into the system (Munn et al., 2003). The defect in cholesterol transport can also be corrected when NRel-4 cells were transfected with a complementary DNA encoding the missing enzyme, dihydroxyacetone phosphate acyltransferase. Collective evidence suggests that plasmalogens play a very important role in cholesterol transport (Munn et al., 2003). It is also reported that ethanolamine plasmalogens reduce the rate of cholesterol oxidation in a dose-dependent manner (Maeba and Ueta, 2003, 2004). Based on oxidizability of cholesterol in phospholipids bilayer using large unilamellar vesicles and water-soluble radical initiator, 2,2′-azobis(2-amidinopropane) dihydrochloride, it is suggested that ethanolamine plasmalogens reduce the total membrane oxidizability more efficiently in the presence of cholesterol in large unilamellar vesicles (Maeba and Ueta, 2004).

5.2.8 Plasmalogens and Their Antioxidant Activity

Plasmalogens protect biological structures against free radical attack (Engelmann, 2004; Maeba and Ueta, 2004). Thus plasmalogens protect cellular membranes of Chinese hamster ovary cells and LDL particles against oxidative stress (Zoeller et al., 1988; Engelmann et al., 1994). In neural membranes, transition metal ions (copper and iron) initiate lipid peroxidation by generating peroxyl and alkoxyl radicals from the decomposition of lipid hydroperoxides according to the following reactions.

$$LOOH + Cu^{2+}/Fe^{3+} \rightarrow LOO + H^{+} + Cu^{+}/Fe^{2+} \qquad (1)$$

$$LOOH + Cu^{2+}/Fe^{3+} \rightarrow LO + OH^{-} + Cu^{+}/Fe^{3+} \quad (2)$$

Plasmalogen-containing liposomes have a strong ability to chelate transition metal ions and thereby prevent the formation of peroxyl and alkoxyl radicals (Zommara et al., 1995; Sindelar et al., 1999). Although the molecular mechanism of plasmalogens-mediated protection is not known, there are two possibilities: (a) the inhibition of lipid peroxidation through plasmalogen-mediated chelation of metal ions (iron and copper) and (b) consumption of plasmalogens during membrane protection. This is because vinyl ether bond of plasmalogen is also a substrate for free radicals. Thus, one vinyl ether double bond in plasmalogens scavenges two peroxy radicals (Reiss et al., 1997). The levels of plasmalogen in biomembranes are 25–100 times higher than vitamin E (Calzada et al., 1997; Hahnel et al., 1999). Both molecules are colocalized in lipoproteins and cellular membranes, and are capable of scavenging peroxy radicals. However, vitamin E scavenges peroxy radicals with 20- to 25-fold higher efficiency than do the plasmalogens (Hahnel et al., 1999), suggesting that vitamin E is the first line of cellular defense against oxidative stress. In contrast, studies on the oxidative degradation of phospholipids in the presence and absence of plasmalogens in human platelet have indicated that plasmalogens block peroxidation of polyunsaturated fatty acids and that vitamin E has no effect on the time course and quantities and composition of the phospholipids, even at a molar ratio of vitamin E to phospholipids four times higher than in platelet membranes (Leray et al., 2002). Based on the effect of iron on liposome with and without plasmalogens and the experiments with lipid- and water-soluble azo compounds, it is suggested that plasmalogens interfere with the propagation rather than the initiation of lipid peroxidation. Their participation in lipid peroxidation does not have a lag phase (Ernster et al., 1992). In contrast, vitamin E involvement in preventing lipid peroxidation shows a lag phase that can be prolonged by increasing concentrations of ascorbate (Sindelar et al., 1999). This behavior of plasmalogen during iron-mediated lipid peroxidation is different from vitamin E.

In the absence of plasmalogens, the degradation of the four glycerophospholipids increases in the order PtdSer < PtdCho < PtdEtn < PtdIns, which is not their unsaturation (PtdCho < PtdSer < PtdEtn < PtdIns. Based on this observation, it is proposed that both unsaturation and head group properties play an important role in peroxidative susceptibility (Leray et al., 2002). When bovine brain glycerophosphoethanolamine is treated with Cu^{2+}/H_2O_2, a remarkable loss of specific plasmalogen molecular species is observed, compared to diacylglycerophospholipids (Khaselev and Murphy, 1999). As stated earlier, the oxidation of plasmalogen results in the generation of 1-lysoglycerophospholipid. This supports the view that vinyl ether double bond of plasmalogen is more vulnerable to oxidation than the double bond found in polyunsaturated fatty acid at the *sn*-2 position. It is stated that the presence of one vinyl ether double bond blocks the oxidation of four double bonds of *sn*-2-located arachidonic acid. This results in oxidative-stress-mediated downregulation of plasmalogen contents (Engelmann, 2004). Decrease in plasmalogen may also be caused by the stimulation of plasmalogen-selective PLA_2, resulting in a decrease in plasmalogen content (McHowat et al., 1998). These studies indicate that more studies are required to determine the relationship between nonenzymic and enzymic decrease in plasmalogen in brain tissue.

Another mechanism that may explain the decrease in plasmalogen levels is that neural membranes contain a plasmalogen redox cycling system mediated by vitamin C and vitamin E (Yavin and Gatt, 1972). This system plays an important role not only in defending neural membranes from the oxidative stress (Engelmann, 2004), but also in protecting LDL from oxidative damage from neural and nonneural tissues.

The reactive brominating species generated by myeloperoxidase, as well as by activated neutrophils also attack the vinyl ether bond of plasmalogens (Albert et al., 2002, Albert et al., 2003). This process results in production of an α-bromofatty aldehyde and lysophosphatidylcholine. The subsequent conversion of the initially generated lysophosphatidylcholine depends on the fatty acid residue in the *sn*-2 position. Matrix- assisted laser desorption and mass spectrometric and ^{31}P NMR spectroscopic studies indicate that plasmalogens containing an oleic acid residue at the *sn*-2 position are converted primarily to 1-lyso-2-oleoyl-*sn*-glycero-3-phosphocholine and corresponding chlorohydrin species by moderate amounts of HOCl (Lessig et al., 2007). In contrast, plasmalogens containing highly unsaturated docosahexaenoic acid at the *sn*-2 position upon HOCl treatment produces 1-lyso-2-docosahexaenoyl-glycerophosphocholine and glycerophosphocholine. The formation of the latter product suggests a novel pathway for the action of HOCl on plasmalogens (Lessig et al., 2007).

The generation of these products in vivo may have a profound effect on the host cell protein kinases and inhibits membrane transport proteins (Sasaki et al., 1993). Thus, plasmalogens may serve as protective agents for the host cells by quenching ROS and hypohalous acids, and thus preventing them from interacting with other targets such as proteins and nucleic acids (Albert et al., 2002, 2003). Collective evidence suggests that plasmalogens represent the principal pool of antioxidant lipids in neural and nonneural membranes and are targeted by oxidants. It is likely that an intramolecular competition occurs between the enol ether double bond and fatty acid double bond for reaction with oxidants (Berry and Murphy, 2005).

In contrast to the above-mentioned view, studies based on the effect of menadione, an intracellular reactive oxygen species generator, on plasmalogen-deficient fibroblasts (Jansen and Wanders, 1997) and lactic acid on astrocytic cultures suggest that plasmalogens do not play a major role in the protection of cells against superoxide anion radicals and lactic-acid-induced oxidative stress (Fauconneau et al., 2001). The fatty aldehyde released from oxidized plasmalogens forms Schiff base adducts with PtdEtn (Stadelmann-Ingrand et al., 2004). Schiff base adducts may have a deleterious effect on neural cell membranes.

Studies described earlier are based on in vitro experiments (Engelmann, 2004; Maeba and Ueta, 2004; Zommara et al., 1995; Sindelar et al., 1999). Information on the in vivo role of plasmalogens during oxidation process would require not only the identification of genes encoding for plasmalogen-synthesizing and plasmalogen-degrading enzymes, but also factors that turn on and off those genes in brain. Activation of multiple intracellular signaling cascades in response to plasmalogen-derived products may involve the participation of nonenzymic and receptor-mediated mechanisms. It is becoming increasingly evident that docosahexaenoic acid stimulates the synthesis of plasmalogens, and modulates genes associated with oxidative stress (Horrocks and Farooqui, 2004). Recently discovered ether lipid deficient mice can be used for obtaining this useful information on the involvement of plasmalogens in antioxidative properties in vivo (Gorgas et al., 2006).

5.2.9 Plasmalogens and Generation of Long-Chain Aldehydes

Plasmalogens are highly sensitive to oxidative attack at the enol ether group at the *sn*-1 position (Yavin and Gatt, 1972). Arachidonic and docosahexaenoic acids at the *sn*-2 position of plasmalogens also undergo oxidation (Berry and Murphy, 2005), resulting in the generation of ω-aldehyde and γ-hydroxy-α,β-unsaturated aldehydes, which are neurotoxic and deleterious for nerve cells. Plasmalogens also undergo epoxidation. Epoxide formation is a biphasic process with stimulation at low concentrations, and inhibition at high concentrations. In contrast, α-hydroxyaldehyde formation is an exclusively inhibitory process. It is proposed that the inhibitory effects of α-hydroxyaldehyde may be related to the inhibition of NADH-oxidase reaction (Heinle et al., 2000). The quotient of plasmalogen epoxide to plasmalogen increases with age, indicating that both epoxidation and lipid peroxidation may be involved in decreasing the plasmalogen content in the aging brain (Weisser et al., 1997). Alterations in receptor function and ion channel activity in aged individuals may reflect this decrease. This may be caused by the reactivity of α-hydroxy aldehyde with membrane-bound proteins that are associated with optimal functioning of receptors and ion channels in neural membranes. Thus, plasmalogens can not only be regarded as antioxidants, but their epoxidation and oxidative degradation products may induce ROS generation in macrophages during phagocytosis (Heinle et al., 2000). Collectively, these studies suggest that oxidative-stress-mediated breakdown of plasmalogens under pathological situations may modulate macrophage function associated with inflammation and atherogenesis.

At this stage, it is difficult to separate the consequences of lower plasmalogen levels in brain that occur during normal aging from the decrease of plasmalogen level that occurs in neurological disorders such as ischemia and Alzheimer disease (Farooqui and Horrocks, 2001). However, both types of decrease in plasmalogen levels may compromise optimal membrane function by affecting fluidity, permeability, and other biophysical properties.

5.2.10 Plasmalogens in Differentiation

Plasmalogens and their metabolizing enzymes are present in the nucleus (Albi et al., 2004). Nuclear plasmalogen-selective PLA_2 is stimulated by retinoic acid (Antony et al., 2001). Retinoic acid treatment produces neuritic outgrowth in LA-N-1 cells. The pan retinoic acid receptor antagonist BMS493 not only inhibits

PlsEtn-PLA$_2$ activity but also blocks the formation of neuritic processes (Antony et al., 2003; Farooqui et al., 2004).

Studies on the incorporation of [^{3}H]ethanolamine in progenitor cell glycerophospholipids have indicated that the ratio between [^{3}H]PlsEtn to [^{3}HPtdEtn] was 1.3. When the progenitor cells are differentiated in the presence of calf serum, the ratio of [^{3}H]PlsEtn to [^{3}H]PtdEtn increased to 2.3 at the second day of differentiation, and remained at elevated levels (2.3–2.7) through 6 days of differentiation. Although the ratio of [^{3}H]PlsEtn to [^{3}H]PtdEtn is decreased after 6 days of differentiation, 1.8 times as much [^{3}H]ethanolamine was still incorporated into [^{3}H]PlsEtn than into [^{3}H]PtdEtn at the ninth day of differentiation (Bichenkov and Ellingson, 1999), suggesting that plasmalogens and plasmalogen-catabolizing enzymes play an important role during differentiation.

Analysis of rat and rabbit myocardial nuclear lipidome by electrospray ionization mass spectrometry has indicated that rabbit myocardial nuclear lipidome contains relatively more plasmenylcholine/phosphatidylcholine molecular species in comparison to that ratio observed in the rat myocardial nuclear lipidome (Albert et al., 2007). The rat myocardial nuclear choline glycerophospholipid pool is enriched with molecular species containing arachidonic acid and docosahexaenoic acid in comparison to that in the rabbit myocardial nuclear choline glycerophospholipid pool. While the ethanolamine glycerophospholipids of the rabbit myocardial nuclei are enriched with arachidonic acid and plasmalogens, the ethanolamine glycerophospholipid profile from rat myocardial nuclei show less plasmalogen and more species containing docosahexaenoic acid. These studies strongly support the occurrence of plasmalogens in the nucleus (Albert et al., 2007).

5.2.11 Plasmalogens in the Ocular Development

Occurrence of ethanolamine plasmalogens in the inner segment of photoreceptors and in the retinal pigment epithelium and ocular abnormalities in dihydroxyacetone-phosphate-acyltransferase-deficient mice suggest that plasmalogens play a crucial role in ocular development and function (Acar et al., 2007). Based on metabolic studies, it is also speculated that the effect of oxidative stress is mediated by plasmalogen-synthesizing enzymes, and PlsEtn-PLA$_2$ may contribute to visual loss observed in patient suffering from peroxisomal disorders (Acar et al., 2007).

5.2.12 Plasmalogens as Precursors for the Platelet-Activating Factor

The action of PlsEtn-PLA$_2$ on 1-alkenyl-2-acyl-*sn*-Gro*P*Etn results in the generation of 1-alkenyl-2-lyso-*sn*-Gro*P*Etn. This is an acceptor for the transfer of arachidonic acid from 1-alkyl-2-arachidonyl-*sn*-Gro*P*Cho. This reaction releases lyso-platelet-activating factor (lyso-PAF), which can be acetylated to PAF (Uemura et al., 1991). Another pathway of lyso-PAF production involves CoA-independent transacylase activity. In this pathway, 1-alkyl-2-acyl-glycerophocholine transfers acyl group to lysoplasmalogen (lyso-PlsEtn) in a CoA-independent manner. CoA-independent transacylation reaction is blocked by ET-18-O-CH_3 (Chilton et al., 1996; Winkler et al., 1996). Thus it seems that in brain PAF synthesis and degradation of plasmalogen are closely linked. This suggestion is supported by studies on the endothelin-mediated stimulation of PlsEtn-PLA_2. It is reported that endothelin-mediated stimulation of brain PlsEtn-PLA_2 decreases PlsEtn levels and evokes the generation of PAF (Collado et al., 2003). This process may contribute to inflammatory processes in the brain tissue.

5.3 Conclusion

Plasmalogens are a subclass of glycerophospholipids with vinyl ether bonds at the sn-1 position of glycerol moiety. These glycerophospholipids are widely distributed in human and animal tissues. Thus brain, heart, lungs, muscle, and red blood cells contain considerable amounts of plasmalogens. Although the role of plasmalogens in mammalian tissues is not fully understood, many in vitro studies indicate that as neural membrane components plasmalogens maintain membrane dynamics. They represent the principal pool for polyunsaturated fatty acids that act as a reservoir for prostaglandins, leukotrienes, thromboxanes, resolvins, and neuroprotectins. Plasmalogen-derived lipid mediators are involved in neural cell proliferation and differentiation. Plasmalogens are endogenous antioxidants. They scavenge free radicals and protect neural cells from oxidative stress. They have antioxidant effects on lipoprotein in serum. Changes in plasmalogen contents have major impact on membrane fluidity and the functioning of a variety of membrane-associated enzymes, ion channels, and receptors. Plasmalogens stabilize neural membranes and modulate HDL-mediated cholesterol efflux, and their deficiency contributes to neural membrane destabilization.

References

Acar N., Gregoire S., Andre A., Juaneda P., Joffre C., Bron A. M., Creuzot-Garcher C. P., and Bretillon L. (2007). Plasmalogens in the retina: *In situ* hybridization of dihydroxyacetone phosphate acyltransferase (DHAP-AT) – the first enzyme involved in their biosynthesis – and comparative study of retinal and retinal pigment epithelial lipid composition. Exp. Eye Res. 84:143–151.

Akbar M., Calderon F., Wen Z. M., and Kim H. Y. (2005). Docosahexaenoic acid: A positive modulator of Akt signaling in neuronal survival. Proc. Natl. Acad. Sci. USA 102:10858–10863.

Albert C. J., Crowley J. R., Hsu F. F., Thukkani A. K., and Ford D. A. (2002). Reactive brominating species produced by myeloperoxidase target the vinyl ether bond of plasmalogens – Disparate

utilization of sodium halides in the production of alpha-halo fatty aldehydes. J. Biol. Chem. 277:4694–4703.
Albert C. J., Thukkani A. K., Heuertz R. M., Slungaard A., Hazen S. L., and Ford D. A. (2003). Eosinophil peroxidase-derived reactive brominating species target the vinyl ether bond of plasmalogens generating a novel chemoattractant, alpha-bromo fatty aldehyde. J. Biol. Chem. 278:8942–8950.
Albert C. J., Anbukumar D. S., Monda J. K., Eckelkamp J. T., and Ford D. A. (2007). Myocardial lipidomics. Developments in myocardial nuclear lipidomics. Front. Biosci. 12:2750–2760.
Albi E., Cataldi S., Magni M. V., and Sartori C. (2004). Plasmalogens in rat liver chromatin: New molecules involved in cell proliferation. J. Cell. Physiol. 201:439–446.
Antony P., Freysz L., Horrocks L. A., and Farooqui A. A. (2001). Effect of retinoic acid on the Ca^{2+}-independent phospholipase A_2 in nuclei of LA-N-1 neuroblastoma cells. Neurochem. Res. 26:83–88.
Antony P., Freysz L., Horrocks L. A., and Farooqui A. A. (2003). Ca^{2+}-independent phospholipases A_2 and production of arachidonic acid in nuclei of LA-N-1 cell cultures: A specific receptor activation mediated with retinoic acid. Mol. Brain Res. 115:187–195.
Basu S. (2004). Isoprostanes: Novel bioactive products of lipid peroxidation. Free Radic. Res. 38:105–122.
Bazan N. G. (2005). Neuroprotectin D1 (NPD1): A DHA-derived mediator that protects brain and retina against cell injury-induced oxidative stress. Brain Pathol. 15:159–166.
Berry K. A. Z. and Murphy R. C. (2005). Free radical oxidation of plasmalogen glycerophosphocholine containing esterified docosahexaenoic acid: Structure determination by mass spectrometry. Antioxidants and Redox Signaling 7:157–169.
Bichenkov E. and Ellingson J. S. (1999). Temporal and quantitative expression of the myelin-associated lipids, ethanolamine plasmalogen, galactocerebroside, and sulfatide, in the differentiating CG-4 glial cell line. Neurochem. Res. 24:1549–1556.
Bick R. J., Youker K. A., Pownall H. J., Van Winkle W. B., and Entman M. L. (1991). Unsaturated aminophospholipids are preferentially retained by the fast skeletal muscle CaATPase during detergent solubilization. Evidence for a specific association between aminophospholipids and the calcium pump protein. Arch. Biochem. Biophys. 286:346–352.
Buddecke E. and Andresen G. (1959). Quantitative bestimmung der acetalphosphatide (plasmalogene) in der aorta des menschen unter berucksichtigung der arteriosklerose. Hoppe-Seyler's Z. Physiol. Chem. 314:38–45.
Calder P. C. (2004). n-3 Fatty acids, inflammation, and immunity – Relevance to postsurgical and critically ill patients. Lipids 39:1147–1161.
Calzada C., Bruckdorfer K. R., and Rice-Evans C. A. (1997). The influence of antioxidant nutrients on platelet function in healthy volunteers. Atherosclerosis 128:97–105.
Chalimoniuk M., King-Pospisil K., Pedersen W. A., Malecki A., Wylegala E., Mattson M. P., Hennig B., and Toborek M. (2004). Arachidonic acid increases choline acetyltransferase activity in spinal cord neurons through a protein kinase C-mediated mechanism. J. Neurochem. 90:629–636.
Chalon S., Delion-Vancassel S., Belzung C., Guilloteau D., Leguisquet A. M., Besnard J. C., and Durand G. (1998). Dietary fish oil affects monoaminergic neurotransmission and behavior in rats. J. Nutr. 128:2512–2519.
Chilton F. H., Fonteh A. N., Surette M. E., Triggiani M., and Winkler J. D. (1996). Control of arachidonate levels within inflammatory cells. Biochim. Biophys. Acta 1299:1–15.
Choi Y. S., Goto S., Ikeda I., and Sugano M. (1989). Effect of dietary n-3 polyunsaturated fatty acids on cholesterol synthesis and degradation in rats of different ages. Lipids 24:45–50.
Clark K. J. and Murray A. W. (1995). Evidence that the bradykinin-induced activation of phospholipase D and of the mitogen-activated protein kinase cascade involve different protein kinase C isoforms. J. Biol. Chem. 270:7097–7103.
Coleman R. A., Smith W. L., and Narumiya S. (1994). International Union of Pharmacology classification of prostanoid receptors: Properties, distribution, and structure of the receptors and their subtypes. Pharmacol. Rev. 46:205–229.

Collado A. P., Latorre E., Fernández I., Aragonés A. D., and Catalán R. E. (2003). Endothelin-1 decreases ethanolamine plasmalogen levels and evokes PAF production in brain microvessels. Microvasc. Res. 66:197–203.

Daniel L. W., Small G. W., and Schmitt J. D. (1988). Alkyl-linked diglycerides inhibit protein kinase C activation by diacylglycerols. Biochem. Biophys. Res. Commun. 151:291–297.

Davies S. S., Amarnath V., and Roberts II L. J., (2004). Isoketals: Highly reactive γ-ketoaldehydes formed from the H_2-isoprostane pathway. Chem. Phys. Lipids 128:85–99.

Doolan C. M. and Keenan A. K. (1994). Inhibition by fatty acids of cyclic AMP-dependent protein kinase activity in brush border membranes isolated from human placental vesicles. Brit. J. Pharmacol. 111:509–514.

Duhm J., Engelmann B., Schönthier U. M., and Streich S. (1993). Accelerated maximal velocity of the red blood cell Na^+/K^+ pump in hyperlipidemia is related to increase in 1-palmitoyl-2-arachidonoyl-plasmalogen phosphatidylethanolamine. Biochim. Biophys. Acta Biomembr. 1149:185–188.

Duncan R. E., El Sohemy A., and Archer M. C. (2005). Regulation of HMG-CoA reductase in MCF-7 cells by genistein, EPA, and DHA, alone and in combination with mevastatin. Cancer Lett. 224:221–228.

Engelmann B. (2004). Plasmalogens: Targets for oxidants and major lipophilic antioxidants. Biochem. Soc. Trans. 32:147–150.

Engelmann B., Streich S., Schönthier U. M., Richter W. O., and Duhm J. (1992). Changes of membrane phospholipid composition of human erythrocytes in hyperlipidemias. I. Increased phosphatidylcholine and reduced sphingomyelin in patients with elevated levels of triacylglycerol-rich lipoproteins. Biochim. Biophys. Acta Lipids Lipid Metab. 1165:32–37.

Engelmann B., Bräutigam C., and Thiery J. (1994). Plasmalogen phospholipids as potential protectors against lipid peroxidation of low density lipoproteins. Biochem. Biophys. Res. Commun. 204:1235–1242.

Ernster L., Forsmark P., and Nordenbrand K. (1992). The mode of action of lipid-soluble antioxidants in biological membranes: Relationship between the effects of ubiquinol and vitamin E as inhibitors of lipid peroxidation in submitochondrial particles. BioFactors 3:241–248.

Esterbauer H., Schaur R. J., and Zollner H. (1991). Chemistry and biochemistry of 4-hydroxynonenal, malonaldehyde and related aldehydes. Free Radic. Biol. Med. 11:81–128.

Fam S. S. and Morrow J. D. (2003). The isoprostanes: Unique products of arachidonic acid oxidation – A review. Curr. Med. Chem. 10:1723–1740.

Farkas T., Kitajka K., Fodor E., Csengeri I., Lahdes E., Yeo Y. K., Krasznai Z., and Halver J. E. (2000). Docosahexaenoic acid-containing phospholipid molecular species in brains of vertebrates. Proc. Natl. Acad. Sci. USA 97:6362–6366.

Farooqui A. A. and Horrocks L. A. (2001). Plasmalogens: Workhorse lipids of membranes in normal and injured neurons and glia. Neuroscientist 7:232–245.

Farooqui A. A. and Horrocks L. A. (2006). Phospholipase A_2-generated lipid mediators in the brain: The good, the bad, and the ugly. Neuroscientist 12:245–260.

Farooqui A. A., Rosenberger T. A., and Horrocks L. A. (1997). Arachidonic acid, neurotrauma, and neurodegenerative diseases. In: Yehuda S. and Mostofsky D. I. (Eds.), *Handbook of Essential Fatty Acid Biology*. Humana, Totowa, NJ, pp. 277–295.

Farooqui A. A., Horrocks L. A., and Farooqui T. (2000). Glycerophospholipids in brain: Their metabolism, incorporation into membranes, functions, and involvement in neurological disorders. Chem. Phys. Lipids 106:1–29.

Farooqui A. A., Antony P., Ong W. Y., Horrocks L. A., and Freysz L. (2004). Retinoic acid-mediated phospholipase A_2 signaling in the nucleus. Brain Res. Rev. 45:179–195.

Farooqui A. A., Ong W. Y., and Horrocks L. A. (2006). Inhibitors of brain phospholipase A_2 activity: Their neuropharmacological effects and therapeutic importance for the treatment of neurologic disorders. Pharmacol. Rev. 58:591–620.

Farooqui A. A., Horrocks L. A., and Farooqui T. (2007a). Interactions between neural membrane glycerophospholipid and sphingolipid mediators: A recipe for neural cell survival or suicide. J. Neurosci. Res. 85:1834–1850.

Farooqui A. A., Horrocks L. A., and Farooqui T. (2007b). Modulation of inflammation in brain: A matter of fat. J. Neurochem. 101:577–599.

Fauconneau B., Stadelmann-Ingrand S., Favrelière S., Baudouin J., Renaud L., Piriou A., and Tallineau C. (2001). Evidence against a major role of plasmalogens in the resistance of astrocytes in lactic acid-induced oxidative stress in vitro. Arch. Toxicol. 74:695–701.

Fernstrom J. D. (1999). Effects of dietary polyunsaturated fatty acids on neuronal function. Lipids 34:161–169.

Ford D. A. and Hale C. C. (1996). Plasmalogen and anionic phospholipid dependence of the cardiac sarcolemmal sodium-calcium exchanger. FEBS Lett. 394:99–102.

Fujimoto K., Yao K., Miyazaki T., Hirano H., Nishikawa M., Kimura S., Murayama K., and Nonaka M. (1989). The effect of dietary docosahexaenoate on the learning ability of rats. In: Chandra R. K. (Ed.), *Health Effects of Fish and Fish Oils*. ARTS Biomedical, The Netherlands, pp. 275–284.

Fujita S., Ikegaya Y., Nishikawa M., Nishiyama N., and Matsuki N. (2001). Docosahexaenoic acid improves long-term potentiation attenuated by phospholipase A_2 inhibitor in rat hippocampal slices. Brit. J. Pharmacol. 132:1417–1422.

Gamberucci A., Fulceri R., Bygrave F. L., and Benedetti A. (1997). Unsaturated fatty acids mobilize intracellular calcium independent of IP3 generation and VIA insertion at the plasma membrane. Biochem. Biophys. Res. Commun. 241:312–316.

Garrido R., Mattson M. P., Hennig B., and Toborek M. (2001). Nicotine protects against arachidonic-acid-induced caspase activation, cytochrome c release and apoptosis of cultured spinal cord neurons. J. Neurochem. 76:1395–1403.

Glaser P. E. and Gross R. W. (1995). Rapid plasmenylethanolamine-selective fusion of membrane bilayers catalyzed by an isoform of glyceraldehyde-3-phosphate dehydrogenase: Discrimination between glycolytic and fusogenic roles of individual isoforms. Biochemistry 34:12193–12203.

Gorgas K., Teigler A., Komljenovic D., and Just W. W. (2006). The ether lipid-deficient mouse: Tracking down plasmalogen functions. Biochim. Biophys. Acta Mol. Cell Res. 1763:1511–1526.

Gross R. W. (1985). Identification of plasmalogen as the major phospholipid constituent of cardiac sarcoplasmic reticulum. Biochemistry 24:1662–1668.

Habib A. and Badr K. F. (2004). Molecular pharmacology of isoprostanes in vascular smooth muscle. Chem. Phys. Lipids 128:69–73.

Hahnel D., Thiery J., Brosche T., and Engelmann B. (1999). Role of plasmalogens in the enhanced resistance of LDL to copper-induced oxidation after LDL apheresis. Arterioscler. Thromb. Vasc. Biol. 19:2431–2438.

Heinle H., Gugeler N., Felde R., Okech D., and Spiteller G. (2000). Oxidation of plasmalogens produces highly effective modulators of macrophage function. Z. Naturforsch. [C] 55:115–120.

Hirafuji M., Machida T., Hamaue N., and Minami M. (2003). Cardiovascular protective effects of n-3 polyunsaturated fatty acids with special emphasis on docosahexaenoic acid. J. Pharmacol. Sci. 92:308–316.

Hofer G., Lichtenberg D., Kostner G. M., and Hermetter A. (1996). Oxidation of fluorescent glycero- and sphingophospholipids in human plasma lipoproteins: Alkenylacyl subclasses are preferred targets. Clin. Biochem. 29:445–450.

Högyes E., Nyakas C., Kiliaan A., Farkas T., Penke B., and Luiten P. G. (2003). Neuroprotective effect of developmental docosahexaenoic acid supplement against excitotoxic brain damage in infant rats. Neuroscience 119:999–1012.

Horrocks L. A. and Farooqui A. A. (2004). Docosahexaenoic acid in the diet: Its importance in maintenance and restoration of neural membrane function. Prostaglandins Leukot. Essent. Fatty Acids 70:361–372.

Horrocks L. A. and Sharma M. (1982). Plasmalogens and O-alkyl glycerophospholipids. In: Hawthorne J. N. and Ansell G. B. (Eds.), *Phospholipids, New Comprehensive Biochemistry, Vol. 4*. Elsevier Biomedical, Amsterdam, pp. 51–93.

Hughson F. M. (1995). Molecular mechanisms of protein-mediated membrane fusion. Curr. Opin. Struct. Biol. 5:507–513.

Jansen G. A. and Wanders R. J. A. (1997). Plasmalogens and oxidative stress: Evidence against a major role of plasmalogens in protection against the superoxide anion radical. J. Inherit. Metab. Dis. 20:85–94.

Jump D. B., Clarke S. D., Thelen A., and Liimatta M. (1994). Coordinate regulation of glycolytic and lipogenic gene expression by polyunsaturated fatty acids. J. Lipid Res. 35:1076–1084.

Katsuki H. and Okuda S. (1995). Arachidonic acid as a neurotoxic and neurotrophic substance. Prog. Neurobiol. 46:607–636.

Keller J. N. and Mattson M. P. (1998). Roles of lipid peroxidation in modulation of cellular signaling pathways, cell dysfunction, and death in the nervous system. Rev. Neurosci. 9:105–116.

Khaselev N. and Murphy R. C. (1999). Susceptibility of plasmenyl glycerophosphoethanolamine lipids containing arachidonate to oxidative degradation. Free Radic. Biol. Med. 26:275–284.

Lahaie I., Hardy P., Hou X., Hassessian H., Asselin P., Lachapelle P., Almazan G., Varma D. R., Morrow J. D., Roberts II L. J., and Chemtob S. (1998). A novel mechanism for vasoconstrictor action of 8-isoprostaglandin $F_{2\alpha}$ on retinal vessels. Am. J. Physiol. 274:R1406–R1416.

Lee T. C. (1998). Biosynthesis and possible biological functions of plasmalogens. Biochim. Biophys. Acta Lipids Lipid Metab. 1394:129–145.

Leray C., Cazenave J. P., and Gachet C. (2002). Platelet phospholipids are differentially protected against oxidative degradation by plasmalogens. Lipids 37:285–290.

Lessig J., Schiller J., Arnhold J., and Fuchs B. (2007). Hypochlorous acid-mediated generation of glycerophosphocholine from unsaturated plasmalogen glycerophosphocholine lipids. J. Lipid Res. 48:1316–1324.

Li Q. R., Zhang Q., Wang M., Zhao S. M., Ma J., Luo N., Li N., Li Y. S., Xu G. W., and Li J. S. (2007). Eicosapentaenoic acid modifies lipid composition in caveolae and induces translocation of endothelial nitric oxide synthase. Biochimie 89:169–177.

Lin D., Lee H. G., Liu Q., Perry G., Smith M. A., and Sayre L. M. (2005). 4-Oxo-2-nonenal is both more neurotoxic and more protein reactive than 4-hydroxy-2-nonenal. Chem. Res. Toxicol. 18:1219–1231.

Liu D. X., Li L. P., and Augustus L. (2001). Prostaglandin release by spinal cord injury mediates production of hydroxyl radical, malondialdehyde and cell death: A site of the neuroprotective action of methylprednisolone. J. Neurochem. 77:1036–1047.

Lohner K. (1996). Is the high propensity of ethanolamine plasmalogens to form non-lamellar lipid structures manifested in the properties of biomembranes? Chem. Phys. Lipids 81:167–184.

Lohner K., Balgavy P., Hermetter A., Paltauf F., and Laggner P. (1991). Stabilization of non-bilayer structures by the etherlipid ethanolamine plasmalogen. Biochim. Biophys. Acta 1061:132–140.

Maeba R. and Ueta N. (2003). Ethanolamine plasmalogens prevent the oxidation of cholesterol by reducing the oxidizability of cholesterol in phospholipid bilayers. J. Lipid Res. 44:164–171.

Maeba R. and Ueta N. (2004). A novel antioxidant action of ethanolamine plasmalogens in lowering the oxidizability of membranes. Biochem. Soc. Trans. 32:141–143.

Maeba R., Maeda T., Kinoshita M., Takao K., Takenaka H., Kusano J., Yoshimura N., Takeoka Y., Yasuda D., Okazaki T., and Teramoto T. (2007). Plasmalogens in human serum positively correlate with high-density lipoprotein and decrease with aging. J. Atheroscler. Thromb. 14:12–18.

Mandel H., Sharf R., Berant M., Wanders R. J. A., Vreken P., and Aviram M. (1998). Plasmalogen phospholipids are involved in HDL-mediated cholesterol efflux: Insights from investigations with plasmalogen-deficient cells. Biochem. Biophys. Res. Commun. 250:369–373.

McHowat J., Liu S., and Creer M. H. (1998). Selective hydrolysis of plasmalogen phospholipids by Ca^{2+}-independent PLA_2 in hypoxic ventricular myocytes. Am. J. Physiol. Cell Physiol. 274: C1727–C1737.

Morrow J. D., Awad J. A., Wu A., Zackert W. E., Daniel V. C., and Roberts II L. J. (1996). Nonenzymatic free radical-catalyzed generation of thromboxane-like compounds (isothromboxanes) in vivo. J. Biol. Chem. 271:23185–23190.

Morrow J. D., Tapper A. R., Zackert W. E., Yang J., Sanchez S. C., Montine T. J., and Roberts II L. J. (1999). Formation of novel isoprostane-like compounds from docosahexaenoic acid. Adv. Exp. Med. Biol. 469:343–347.
Munn N. J., Arnio E., Liu D., Zoeller R. A., and Liscum L. (2003). Deficiency in ethanolamine plasmalogen leads to altered cholesterol transport. J. Lipid Res. 44:182–192.
Nagan N. and Zoeller R. A. (2001). Plasmalogens: Biosynthesis and functions. Prog. Lipid Res. 40:199–229.
Nakanishi K., Yasugi E., Morita H., Dohi T., and Oshima M. (1994). Plasmenylethanolamine in human intestinal mucosa detected by an improved method for analysis of phospholipid. Biochem. Mol. Biol. Int. 33:457–462.
Nishikawa M., Kimura S., and Akaike N. (1994). Facilitatory effect of docosahexaenoic acid on *N*-methyl-D-aspartate response in pyramidal neurones of rat cerebral cortex. J. Physiol. (Lond.) 475:83–93.
Nourooz-Zadeh J., Liu E. H. C., Yhlen B., Änggård E. E., and Halliwell B. (1999). F_4-isoprostanes as specific marker of docosahexaenoic acid peroxidation in Alzheimer's disease. J. Neurochem. 72:734–740.
Opere C. A., Zheng W. D., Huang J. F., Adewale A., Kruglet M., and Ohia S. E. (2005). Dual effect of isoprostanes on the release of [^{3}H]D-aspartate from isolated bovine retinae: Role of arachidonic acid metabolites. Neurochem. Res. 30:129–137.
Page S., Fischer C., Baumgartner B., Haas M., Kreusel U., Loidl G., Hayn M., Ziegler-Heitbrock H. W., Neumeier D., and Brand K. (1999). 4-Hydroxynonenal prevents NF-κB activation and tumor necrosis factor expression by inhibiting IκB phosphorylation and subsequent proteolysis. J. Biol. Chem. 274:11611–11618.
Parlati F., McNew J. A., Fukuda R., Miller R., Sollner T. H., and Rothman J. E. (2000). Topological restriction of SNARE-dependent membrane fusion. Nature 407:194–198.
Phillis J. W., Horrocks L. A., and Farooqui A. A. (2006). Cyclooxygenases, lipoxygenases, and epoxygenases in CNS: Their role and involvement in neurological disorders. Brain Res. Rev. 52:201–243.
Pike L. J., Han X., Chung K. N., and Gross R. W. (2002). Lipid rafts are enriched in arachidonic acid and plasmenylethanolamine and their composition is independent of caveolin-1 expression: A quantitative electrospray ionization/mass spectrometric analysis. Biochemistry 41:2075–2088.
Portilla D. and Dai G. (1996). Purification of a novel calcium-independent phospholipase A_2 from rabbit kidney. J. Biol. Chem. 271:15451–15457.
Rao K. V., Vaidyanathan V. V., and Sastry P. S. (1994). Diacylglycerol kinase is stimulated by arachidonic acid in neural membranes. J. Neurochem. 63:1454–1459.
Rapoport S. I. (1999). In vivo fatty acid incorporation into brain phospholipids in relation to signal transduction and membrane remodeling. Neurochem. Res. 24:1403–1415.
Reiss D., Beyer K., and Engelmann B. (1997). Delayed oxidative degradation of polyunsaturated diacyl phospholipids in the presence of plasmalogen phospholipids in vitro. Biochem. J. 323:807–814.
Roberts II L. J. and Fessel J. P. (2004). The biochemistry of the isoprostane, neuroprostane, and isofuran pathways of lipid peroxidation. Chem. Phys. Lipids 128:173–186.
Roberts II L. J., Montine T. J., Markesbery W. R., Tapper A. R., Hardy P., Chemtob S., Dettbarn W. D., and Morrow J. D. (1998). Formation of isoprostane-like compounds (neuroprostanes) *in vivo* from docosahexaenoic acid. J. Biol. Chem. 273:13605–13612.
Rossi M. A., Di Mauro C., and Dianzani M. U. (1993). Action of lipid peroxidation products on phosphoinositide specific phospholipase C. Mol. Aspects Med. 14:273–279.
Sakata A., Ida E., Tominaga M., and Onoue K. (1987). Arachidonic acid acts as an intracellular activator of NADPH-oxidase in Fc gamma receptor-mediated superoxide generation in macrophages. J. Immunol. 138:4353–4359.
Sasaki Y., Asaoka Y., and Nishizuka Y. (1993). Potentiation of diacylglycerol-induced activation of protein kinase C by lysophospholipids. FEBS Lett. 320:47–51.

Sawyer D. B. and Andersen O. S. (1989). Platelet-activating factor is a general membrane perturbant. Biochim. Biophys. Acta 987:129–132.

Serhan C. N. (2004). A search for endogenous mechanisms of anti-inflammation uncovers novel chemical mediators: Missing links to resolution. Histochem. Cell Biol. 122:305–321.

Serhan C. N. (2005). Novel ω-3-derived local mediators in anti-inflammation and resolution. Pharmacol. Ther. 105:7–21.

Sindelar P. J., Guan Z. Z., Dallner G., and Ernster L. (1999). The protective role of plasmalogens in iron-induced lipid peroxidation. Free Radic. Biol. Med. 26:318–324.

Songur A., Sarsilmaz M., Sogut S., Ozyurt B., Ozyurt H., Zararsiz I., and Turkoglu A. O. (2004). Hypothalamic superoxide dismutase, xanthine oxidase, nitric oxide, and malondialdehyde in rats fed with fish ω-3 fatty acids. Prog. Neuropsychopharmacol. Biol. Psychiatry 28:693–698.

Stadelmann-Ingrand S., Pontcharraud R., and Fauconneau B. (2004). Evidence for the reactivity of fatty aldehydes released from oxidized plasmalogens with phosphatidylethanolamine to form Schiff base adducts in rat brain homogenates. Chem. Phys. Lipids 131:93–105.

Sun G. Y., Horrocks L. A., and Farooqui A. A. (2007). The role of NADPH oxidase and phospholipases A_2 in mediating oxidative and inflammatory responses in neurodegenerative diseases. J. Neurochem. (In press)

Takahashi K., Nammour T. M., Fukunaga M., Ebert J., Morrow J. D., Roberts L. J., Hoover R. L., and Badr K. F. (1992). Glomerular actions of a free radical-generated novel prostaglandin, 8-epi-prostaglandin $F_{2\alpha}$, in the rat. Evidence for interaction with thromboxane A_2 receptors. J. Clin. Invest. 90:136–141.

Tanford C. (1980). *The Hydrophobic Effects: Formation of Micelles and Biological Membranes*. Wiley, New York.

Toborek M., Garrido R., Malecki A., Kaiser S., Mattson M. P., Hennig B., and Young B. (2000). Nicotine attenuates arachidonic acid-induced overexpression of nitric oxide synthase in cultured spinal cord neurons. Exp. Neurol. 161:609–620.

Uemura Y., Lee T. C., and Snyder F. (1991). A coenzyme A-independent transacylase is linked to the formation of platelet-activating factor (PAF) by generating the lyso-PAF intermediate in the remodeling pathway. J. Biol. Chem. 266:8268–8272.

Weisser M., Vieth M., Stolte M., Riederer P., Pfeuffer R., Leblhuber F., and Spiteller G. (1997). Dramatic increase of alpha-hydroxyaldehydes derived from plasmalogens in the aged human brain. Chem. Phys. Lipids 90:135–142.

West J. D. and Marnett L. J. (2005). Alterations in gene expression induced by the lipid peroxidation product, 4-hydroxy-2-nonenal. Chem. Res. Toxicol. 18:1642–1653.

Wilschut J., Duzgunes N., Hoekstra D., and Papahadjopoulos D. (1985). Modulation of membrane fusion by membrane fluidity: Temperature dependence of divalent cation induced fusion of phosphatidylserine vesicles. Biochemistry 24:8–14.

Winkler J. D., Eris T., Sung C. M., Chabot-Fletcher M., Mayer R. J., Surette M. E., and Chilton F. H. (1996). Inhibitors of coenzyme A-independent transacylase induce apoptosis in human HL-60 cells. J. Pharmacol. Exp. Ther. 279:956–966.

Wu M., Harvey K. A., Ruzmetov N., Welch Z. R., Sech L., Jackson K., Stillwell W., Zaloga G. P., and Siddiqui R. A. (2005). Omega-3 polyunsaturated fatty acids attenuate breast cancer growth through activation of a neutral sphingomyelinase-mediated pathway. Int. J. Cancer 117:340–348.

Xiao Y. F. and Li X. Y. (1999). Polyunsaturated fatty acids modify mouse hippocampal neuronal excitability during excitotoxic or convulsant stimulation. Brain Res. 846:112–121.

Yavin E. and Gatt S. (1972). Oxygen-dependent cleavage of the vinyl ether linkage of plasmalogens. 2. Identification of the low molecular weight active component and the reaction mechanism. Eur. J. Biochem. 25:437–446.

Yehuda S., Rabinovitz S., Carasso R. L., and Mostofsky D. I. (2002). The role of polyunsaturated fatty acids in restoring the aging neuronal membrane. Neurobiol. Aging 23:843–853.

Yin H. Y., Musiek E. S., Gao L., Porter N. A., and Morrow J. D. (2005). Regiochemistry of neuroprostanes generated from the peroxidation of docosahexaenoic acid in vitro and in vivo. J. Biol. Chem. 280:26600–26611.

Young C., Gean P. W., Chiou L. C., and Shen Y. Z. (2000). Docosahexaenoic acid inhibits synaptic transmission and epileptiform activity in the rat hippocampus. Synapse 37:90–94.

Zarrouki B., Soares A. F., Guichardant M., Lagarde M., and Geloen A. (2007). The lipid peroxidation end-product 4-HNE induces COX-2 expression through p38MAPK activation in 3T3-L1 adipose cell. FEBS Lett. 581:2394–2400.

Zimmer L., Delion-Vancassel S., Durand G., Guilloteau D., Bodard S., Besnard J. C., and Chalon S. (2000). Modification of dopamine neurotransmission in the nucleus accumbens of rats deficient in n-3 polyunsaturated fatty acids. J. Lipid Res. 41:32–40.

Zoeller R. A., Morand O. H., and Raetz C. R. H. (1988). A possible role for plasmalogens in protecting animal cells against photosensitized killing. J. Biol. Chem. 263:11590–11596.

Zommara M., Tachibana N., Mitsui K., Nakatani N., Sakono M., Ikeda I., and Imaizumi K. (1995). Inhibitory effect of ethanolamine plasmalogen on iron- and copper-dependent lipid peroxidation. Free Radic. Biol. Med. 18:599–602.

Chapter 6
Involvement of Plasmalogens in Neurological Disorders

6.1 Introduction

Neural membranes are complex, well-organized, and highly specialized structures involved in receiving, processing, transporting, and transmitting information, not only from the plasma membrane to the nucleus, but also from one cell to another through chemical mediators generated during the catabolism of various glycerophospholipids (Guan et al., 1999; Farooqui and Horrocks, 2007). Neural membranes are highly interactive and dynamic. These properties facilitate optimal interactions of lipid mediators with transmembrane proteins, receptors, and ion channels and maintain normal brain function and adaptive responses (Farooqui et al., 1995; Farooqui and Horrocks, 2004). Although very little is known about the regulation of lipid dynamics in neural membranes, this process has been reported to link with the biosynthesis, metabolism, and transport of individual molecular species of glycerophospholipid (Farooqui and Horrocks, 2007). The catabolism of neural membrane glycerophospholipids, including plasmalogens, involves phospholipases, whose activities are modulated by receptors and ion channels. Plasmalogens provide neural membranes with suitable stability, fluidity, and permeability. They serve as storage depot and precursors for eicosanoids, docosanoids, and platelet activating factor.

In neural membranes, the maintenance of lipid asymmetry requires up to 20–26% consumption of ATP (Purdon et al., 2002; Purdon and Rapoport, 2007). This high rate of ATP consumption includes 1.4% of net brain ATP consumption for de novo synthesis of ether lipids, 5% for recycling of fatty acids within glycerophospholipid, 7.7% for maintaining membrane asymmetries of charged aminophospholipids, and about 12% for maintaining the phosphorylation state and de novo synthesis of inositol containing phospholipids involving phosphatidylinositol signaling (Purdon and Rapoport, 2007). Much of the remaining ATP maintains the distribution and transport of ions and activities of membrane-bound enzymes and ion channels. The high rate of ATP consumption is consistent with the role of glycerophospholipids in neural cell signaling, apoptosis, and membrane-associated processes such as membrane fusion, anchoring, and recycling (Purdon and Rapoport, 2007). At present, no information is available on ATP consumption during trafficking and sorting of various glycerophospholipids in neurons, astrocytes, oligodendrocytes, and microglial glial cells. The situation on ATP consumption during glycerophospholipid trafficking and sorting becomes more complex at the subcellular level (endoplasmic reticulum, Golgi apparatus, nucleus, etc.) of various cell types of neural cells in normal brain and brains from patients with neural trauma and neurodegenerative diseases.

In recent years, lipidomics has emerged as an important technology for the characterization of plasmalogen molecular species found in developing, adult, and aging brain (Piomelli, 2005; Lee et al., 2005; Berry and Murphy, 2005). Synaptic function and signal transduction processes depend on the temporal and spatially coordinated interactions between lipid mediators and their receptors and on their interrelationships with the organization of lipid mediator network (Gross et al., 2005). With the development of proteomics technology and DNA microarray analysis, it is possible to identify not only specific genes involved in the regulation of biosynthesis of individual molecular species of plasmalogen, but also genes related to its sorting and transport (Vreken et al., 2000; Voelker, 2003; Lee et al., 2005; Forrester et al., 2004). The oxidation of glycerophospholipids in neural membranes has been implicated in a variety of neural and nonneural human disorders such as ischemia, neurodegenerative diseases, atherosclerosis, and tumorogenesis.

6.2 Plasmalogens in Neurological Disorders

In neural membranes, homeostasis of normal plasmalogens is based on a balance between their catabolism, resynthesis in the reacylation/deacylation cycle, and de novo synthesis pathways (Porcellati, 1983). These processes not only result in the maintenance of optimal levels of plasmalogens, but also facilitate the inner and outer leaflets lipid asymmetry. Alterations in plasmalogen content occur in many neurological disorders (Fig. 6.1) (Table 6.1). This may be due either to a decrease in their synthesis or to an increase in their degradation. Peroxisomal disorders are characterized by the deficiency of plasmalogen-synthesizing enzymes (dihydroxyacetone phosphate acyltransferase and alkyldihydroxyacetone phosphate synthase). In contrast, decreases in plasmalogens in neurotrauma and neurodegenerative diseases may be caused by the stimulation of plasmalogen-selective phospholipase A_2 (Farooqui et al., 1997a, 2006). Beside peroxisomal disorders, neural trauma, and Alzheimer disease (AD), rats with experimental autoimmune encephalomyelitis show a downregulation in dihydroxyacetone phosphate acyltransferase (DHAP-AT) activity (66% of control). This downregulation in enzymic activity produces a decrease (16–30%) in the levels of plasmalogens (Singh et al., 2004).

In brain traumatic injury or neurodegenerative disease, the stimulation of glutamate receptors, upregulation of phospholipases (cPLA$_2$, PlsEtn-PLA$_2$, PLC, and D), and the release of high levels of arachidonic acid, docosahexaenoic acid (AA and DHA), platelet activating factor, diacylglycerol, and inositol trisphosphate result in

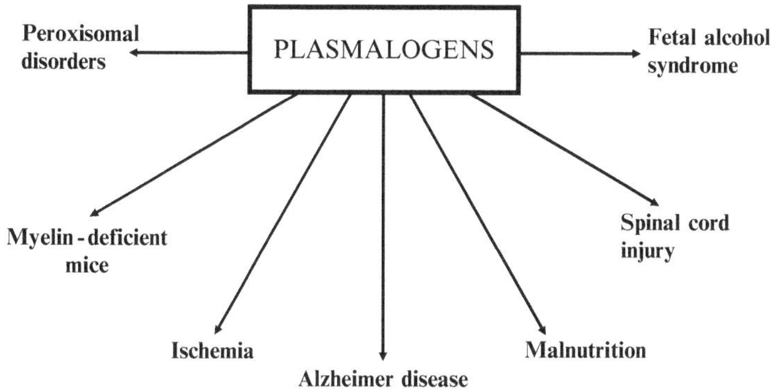

Fig. 6.1 Neurological disorders associated with alterations in plasmalogen levels in neural membranes.

Table 6.1 Alterations in plasmalogen levels in neurological disorders.

Disorder	Plasmalogen level	Reference
Ischemia	Decreased	Zhang and Sun, 1995
Alzheimer disease	Decreased	Wells et al., 1995; Daniel et al., 1999; Guan et al., 1999b
Alzheimer disease rat model	Decreased	Hashimoto et al., 2002
Spinal cord injury	Decreased	Demediuk et al., 1985
Zellweger syndrome	Decreased	Datta et al., 1984; Martínez, 1990; Martínez et al., 2000
Zellweger syndrome mouse model	Decreased	Janssen et al., 2000
Experimental allergic encephalomyelitis	Decreased	Jagannatha and Sastry, 1981
Rhizomelic chondrodysplasia punctata	Decreased	Poulos et al., 1991; Beams-Mengerink et al., 2006; de Vet and van den Bosch, 2000
Bronchopulmonary dysplasia	Decreased	Rüdiger et al., 2000

neural injury (Farooqui et al., 2006; Farooqui and Horrocks, 2007). Multiple forms of phospholipases form a network that generates common second messengers. The cross talk among various receptors through second messengers is essential for maintaining normal neuronal and glial cell growth (Farooqui et al., 1992; Ong et al., 2005). The occurrence of molecular species of plasmalogens, multiplicity of PLA_2, and cross talk among various phospholipases provide diversity in their function and specificity of various isoforms in the regulation of enzymic activity in response to a wide range of extracellular signals. However, at the same time it complicates the analysis of function of various phospholipases in the brain tissue (Farooqui and Horrocks, 2005).

6.2.1 *Plasmalogens in Ischemic Injury*

There are considerable differences in plasmalogens content of different regions of gerbil brain. Thus, the levels of PlsEtn in the hippocampal CA1 region are 5.65 ± 0.34 mmol kg^{-1} wet tissue, which is 2.6- and 2.7-fold higher than in the hippocampal CA3 region (2.45 ± 0.17 mmol kg^{-1}) and in the cerebral cortex (2.09 ± 0.22 mmol kg^{-1}), respectively. Higher PlsEtn content may explain the selective vulnerability of CA1 subfield of the hippocampus to ischemic injury (Kubota et al., 2001). Ischemic injury results in a marked decrease in plasmalogen content of neural membranes (Viani et al., 1995; Zhang and Sun, 1995). Thus, levels of PlsEtn are markedly decreased in endothelin-1-induced ischemia in synaptosomal membranes from rat striatum. This may be due to the stimulation of plasmalogen-selective PLA_2. In rabbit myocardium, microsomal plasmalogen-selective PLA_2 activity is increased tenfold during ischemic injury (Hazen et al., 1991). The activation of this enzyme is accompanied by the accumulation of lysoplasmalogen in critical subcellular loci, leading to alterations in membrane dynamics.

A marked stimulation of plasmalogen-selective PLA_2 also occurs during hypoxic injury to rabbit proximal tubules (Portilla et al., 1994). In contrast to myocardial plasmalogen-selective PLA_2, in rabbit kidney, the stimulation of PLA_2 after hypoxic injury is associated with the cytosolic fraction. The activation of the plasmalogen-selective PLA_2 can be abolished by pretreatment with bromoenol lactone, a specific inhibitor of plasmalogen-selective PLA_2 (Hazen et al., 1991). Pretreatment with bromoenol lactone also reduces total lactate dehydrogenase release and free arachidonate release from rabbit proximal tubule cells after hypoxic injury, suggesting that plasmalogen-selective PLA_2 plays a major role in arachidonate release at the initial stage of cell injury during hypoxic and ischemic injury.

Under normal conditions, plasmalogen-selective PLA_2 may be involved in the synthesis of arachidonic acid, eicosanoids, and PAF. Lysoplasmalogens, the other product of the plasmalogen-selective PLA_2-catalyzed reaction, are rapidly reacylated and normal levels of plasmalogens are maintained in the neural membranes. However, under pathological situations such as ischemia, high levels of lysoplasmalogens may act as detergents, and affect the integrity of neural membranes by interacting with individual proteins, or by affecting the biophysical properties of membranes. Lysoplasmalogens increase membrane fluidity correlating with an increased permeability for a number of solutes (Han and Gross, 1991). In ischemic heart tissue, high levels of lysoplasmalogens may produce spontaneous contraction, which may cause arrhythmia (Caldwell and Baumgarten, 1998).

Ischemic injury is also accompanied by the stimulation of the Ca^{2+}-dependent PLA_2 hydrolyzing PtdEtn (Edger et al., 1982). This stimulation of cytosolic Ca^{2+}-dependent PLA_2 may be due to increased gene expression or covalent modification of this enzyme (Edger et al., 1982; Owada et al., 1994). The stimulation of phospholipases A_2 induces membrane dysfunction, which may lead to cellular injury (Farooqui and Horrocks, 1991; Ray et al., 1994; Farooqui et al., 1997c). The stimulation of plasmalogen-selective PLA_2 may occupy a proximal position in the injury pathway, initiating cell injury, whereas participation of Ca^{2+}-dependent PLA_2 that hydrolyzes PtdCho may result in amplification of the injury process. The Ca^{2+}-dependent PLA_2 hydrolyzing PtdCho is cleaved and stimulated by caspase-3, an enzyme involved in apoptotic cell death (Wissing et al., 1997; Cummings et al., 2000). This type of cell death occurs in the brain tissue after ischemic injury, and phospholipases A_2 are known to play an important role in apoptotic cell death by altering membrane phospholipid composition (including a decrease in plasmalogens), permeability, and fluidity (Farooqui and Horrocks, 1991; Farooqui et al., 1997b; Sapirstein and Bonventre, 2000; Cummings et al., 2000).

6.2.2 *Plasmalogens in Alzheimer Disease*

Levels of ethanolamine plasmalogens are markedly decreased in autopsy brain samples from AD patients, compared with age-matched controls (Wells et al., 1995; Ginsberg et al., 1995; Guan et al., 1999; Farooqui et al., 1997a; Han et al., 2001; Pettegrew et al., 2001). The decrease in ethanolamine plasmalogen in AD is accompanied by a marked elevation not only in phosphatidyserine and phosphoethanolamine (Wells et al., 1995) (Fig. 6.2), but also in the generation of prostaglandins and lipid peroxides (Farooqui et al., 2003a, b). Determination of PlsEtn-PLA_2 activity in the cytosolic fraction prepared from different regions of brains from normal individuals and AD patients has indicated a three- to fourfold increase in PlsEtn-PLA_2 activity in the nucleus basalis, and a twofold increase in the hippocampus in AD brain, compared with control human brain (Fig. 6.3). It is proposed that elevation in PlsEtn-PLA_2 activity may cause a deficiency of plasmalogens and loss of synapses in AD brain (Wells et al., 1995; Ginsberg et al., 1995; Ginsberg et al., 1998; Guan et al., 1999; Han et al., 2001; Pettegrew et al., 2001). This deficiency of plasmalogens in the AD patient's neural membranes may result not only in abnormal signal transduction but also in accumulation of eicosanoids and lipid peroxides, including 4-hydroxynonenal (Farooqui and Horrocks, 2006). It is worth mentioning here that a reduction in ethanolamine plasmalogen content, compared to age-matched controls, is found only in AD. Other neurodegenerative diseases do not show the loss of plasmalogens from the brain tissue (Fig. 6.4).

The cause of the increased PlsEtn-PLA_2 activity is not fully understood. However, it is well known that translocation of PLA_2 from the cytosol to plasma and nuclear membranes, and generation of ceramide and cytokines stimulate various isoforms of PLA_2. So the increase in activity of PlsEtn-PLA_2 may be due to increases in levels of ceramide and cytokines (Farooqui and Horrocks, 2007). It remains to be seen whether the increase in activity of PlsEtn-PLA_2 correlates with the number of senile plaques or neurofibrillary tangles. It must be admitted that at this stage, we do not know whether changes in PlsEtn-PLA_2 are the cause or the consequence of neurodegenerative processes or whether changes in this PlsEtn-PLA_2 enzyme are primary or secondary. Furthermore, it is not known whether changes in PlsEtn-PLA_2 are specific for AD or if other neurodegenerative diseases also show similar increases in the activity of this enzyme (Farooqui and Horrocks, 2007).

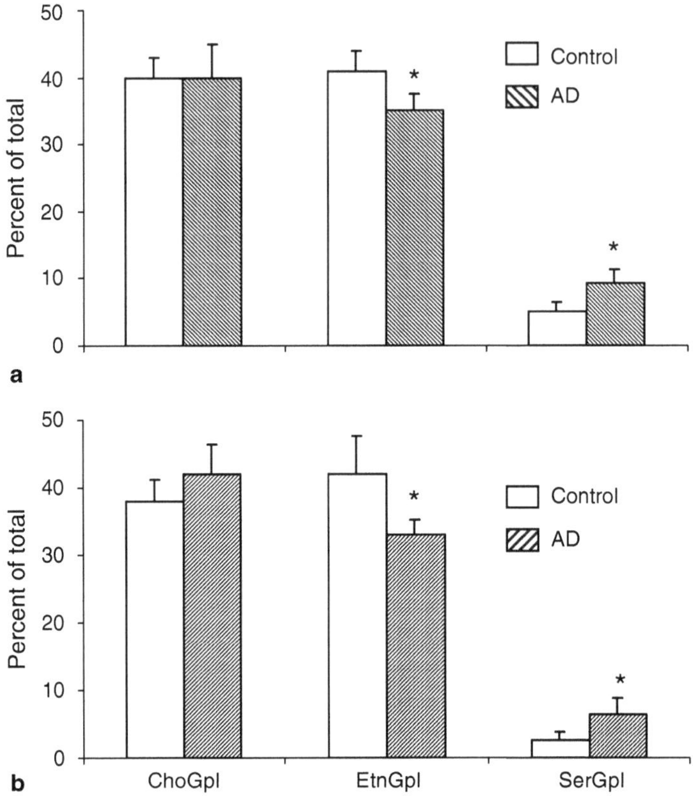

Fig. 6.2 Proportions of choline, ethanolamine, and serine glycerophospholipids in the plasma membrane (**a**) and synaptosomal plasma membrane (**b**) fractions from cerebral cortex of control subjects and AD patients. EtnGpl ($p = 0.0001$) and SerGpl ($p = 0.0001$). These values differed significantly. Data modified from Wells et al. (1995).

A deficiency of ethanolamine plasmalogen in neural membranes from AD patients may lead to neural membrane destabilization due to changes in the critical temperature necessary for maintaining the stability of the lipid bilayer (Ginsberg et al., 1998). A deficiency of plasmalogens also produces impairment of muscarinic cholinergic signals and abnormal amyloid precursor processing. These are characteristic features of AD (Périchon et al., 1998; Roth et al., 1995). Neuronal membranes are highly interactive and dynamic structures. They are involved in receiving, processing, transporting, and transmitting information from one cell type to another in the brain tissue. Their function depends not only upon receptors, ion channels, and enzymes that are embedded in the bilayer, but also on a delicate balance in lipid composition. Another effect of plasmalogen deficiency in AD is the inhibition of free cholesterol esterification in the plasma membrane (Munn et al., 2003). This may not only modulate intracellular cholesterol signaling but also reverse extracellular cholesterol transport (Mandel et al., 1998).

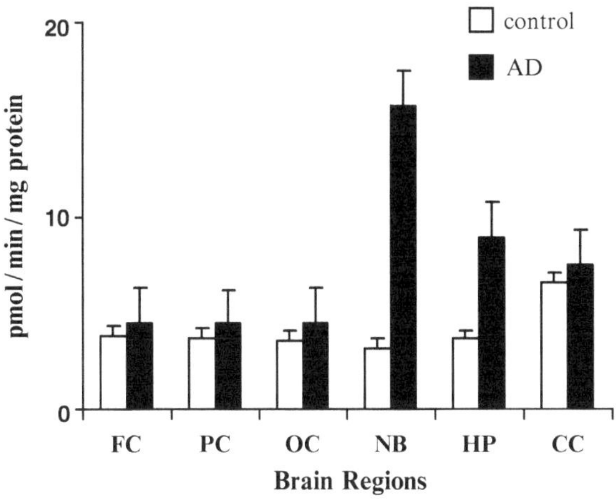

Fig. 6.3 Activity of PlsEtn-PLA_2 in different regions of brain from normal subjects and Alzheimer disease patients. Age-matched autopsy brains from normal subjects and Alzheimer disease patients were obtained 10–12h after death and used for the determination of enzymic activity. *FC* frontal cortex, *PC* parietal cortex, *OC* occipital cortex, *NB* nucleus basalis, *HP* hippocampus, and *CC* carpus callosum. Modified from Farooqui et al. (2003a).

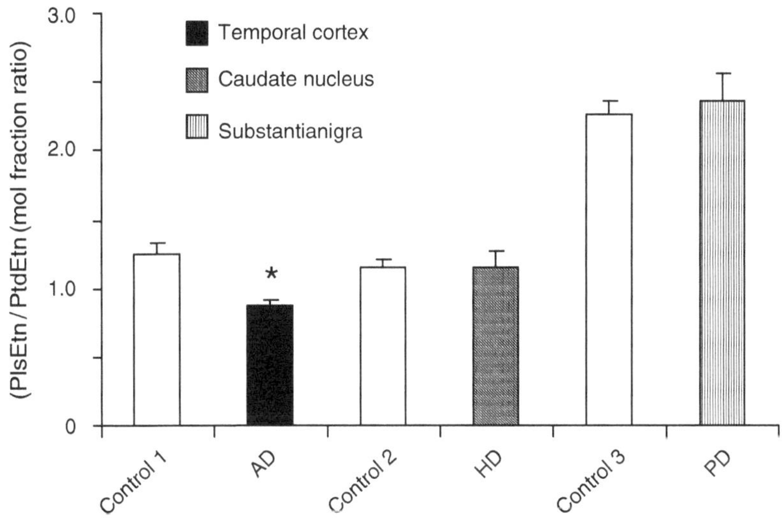

Fig. 6.4 Mole fraction ratio of PlsEtn to PtdEtn in control and Alzheimer disease (AD) midtemporal cortex, control and Huntington disease (HD) caudate nucleus, and control and Parkinson disease (PD) substantia nigra. Results are expressed as mean ± SEM. Significance relative to control by the Mann–Whitney *U* test is $p = 0.0009$, $p = 0.508$, and $p = 0.69$. Modified from Farooqui et al. (1997a).

Normally, in the brain tissue, the damage caused by lipid peroxidation is balanced by antioxidant defense mechanisms (Halliwell, 1994; Farooqui et al., 2008). During normal aging, neural cells can tolerate mild stress by upregulating the synthesis of antioxidant defense systems in an attempt to restore the balance. However, severe oxidative stress, as in AD, can produce major interdependent derangement of cell metabolic processes, including decrease in plasmalogens, DNA strand breakage, and damage to membrane ion transporters. Plasmalogens are known to act as antioxidant and protect polyunsaturated fatty acid from iron-induced lipid peroxidation (Sindelar et al., 1999; Guan et al., 1999). Their deficiency may cause impairment of the antioxidant defense system, resulting in oxidative-stress-mediated neuronal injury in AD. This is tempting to speculate that more studies are required on the involvement of plasmalogens in neural membrane destabilization/stabilization, signal transduction, and modulation of membrane function in AD and its animal models.

6.2.3 *Plasmalogens in Spinal Cord Injury*

In the spinal cord, plasmalogens account for about one-third of the total glycerophospholipids. Much of the PlsEtn are found in the myelin sheath. Degradation of the PlsEtn is similar in gray and white matter during and after compression spinal cord trauma (Demediuk et al., 1985). About 10% of plasmalogens are lost during the first minute of compression, with an overall loss of 18% found at 30 min after compression injury (Fig. 6.5). In another model of spinal cord injury in rabbits (Lukácová et al., 1996), the increased levels of thiobarbituric acid reactive substance

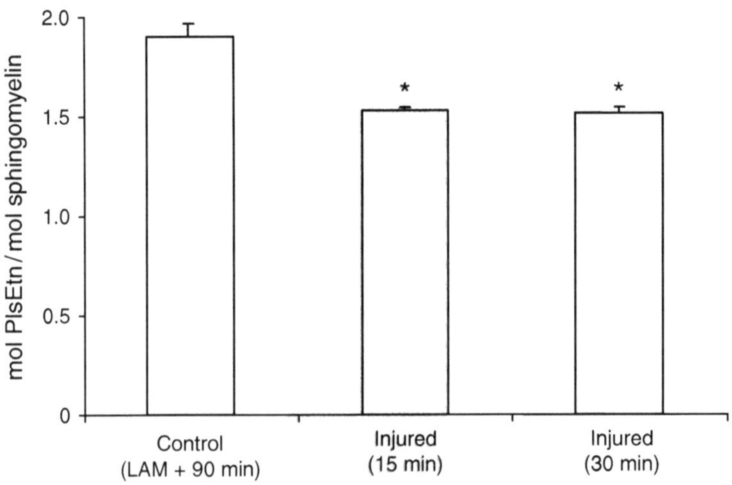

Fig. 6.5 Decrease in levels of ethanolamine plasmalogen after compression spinal cord injury. After laminectomy (LAM), all animals are allowed to recover 90 min before compression trauma. Samples from control (LAM + 90 min), compression for 15 min, and 30 min were removed and phospholipid analysis was performed. Values are means ± SEM, $n = 3$. Modified from Demediuk et al. (1985).

is reported to be associated with a decrease in PlsEtn levels. The loss of plasmalogens after compression and ischemic injuries can be explained by the stimulation of plasmalogen-selective PLA_2 due to shear stress. It is interesting to note that shear stress is known to stimulate another form of PLA_2, namely cytosolic PLA_2 (molecular mass, 85 kDa), which acts on phosphatidylcholine. Stimulation of plasmalogen-selective PLA_2 in spinal cord injury may result in changes in membrane fluidity and permeability, resulting in increased Ca^{2+} influx, impaired mitochondrial function, and the subsequent formation of reactive oxygen species (ROS), leading to lipid peroxidation. Each of these processes may contribute to neurodegeneration in spinal cord injury.

6.2.4 Plasmalogens in Peroxisomal Disorders

Zellweger syndrome or cerebrohepatorenal syndrome is an autosomal recessive disease characterized by the absence of peroxisomes (Heymans et al., 1983). The absence of peroxisomes results in deficiencies of peroxisomal enzymes, causing multiple neurochemical abnormalities associated with lipid metabolism, including the deficiency of plasmalogens and increase in very long chain fatty acids. The deficiency of plasmalogens is caused by the decrease in activities of dihydroxyacetone phosphate acyltransferase and alkyldihydroxyacetone phosphate synthase (Datta et al., 1984). The incorporation of [^{14}C]hexadecanol into the alk-1-enyl moiety of plasmalogens is markedly decreased in Zellweger patients, compared with controls. In contrast, the incorporation of [^{3}H]alkylglycerol into plasmalogens occurs with the same efficiency in Zellweger syndrome patients as in controls (van den Bosch et al., 1993), suggesting that only the reaction(s) involved in the introduction of the alkyl side-chain during biosynthesis of plasmalogens are deficient in Zellweger patients. These results have been recently confirmed by magnetic resonance imaging of Zellweger syndrome patients' fetuses (Mochel et al., 2006). The high level resolution of magnetic resonance imaging, which allows analysis of cerebral gyration and myelination, facilitates the prenatal diagnosis of complex polymalformative syndromes such as Zellweger syndrome. It is stated that dihydroxyacetone phosphate acyltransferase activity and C26:0 β-oxidation are the best markers in predicting life expectancy of patients with borderline personality disorder. Combination of both markers may give an even better prediction. Plasmalogens are essential for the myelination (Horrocks and Sharma, 1982), and so their deficiency in Zellweger syndrome may be responsible for many neurological as well as retinal problems in these patients.

Rhizomelic chondrodysplasia punctata (RCDP) is another peroxisomal disorder. The disorder is caused by mutations in the *PEX7* gene, which encodes the receptor for a class of peroxisomal matrix enzymes (Bams-Mengerink et al., 2006; Gorgas et al., 2006). Clinically the disease is characterized by symmetrical shortening of the proximal limbs, contractures of joints, a characteristic dysmorphic face, and cataracts. Although some RCDP patients have single enzyme deficiency, the majority of RCDP patients (86%) belong to a single complementation group (CG11).

Cells from CG11 show many biochemical abnormalities, including a deficiency of (a) DHAP-AT, (b) alkyldihydroxyacetonephosphate synthase, and (c) phytanic acid α-oxidation, and (d) inability to import peroxisomal thiolase (Motley et al., 1997). It is reported that all RCDP patients from CG11 group with detectable *PEX7* mRNA have mutations in *PEX7*, a gene encoding peroxin 7, the cytosolic PTS2-receptor protein required for targeting a subset of enzymes to peroxisomes (Motley et al., 1997). A mutation resulting in C-terminal truncation of *PEX7* cosegregates with the disease, and expression of *PEX7* in RCDP fibroblasts from CG11 rescues the PTS2 protein import deficiency.

In many RCDP patients, plasmalogen deficiency is accompanied by death in early childhood. Several variant forms of this peroxisomal disorder are known to occur in human population. Some variants are characterized by the deficiency of DHAP-AT, whereas others are associated with the deficiency of alkyl-dihydroxyacetonephosphate synthase activity in fibroblasts from patients of RCDP (de Vet and van den Bosch, 2000). Based on metabolic and kinetic studies, it is suggested that that the activity of DHAP-AT is dependent on the presence of alkyl-dihydroxyacetonephosphate synthase protein. The deficiency of DHAP-AT in the classic form of RCDP is a consequence of the absence of the alkyl-dihydroxyacetonephosphate synthase protein (de Vet and van den Bosch, 2000). Some variants of RCDP are also characterized by the deficiency of 3-oxoacyl-coenzyme A thiolase in peroxisomes with impaired processing of this enzyme (Heikoop et al., 1990). The reduction of 3-oxoacyl-CoA thiolase activity results in a decrease in the rate of peroxisomal β-oxidation of palmitoyl-CoA. However, the capacity of the peroxisomes to oxidize very long chain fatty acids is sufficient to prevent excessive accumulation of these compounds in vivo.

A mouse model of RCDP has been recently developed (Rodemer et al., 2003). This mouse is characterized by a targeted disruption of the DHAP-AT gene. The mutant mouse shows multiple abnormalities, such as male infertility, defects in eye development, cataract, and optic nerve hypoplasia, some of which are also observed in RCDP. Plasmalogens are completely absent and the concentration of brain DHAis decreased in mutant mouse. The marker proteins such as flotillin-1 and F3/contactin are found in brain lipid raft or microdomains in reduced concentrations. It is proposed that this mouse can be used to obtain more information on the pathogenesis of RCDP (Rodemer et al., 2003).

Dietary supplementation of plasmalogen has failed to restore normal levels of erythrocyte plasmalogen in patients with peroxisomal disorders such as RCDP (Heymans et al., 1985). In contrast, Zellweger syndrome patients (20 weeks or younger) had erythrocyte plasmalogen levels lower than those in older patients or normal controls (Wanders et al., 1986), indicating that an increase in erythrocyte plasmalogen levels in the older Zellweger patients has occurred during aging. The molecular mechanism of this process remains unknown. However, the increase in plasmalogen levels can be attributed to exchange of phospholipids with plasmalogens from circulating lipoproteins. Supplementation of diet with plasmalogens results in the absorption of plasmalogens in rat intestine, contributing to a significant increase in plasmalogen levels in blood plasma (Nishimukai et al., 2003). Significant amount of plasmalogen is also absorbed into the lymph (Hara et al., 2003). This increase in plasmalogen levels in blood may provide increased protection against lipoprotein oxidation. Similarly, studies on human endothelial cells have indicated that supplementation of cell cultures with *sn*-1-O-hexadecylglycerol produces a twofold increase in cellular levels of plasmalogen (Zoeller et al., 2002). These supplemented endothelial cells with increased plasmalogen levels have been shown to be resistant to hypoxic injury with no evidence of an increase in ROS (Zoeller et al., 2002). In contrast, exposure of nonsupplemented endothelial cells to hypoxic injury decreases plasmalogen levels and increases cellular ROS (Viani et al., 1995; Zoeller et al., 2002). This observation strongly suggests that it is possible to increase plasmalogen levels in cultured endothelial cells by supplementing cultures with *sn*-1-O-hexadecylglycerol. These results are supported by earlier studies indicating that erythrocyte plasmalogen levels are increased in patients with peroxisomal disorders after oral feeding of batyl alcohol for a few months (Das et al., 1992). Plasmalogen levels in brain tissue of these patients were not determined. Pregnant rats ate powdered food containing 0.5% (w/w) 1-O-heptadecyl-*sn*-glycerol from the 5th day before delivery until the 14th day after the pups were born. This resulted in considerable incorporation of 17:0 in alkenyl chains of ethanolamine plasmalogen in liver (28%) and kidney (17%), but the total concentration of plasmalogen in these tissues was not altered. In brain, a low (2%) incorporation of 17:0 alkenyl chains was observed in rat pups (Das et al., 1992). The reason for the low incorporation of the ether lipid precursor in brain tissue is not fully understood. However, it may be due either to the blood brain barrier or to the low consumption of heptadecylglycerol by rat pups. It still remains to be seen whether a similar increase occurs in cultures of neuronal and glial origin. It is an open question whether *sn*-1-O-hexadecylglycerol can be used to restore plasmalogen levels in Zellweger syndrome patients, and whether this restoration in plasmalogen levels can provide actual benefit to the Zellweger syndrome patients.

Alterations in the incorporation of [^{14}C]hexadecanol and [^{3}H]alkylglycerol are found in fibroblasts from patients with other peroxisomal disorders such as RCDP, neonatal adrenoleukodystrophy, and infantile Refsum disease (van den Bosch et al., 1993). A deficiency of ethanolamine plasmalogen is also observed in Niemann-Pick type C disease (Schedin et al., 1997). The deficiency of plasmalogens in Zellweger syndrome and Niemann-Pick type C disease suggests the involvement of peroxisomes in the synthesis of ether lipids. The pathophysiological events associated with peroxisomal disorders may not reflect the effect of plasmalogen insufficiency, but may be the consequences of the loss of other peroxisomal function, such as a defect in the degradation of very long chain fatty acids (Lee, 1998). In most peroxisomal disorders, the decrease in plasmalogen levels is in parallel with the DHA levels. This results in disturbances in brain and spinal cord function (Roels et al., 1993; Martínez, 1992). A deficit in DHA-containing plasmalogens may not only affect the integrity of neural membranes, but also alter the activities of membrane-bound enzymes and receptors and metabolic reactions such as superoxide dismutase and nitric oxide synthase (Yehuda et al., 1999, 2002).

Treatment of Zellweger syndrome patients with purified DHA partially improves visual function, increases levels of plasmalogens, and reduces levels of saturated very long chain free fatty acids (Martínez et al., 2000). The treatment with DHA is more effective when started very early in life. In contrast, recent studies in a mouse model of Zellweger syndrome, peroxisome-deficient mice, indicate that normalization of DHA levels in Zellweger pups does not improve symptoms of Zellweger syndrome. The neuronal migration defect was unaltered, indicating that a DHA deficit is not a major pathogenic factor in these newborn Zellweger mice (Janssen et al., 2000; Infante and Huszagh, 2001). The reason for this discrepancy between human and mouse is not known. Apparently, the peroxisome-deficient mouse is not an exact model for the human condition. The discrepancy may be due to differences in signal transduction between the human disease and its mouse model.

DHA supplementation reduces stress-induced aggression in students, indicating that DHA may have psychotropic effects (Hamazaki et al., 1996). Thus, the supplementation of DHA in diet may protect brain tissue against schizophrenia, depression, hyperactivity, stroke, and AD by stimulating the synthesis of plasmalogens, correcting abnormal signal transduction processes, and restoring neural membrane integrity. Recent studies also indicate that plasmalogen levels in rat brain are increased by the administration of *myo*-inositol, a six-carbon sugar that has been commonly used for the treatment of depression, panic disorder, and obsessive-compulsive disorders (Pettegrew et al., 2001).

DHA and plasmalogens turn over rapidly in the brain (Jones et al., 1997; Rapoport, 1999; Rosenberger et al., 2002; Farooqui and Horrocks, 2001), and so a continued supply of a DHA-enriched diet is necessary throughout life (Horrocks and Yeo, 1999). Multicenter controlled studies are required for testing the ability of DHA-enriched diets to treat not only peroxisomal disorders, but also learning and psychotic changes in schizophrenia, depression, hyperactivity, stroke, and AD.

6.2.5 Plasmalogens in Sjogren-Larsson Syndrome

Sjögren-Larsson syndrome (SLS) is an inherited disorder characterized by ichthyosis, spastic di- or tetraplegia, and mental retardation. It is associated with reduced oxidation of long-chain aliphatic alcohols due to deficient activity of fatty alcohol: NAD^+ oxidoreductase (FAO) (Rizzo and Craft, 1991). FAO is a complex enzyme, which consists of two separate proteins, namely, fatty aldehyde dehydrogenase (FALDH) and fatty alcohol dehydrogenase. These enzymes catalyze the oxidation of fatty alcohol to fatty aldehyde and fatty acid, respectively. Determination of FAO activity in SLS fibroblast has indicated that FALDH component of FAO is selectively deficient in SLS, and fatty alcohol dehydrogenase shows normal activity (Rizzo and Craft, 1991). Accumulation of plasmalogen-derived fatty aldehydes in fibroblasts from SLS has also been reported (Rizzo et al., 2000). Accumulated fatty aldehyde forms a Schiff base with PtdEtn (Stadelmann-Ingrand et al., 2004). Levels of Schiff base adduct are fourfold higher in SLS fibroblasts, compared with those in normal controls. Collective evidence implicates FALDH in the oxidation of ether-derived fatty aldehyde in human and rodent cells. It is proposed that abnormal metabolism of ether glycerophospholipids may contribute to the pathogenesis of SLS (Rizzo et al., 2000).

6.2.6 *Plasmalogens in Malnutrition*

Neonatal undernutrition affects the phospholipid content of brain tissue (Reddy and Horrocks, 1982). The plasmalogen content of white matter of undernourished brain is 36% lower. The lower plasmalogen content correlates with the decrease in the number of synapses per unit area of neonatal undernourished brain (Shoemaker and Bloom, 1977). The amount of myelin recovered from malnourished rat brain is lower than from control rat brain. Nutritional rehabilitation restores the phospholipid content and reverses morphological changes (Reddy et al., 1982).

6.2.7 *Plasmalogens in Fetal Alcohol Syndrome*

Ethanol effects the incorporation of [^{3}H]ethanolamine into plasmalogens in the differentiating CG-4 oligodendrocytes cell line. Thus, ethanol inhibits the increased labeling of plasmalogens if it is present for the first 48 h of differentiation (Bichenkov and Ellingson, 1999). This may be due either to the decrease in activities of PlsEtn-synthesizing enzymes or to increased activity of plasmalogen-selective PLA_2. Chronic maternal ethanol consumption during pregnancy is known to cause fetal alcohol syndrome. Infants born with this syndrome have impaired intelligence and motor function, and hyperactivity (Clarren and Smith, 1978). The neurochemical mechanism of this type of brain damage remains unknown. However, it has been hypothesized that ethanol consumption impairs the accumulation of DHA, 22:6n-3, during brain development. This fatty acid is needed for the biosynthesis of plasmalogens in peroxisomes (Burdge, 1998). This hypothesis is supported by studies on ethanol exposure. Lower plasmalogen and DHA contents of neural membranes are found in developing and adult brain (Hofteig et al., 1985; Wing et al., 1982). The lower DHA content in brain induced by ethanol can be corrected by increasing maternal dietary DHA intake (Burdge, 1998). Plasmalogen levels in these brains were not reported. However, it has been suggested that some harmful effects of ethanol on brain can be reduced or reversed by nutritional intervention with DHA-enriched diet (Burdge, 1998).

6.2.8 *Plasmalogens in Diabetic Heart*

Plasmalogens are the predominant glycerophospholipids of sarcolemma and sarcoplasmic reticulum of myocardial cells (Gross, 1985), and majority of plasmalogen-selective

PLA_2 is localized in surface membrane of isolated myocytes. A significant increase in plasmalogen-selective PLA_2 is reported in myocardial membranes of diabetic heart. This diabetes-induced increase in plasmalogen-selective PLA_2 can be reversed by treating diabetic animals with insulin (McHowat et al., 2000). Diabetic condition has no effect on cytosolic PLA_2 activity of heart tissue. The increase in plasmalogen-selective PLA_2 activity in diabetes is accompanied with increase in lysoplasmalogen content. This may contribute to arrhythmias vulnerability in diabetic animals. It is also reported that lysoplasmalogen at low concentration (0.3–1.0 μM) is a potent modulator of multiple membrane ionic currents (Liu et al., 1999), which are another factor responsible for arrhythmias. In addition, plasmalogen levels are significantly decreased in patients with acute coronary syndrome or supraventricular tachycardias (Brosche et al., 2007). This decrease in plasmalogen may be due to an increase in PlsEtn-PLA_2 activity.

6.2.9 Plasmalogens in Other Neurological Disorders

A deficiency of ethanolamine plasmalogens also occurs in Niemann-Pick type C disease (Schedin et al., 1997), a lysosomal storage disease characterized not only by an accumulation of cholesterol and sphingomyelin but also by abnormal cellular trafficking of cholesterol. It is stated that plasmalogen deficiency may not be the primary defect, but a contributing factor to the progressive neurological dysfunction observed in this disorder. Deficiency of plasmalogens also occurs in hyperlipidemia (Engelmann et al., 1992, 1994). Analyses of autopsy brain samples from older Down syndrome and multiple sclerosis patients indicate reduced levels of plasmalogens in this disorder (Murphy, 2001; Yanagihara and Cumings, 1969). This suggests that plasmalogens are necessary for proper brain development, and the loss of plasmalogen may be an indication of demyelination and oxidative stress (Farooqui and Horrocks, 2001).

6.3 Plasmalogens in Uremic Patients

Determination of fatty aldehyde dimethyl acetals (16:0 DMA and 18:0 DMA, representing derivatives of plasmalogens) in fasting serum phospholipids of 30 patients with chronic renal failure receiving repeated ambulatory hemodialysis, compared with 99 normal control subjects, has indicated reduced levels of plasmalogens in uremic patients (Brosche et al., 2002). Thus, the reduced content of serum plasmalogen in uremic patients undergoing hemodialysis suggests an increased oxidative stress. Enrichment of lipoproteins with plasmalogens may increase oxidative resistance in uremic patients.

6.4 Plasmalogens in Myelin-Deficient Mutant Mice

Plasmalogens are major phospholipid constituents of myelin membranes (Horrocks and Sharma, 1982), and their lack is an indication of dysmyelination or demyelination. Low levels of plasmalogens occur in the myelin-deficient mice, jimpy and quaking (Nussbaum et al., 1969; Dawson and Clarke, 1971). The molecular species of myelin ethanolamine plasmalogen that is most affected in myelin-deficient mice contains 18:1 acyl chain on both positions of the glycerol moiety (Hack and Helmy, 1978). This molecular species is apparently unique to myelin, and has been suggested to play an important role in maintaining myelin structure. Levels of plasmalogens are also decreased in multiple sclerosis (Yanagihara and Cumings, 1969), suggesting the stimulation of plasmalogen-selective PLA_2 during demyelination (Huterer et al., 1995). Autopsy brains from older Down syndrome patients have been reported to show decreased levels of plasmalogens (Murphy et al., 2000). Plasmalogen-deficient mutant cell lines have been developed (Zoeller et al., 1999; Gaposchkin and Zoeller, 1999). These cell lines can be used not only to study the mechanism of restoration of plasmalogen levels with 1-O-hexadecyl-*sn*-glycerol, but to elucidate the functions of plasmalogens.

6.5 Conclusion

The consequences of plasmalogen deficiency due to stimulation of plasmalogen-selective PLA_2 during acute neural trauma, ischemia, and spinal cord injury, and in neurodegenerative diseases (including AD) can be lethal. Generally, under normal conditions, the stability of the cell membrane lipid bilayer depends on the interrelationship between physiologic temperature and membrane lipid composition (Farooqui et al., 2000). A plasmalogen deficiency can lead to neuronal membrane destabilization with alterations in membrane fluidity and permeability (Farooqui et al., 1997a), leading to Ca^{2+} influx, which can initiate a cascade of neurochemical reactions resulting in apoptotic as well as necrotic cell death (Farooqui et al., 2000). A decrease in plasmalogen levels may produce membrane destabilization by changing the critical temperature necessary for maintaining the stability of the lipid bilayer (Ginsberg et al., 1998). Fluorescent probe studies have shown that plasmalogen deficiency increases membrane lipid mobility, resulting in relocation of cholesterol to more hydrophobic areas in the lipid bilayer (Maeba and Ueta, 2004), thereby exposing to peroxidation many phospholipids containing polyunsaturated fatty acids. These processes are closely associated with neural cell injury and neurodegeneration in acute neural trauma and neurodegenerative diseases (Farooqui and Horrocks, 2001). It remains an open question whether plasmalogen deficiency in neurological disorders is the cause, or the result of the disease progression.

References

Bams-Mengerink A. M., Majoie C. B., Duran M., Wanders R. J., Van Hove J., Scheurer C. D., Barth P. G., and Poll-The B. T. (2006). MRI of the brain and cervical spinal cord in rhizomelic chondrodysplasia punctata. Neurology 66:798–803.

Berry K. A. Z. and Murphy R. C. (2005). Free radical oxidation of plasmalogen glycerophosphocholine containing esterified docosahexaenoic acid: Structure determination by mass spectrometry. Antioxidants and Redox Signaling 7:157–169.

Bichenkov E. and Ellingson J. S. (1999). Temporal and quantitative expression of the myelin-associated lipids, ethanolamine plasmalogen, galactocerebroside, and sulfatide, in the differentiating CG-4 glial cell line. Neurochem. Res. 24:1549–1556.

Brosche T., Platt D., and Knopf B. (2002). Decreased concentrations of serum phospholipid plasmalogens indicate oxidative burden of uraemic patients undergoing haemodialysis. Nephron 90:58–63.

Brosche T., Brueckmann M., Haase K. K., Sieber C., and Bertsch T. (2007). Decreased plasmalogen concentration as a surrogate marker of oxidative stress in patients presenting with acute coronary syndromes or supraventricular tachycardias. Clin. Chem. Lab. Med. 45:689–691.

Burdge G. C. (1998). The role of docosahexaenoic acid in brain development and fetal alcohol syndrome. Biochem. Soc. Trans. 26:246–252.

Caldwell R. A. and Baumgarten C. M. (1998). Plasmalogen-derived lysolipid induces a depolarizing cation current in rabbit ventricular myocytes. Circ. Res. 83:533–540.

Clarren S. K. and Smith D. W. (1978). The fetal alcohol syndrome. N. Engl. J. Med. 298:1063–1067.

Cummings B. S., McHowat J., and Schnellmann R. G. (2000). Phospholipase A_2s in cell injury and death. J. Pharmacol. Exp. Ther. 294:793–799.

Daniel L. W., Sciorra V. A., and Ghosh S. (1999). Phospholipase D, tumor promoters, proliferation and prostaglandins. Biochim. Biophys. Acta Mol. Cell Biol. Lipids 1439:265–276.

Das A. K., Holmes R. D., Wilson G. N., and Hajra A. K. (1992). Dietary ether lipid incorporation into tissue plasmalogens of humans and rodents. Lipids 27:401–405.

Datta N. S., Wilson G. N., and Hajra A. K. (1984). Deficiency of enzymes catalyzing the biosynthesis of glycerol-ether lipids in Zellweger syndrome. A new category of metabolic disease involving the absence of peroxisomes. N. Engl. J. Med. 311:1080–1083.

Dawson R. M. and Clarke N. (1971). Cerebral phospholipids in 'quaking' mice. J. Neurochem. 18:1313–1316.

Demediuk P., Saunders R. D., Anderson D. K., Means E. D., and Horrocks L. A. (1985). Membrane lipid changes in laminectomized and traumatized cat spinal cord. Proc. Natl. Acad. Sci. USA 82:7071–7075.

de Vet E. C. and van den Bosch H. (2000). Alkyl-dihydroxyacetonephosphate synthase. Cell Biochem. Biophys. 32:117–121.

Edger A. D., Strosznajder J., Horrocks L. A. (1982). Activation of ethanolamine phospholipase A_2 in Brain during ischemia. J. Neurochem. 39:1111–1116.

Engelmann B., Streich S., Schönthier U. M., Richter W. O., and Duhm J. (1992). Changes of membrane phospholipid composition of human erythrocytes in hyperlipidemias. I. Increased phosphatidylcholine and reduced sphingomyelin in patients with elevated levels of triacylglycerol-rich lipoproteins. Biochim. Biophys. Acta Lipids Lipid Metab. 1165:32–37.

Engelmann B., Bräutigam C., and Thiery J. (1994). Plasmalogen phospholipids as potential protectors against lipid peroxidation of low density lipoproteins. Biochem. Biophys. Res. Commun. 204:1235–1242.

Farooqui A. A. and Horrocks L. A. (1991). Excitatory amino acid receptors, neural membrane phospholipid metabolism and neurological disorders. Brain Res. Rev. 16:171–191.

Farooqui A. A. and Horrocks L. A. (2001). Plasmalogens: Workhorse lipids of membranes in normal and injured neurons and glia. Neuroscientist 7:232–245.

Farooqui A. A. and Horrocks L. A. (2004). Plasmalogens, platelet-activating factor, and other ether lipids. In: Nicolaou A. and Kokotos G. (Eds.), *Bioactive Lipids*. Oily, Bridgwater, England, pp. 107–134.

Farooqui A. A. and Horrocks L. A. (2005). Signaling and interplay mediated by phospholipases A_2, C, and D in LA-N-1 cell nuclei. Reprod. Nutr. Dev. 45:613–631.

Farooqui A. A. and Horrocks L. A. (2006). Phospholipase A_2-generated lipid mediators in the brain: The good, the bad, and the ugly. Neuroscientist 12:245–260.

Farooqui A. A. and Horrocks L. A. (2008). *Glycerophospholipids in the Brain: Phospholipases A_2 in Neurological Disorders*. Springer, New York, pp. 1–290.

Farooqui A. A., Hirashima Y., and Horrocks L. A. (1992). Brain phospholipases and their role in signal transduction. Adv. Exp. Med. Biol. 318:11–25.

Farooqui A. A., Yang H. C., and Horrocks L. A. (1995). Plasmalogens, phospholipases A_2, and signal transduction. Brain Res. Rev. 21:152–161.

Farooqui A. A., Rapoport S. I., and Horrocks L. A. (1997a). Membrane phospholipid alterations in Alzheimer disease: Deficiency of ethanolamine plasmalogens. Neurochem. Res. 22:523–527.

Farooqui A. A., Yang H. C., Rosenberger T. A., and Horrocks L. A. (1997b). Phospholipase A_2 and its role in brain tissue. J. Neurochem. 69:889–901.

Farooqui A. A., Yang H. C., and Horrocks L. A. (1997c). Involvement of phospholipase A_2 in neurodegeneration. Neurochem. Int. 30:517–522.

Farooqui A. A., Horrocks L. A., and Farooqui T. (2000). Glycerophospholipids in brain: Their metabolism, incorporation into membranes, functions, and involvement in neurological disorders. Chem. Phys. Lipids 106:1–29.

Farooqui A. A., Ong W. Y., and Horrocks L. A. (2003a). Plasmalogens, docosahexaenoic acid, and neurological disorders. In: Roels F., Baes M., and de Bies S. (Eds.), *Peroxisomal Disorders and Regulation of Genes*. Kluwer Academic/Plenum, London, pp. 335–354.

Farooqui A. A., Ong W. Y., and Horrocks L. A. (2003b). Stimulation of lipases and phospholipases in Alzheimer disease. In: Szuhaj B. and van Nieuwenhuyzen W. (Eds.), *Nutrition and Biochemistry of Phospholipids*. AOCS, Champaign, IL, pp. 14–29.

Farooqui A. A., Ong W. Y., and Horrocks L. A. (2006). Inhibitors of brain phospholipase A_2 activity: Their neuropharmacological effects and therapeutic importance for the treatment of neurologic disorders. Pharmacol. Rev. 58:591–620.

Farooqui A. A., Ong W. Y., and Horrocks L. A. (2008). *Neurochemical Aspects of Excitotoxicity*. Springer, New York, pp. 1–290.

Forrester J. S., Milne S. B., Ivanova P. T., and Brown H. A. (2004). Computational lipidomics: A multiplexed analysis of dynamic changes in membrane lipid composition during signal transduction. Mol. Pharmacol. 65:813–821.

Gaposchkin D. P. and Zoeller R. A. (1999). Plasmalogen status influences docosahexaenoic acid levels in a macrophage cell line: Insights using ether lipid-deficient variants. J. Lipid Res. 40:495–503.

Ginsberg L., Rafique S., Xuereb J. H., Rapoport S. I., and Gershfeld N. L. (1995). Disease and anatomic specificity of ethanolamine plasmalogen deficiency in Alzheimer's disease brain. Brain Res. 698:223–226.

Ginsberg L., Xuereb J. H., and Gershfeld N. L. (1998). Membrane instability, plasmalogen content, and Alzheimer's disease. J. Neurochem. 70:2533–2538.

Gorgas K., Teigler A., Komljenovic D., and Just W. W. (2006). The ether lipid-deficient mouse: Tracking down plasmalogen functions. Biochim. Biophys. Acta Mol. Cell Res. 1763:1511–1526.

Gross R. W. (1985). Identification of plasmalogen as the major phospholipid constituent of cardiac sarcoplasmic reticulum. Biochemistry 24:1662–1668.

Gross R. W., Jenkins C. M., Yang J. Y., Mancuso D. J., and Han X. L. (2005). Functional lipidomics: The roles of specialized lipids and lipid–protein interactions in modulating neuronal function. Prostaglandins Other Lipid Mediat. 77:52–64.

Guan Z. Z., Wang Y. A., Cairns N. J., Lantos P. L., Dallner G., and Sindelar P. J. (1999). Decrease and structural modifications of phosphatidylethanolamine plasmalogen in the brain with Alzheimer disease. J. Neuropathol. Exp. Neurol. 58:740–747.

Hack M. H. and Helmy F. M. (1978). The diminution of the myelin ethanolamine plasmalogen in brain of the jimpy mouse and brain and spinal cord of the quaking mouse as visualized by thin-layer chromatography. J. Chromatogr. 145:307.

Halliwell B. (1994). Free radicals and antioxidants: A personal view [Review]. Nutr. Rev. 52:253–265.

Hamazaki T., Sawazaki S., Itomura M., Asaoka E., Nagao Y., Nishimura N., Yazawa K., Kuwamori T., and Kobayashi M. (1996). The effect of docosahexaenoic acid on aggression in young adults – A placebo-controlled double-blind study. J. Clin. Invest. 97:1129–1133.

Han X. and Gross R. W. (1991). Alterations in membrane dynamics elicited by amphiphilic compounds are augmented in plasmenylcholine bilayers. Biochim. Biophys. Acta Biomembr. 1069:37–45.

Han X. L., Holtzman D. M., and McKeel D. W., Jr. (2001). Plasmalogen deficiency in early Alzheimer's disease subjects and in animal models: Molecular characterization using electrospray ionization mass spectrometry. J. Neurochem. 77:1168–1180.

Hara H., Wakisaka T., and Aoyama Y. (2003). Lymphatic absorption of plasmalogen in rats. Br. J. Nutr. 90:29–32.

Hashimoto M., Hossain S., Shimada T., Sugioka K., Yamasaki H., Fujii Y., Ishibashi Y., Oka J. I., and Shido O. (2002). Docosahexaenoic acid provides protection from impairment of learning ability in Alzheimer's disease model rats. J. Neurochem. 81:1084–1091.

Hazen S. L., Ford D. A., and Gross R. W. (1991). Activation of a membrane-associated phospholipase A_2 during rabbit myocardial ischemia which is highly selective for plasmalogen substrate. J. Biol. Chem. 266:5629–5633.

Heikoop J. C., van Roermund C. W., Just W. W., Ofman R., Schutgens R. B., Heymans H. S., Wanders R. J., and Tager J. M. (1990). Rhizomelic chondrodysplasia punctata. Deficiency of 3-oxoacyl-coenzyme A thiolase in peroxisomes and impaired processing of the enzyme. J. Clin. Invest. 86:126–130.

Heymans H. S. A., Schutgens R. B. H., Tan R., van den Bosch H., and Borst P. (1983). Severe plasmalogen deficiency in tissues of infants without peroxisomes (Zellweger syndrome). Nature 306:69–70.

Heymans H. S. A., Oorthuys J. W. E., Nelck G., Wanders R. J. A., Dingemans K. P., and Schutgens R. B. H. (1985). Peroxisomal abnormalities in rhizomelic chondrodysplasia punctata. J. Inherit. Metab. Dis. 9:329–331.

Hofteig J. H., Noronha A. B., Druse M. J., and Keresztes-Nagy C. (1985). Synaptic membrane phospholipids: Effects of maternal ethanol consumption. Exp. Neurol. 87:165–171.

Horrocks L. A. and Sharma M. (1982). Plasmalogens and O-alkyl glycerophospholipids. In: Hawthorne J. N. and Ansell G. B. (Eds.), *Phospholipids, New Comprehensive Biochemistry, Vol. 4*. Elsevier Biomedical, Amsterdam, pp. 51–93.

Horrocks L. A. and Yeo Y. K. (1999). Health benefits of docosahexaenoic acid (DHA). Pharmacol. Res. 40:211–225.

Huterer S. J., Tourtellotte W. W., and Wherrett J. R. (1995). Alterations in the activity of phospholipases A_2 in post-mortem white matter from patients with multiple sclerosis. Neurochem. Res. 20:1335–1343.

Infante J. P. and Huszagh V. A. (2001). Zellweger syndrome knockout mouse models challenge putative peroxisomal beta-oxidation involvement in docosahexaenoic acid (22: 6n-3) biosynthesis. Mol. Genet. Metab. 72:1–7.

Jagannatha H. M. and Sastry P. S. (1981). Ethanolamine plasmalogen and cholesterol ester metabolism in experimental allergic encephalomyelitis. Indian J. Biochem. Biophys. 18:411–416.

Janssen A., Baes M., Gressens P., Mannaerts G. P., Declercq P., and Van Veldhoven P. P. (2000). Docosahexaenoic acid deficit is not a major pathogenic factor in peroxisome-deficient mice. Lab. Invest. 80:31–35.

Jones C. R., Arai T., and Rapoport S. I. (1997). Evidence for the involvement of docosahexaenoic acid in cholinergic stimulated signal transduction at the synapse. Neurochem. Res. 22:663–670.

Kubota M., Nakane M., Nakagomi T., Tamura A., Hisaki H., Shimasaki H., and Ueta N. (2001). Regional distribution of ethanolamine plasmalogen in the hippocampal CA1 and CA3 regions and cerebral cortex of the gerbil. Neurosci. Lett. 301:175–178.

Lee S. H., Williams M. V., and Blair I. A. (2005). Targeted chiral lipidomics analysis. Prostaglandins Other Lipid Mediat. 77:141–157.

Lee T. C. (1998). Biosynthesis and possible biological functions of plasmalogens. Biochim. Biophys. Acta Lipids Lipid Metab. 1394:129–145.

Liu S. J., McHowat J., and Creer M. H. (1999). Effects of lysoplasmenylcholine on membrane currents in rabbit ventricular myocytes. J. Mol. Cell Cardiol. 31, 27 (abstract).

Lukácová N., Halát G., Chavko M., and Maršala J. (1996). Ischemia-reperfusion injury in the spinal cord of rabbits strongly enhances lipid peroxidation and modifies phospholipid profiles. Neurochem. Res. 21:869–873.

Maeba R. and Ueta N. (2004). A novel antioxidant action of ethanolamine plasmalogens in lowering the oxidizability of membranes. Biochem. Soc. Trans. 32:141–143.

Mandel H., Sharf R., Berant M., Wanders R. J. A., Vreken P., and Aviram M. (1998). Plasmalogen phospholipids are involved in HDL-mediated cholesterol efflux: Insights from investigations with plasmalogen-deficient cells. Biochem. Biophys. Res. Commun. 250:369–373.

Martínez M. (1990). Severe deficiency of docosahexaenoic acid in peroxisomal disorders: A defect of delta 4 desaturation? Neurology 40:1292–1298.

Martínez M. (1992). Abnormal profiles of polyunsaturated fatty acids in the brain, liver, kidney and retina of patients with peroxisomal disorders. Brain Res. 583:171–182.

Martínez M., Vázquez E., García-Silva M. T., Manzanares J., Bertran J. M., Castelló F., and Mougan I. (2000). Therapeutic effects of docosahexaenoic acid ethyl ester in patients with generalized peroxisomal disorders. Am. J. Clin. Nutr. 71:376S–385S.

McHowat J., Creer M. H., Hicks K. K., Jones J. H., McCrory R., and Kennedy R. H. (2000). Induction of Ca-independent PLA_2 and conservation of plasmalogen polyunsaturated fatty acids in diabetic heart. Am. J. Physiol. Endocrinol. Metab. 279:E25–E32.

Mochel F., Grebille A. G., Benachi A., Martinovic J., Razavi F., Rabier D., Simon I., Boddaert N., Brunelle F., and Sonigo P. (2006). Contribution of fetal MR imaging in the prenatal diagnosis of Zellweger syndrome. AJNR Am. J. Neuroradiol. 27:333–336.

Motley A. M., Hettema E. H., Hogenhout E. M., Brites P., ten Asbroek A. L., Wijburg F. A., Baas F., Heijmans H. S., Tabak H. F., Wanders R. J., and Distel B. (1997). Rhizomelic chondrodysplasia punctata is a peroxisomal protein targeting disease caused by a non-functional PTS2 receptor. Nat. Genet. 15:377–380.

Munn N. J., Arnio E., Liu D., Zoeller R. A., and Liscum L. (2003). Deficiency in ethanolamine plasmalogen leads to altered cholesterol transport. J. Lipid Res. 44:182–192.

Murphy E. J., Schapiro M. B., Rapoport S. I., and Shetty H. U. (2000). Phospholipid composition and levels are altered in Down syndrome brain. Brain Res. 867:9–18.

Murphy R. C. (2001). Free-radical-induced oxidation of arachidonoyl plasmalogen phospholipids: Antioxidant mechanism and precursor pathway for bioactive eicosanoids. Chem. Res. Toxicol. 14:463–472.

Nishimukai M., Wakisaka T., and Hara H. (2003). Ingestion of plasmalogen markedly increased plasmalogen levels of blood plasma in rats. Lipids 38:1227–1235.

Nussbaum J. L., Neskovic N., and Mandel P. (1969). A study of lipid components in brain of the 'Jimpy' mouse, a mutant with myelin deficiency. J. Neurochem. 16:927–934.

Ong W. Y., Ling S. F., Yeo J. F., Chiueh C. C., and Farooqui A. A. (2005). Injury and recovery of pyramidal neurons in the rat hippocampus after a single episode of oxidative stress induced by intracerebroventricular injection of ferrous ammonium citrate. Reprod. Nutr. Dev. 45:647–662.

Owada Y., Tominaga T., Yoshimoto T., and Kondo H. (1994). Molecular cloning of rat cDNA for cytosolic phospholipase A2 and the increased gene expression in the dentate gyrus following transient forebrain ischemia. Brain Res Mol Brain Res. 25:364–368.

Périchon R., Moser A. B., Wallace W. C., Cunningham S. C., Roth G. S., and Moser H. W. (1998). Peroxisomal disease cell lines with cellular plasmalogen deficiency have impaired muscarinic cholinergic signal transduction activity and amyloid precursor protein secretion. Biochem. Biophys. Res. Commun. 248:57–61.

Pettegrew J. W., Panchalingam K., Hamilton R. L., and McClure R. J. (2001). Brain membrane phospholipid alterations in Alzheimer's disease. Neurochem. Res. 26:771–782.

Piomelli D. (2005). The challenge of brain lipidomics. Prostaglandins Other Lipid Mediat. 77:23–34.

Porcellati G. (1983). Phospholipid metabolism in neural membranes. In: Sun G. Y., Bazan N., Wu J. Y., Porcellati G., and Sun A. Y. (Eds.), *Neural Membranes*. Humana, New York, pp. 3–35.

Portilla D., Shah S. V., Lehman P. A., and Creer M. H. (1994). Role of cytosolic calcium-independent plasmalogen-selective phospholipase A_2 in hypoxic injury to rabbit proximal tubules. J. Clin. Invest. 93:1609–1615.

Poulos A., Bankier A., Beckman K., Johnson D., Robertson E. F., Sharp P., Sheffield L., Singh H., Usher S., and Wise G. (1991). Glyceryl ethers in peroxisomal disease. Clin Genet. 39:13–25.

Purdon A. D. and Rapoport S. I. (2007). Energy consumption by phospholipid metabolism in mammalian brain. In: Gibson G. and Dienel G. (Eds.), *Brain Energetics. Integration of Molecular and Cellular Processes*, in Handbook of Neurochemistry and Molecular Neurobiology (Lajtha, A., Ed.). Springer, New York. (In press)

Purdon A. D., Rosenberger T. A., Shetty H. U., and Rapoport S. I. (2002). Energy consumption by phospholipid metabolism in mammalian brain. Neurochem. Res. 27:1641–1647.

Rapoport S. I. (1999). In vivo fatty acid incorporation into brain phospholipids in relation to signal transduction and membrane remodeling. Neurochem. Res. 24:1403–1415.

Ray P., Ray R., Broomfield C. A., and Berman J. D. (1994). Inhibition of bioenergetics alters intracellular calcium, membrane composition, and fluidity in a neuronal cell line. Neurochem. Res. 19:57–63.

Reddy T. S. and Horrocks L. A. (1982). Effects of neonatal undernutrition on the lipid composition of gray matter and white matter in rat brain. J. Neurochem. 38:601–605.

Reddy T. S., Rajalakshmi R., and Ramakrishnan C. V. (1982). Effects of nutritional rehabilitation on the content and lipid composition of brain gray and white matter of neonatally undernourished rats. J. Neurochem. 39:1297–1301.

Rizzo W. B. and Craft D. A. (1991). Sjögren-Larsson syndrome. Deficient activity of the fatty aldehyde dehydrogenase component of fatty alcohol:NAD^+ oxidoreductase in cultured fibroblasts. J. Clin. Invest. 88:1643–1648.

Rizzo W. B., Heinz E., Simon M., and Craft D. A. (2000). Microsomal fatty aldehyde dehydrogenase catalyzes the oxidation of aliphatic aldehyde derived from ether glycerolipid catabolism: Implications for Sjögren-Larsson syndrome. Biochim. Biophys. Acta Mol. Basis Dis. 1535:1–9.

Rodemer C., Thai T. P., Brugger B., Kaercher T., Werner H., Nave K. A., Wieland F., Gorgas K., and Just W. W. (2003). Inactivation of ether lipid biosynthesis causes male infertility, defects in eye development and optic nerve hypoplasia in mice. Hum. Mol. Genet. 12:1881–1895.

Roels F., Fischer S., and Kissling W. (1993). Polyunsaturated fatty acids in peroxisomal disorders: A hypothesis and a proposal for treatment. J. Neurol. Neurosurg. Psychiatry 56:937.

Rosenberger T. A., Oki J., Purdon A. D., Rapoport S. I., and Murphy E. J. (2002). Rapid synthesis and turnover of brain microsomal ether phospholipids in the adult rat. J. Lipid Res. 43:59–68.

Roth G. S., Joseph J. A., and Mason R. P. (1995). Membrane alterations as causes of impaired signal transduction in Alzheimer's disease and aging. Trends Neurosci. 18:203–206.

Rüdiger M., von Baehr A., Haupt R., Wauer R. R., and Rüstow B. (2000). Preterm infants with high polyunsaturated fatty acid and plasmalogen content in tracheal aspirates develop bronchopulmonary dysplasia less often. Crit. Care Med. 28:1572–1577.

Sapirstein A. and Bonventre J. V. (2000). Phospholipases A_2 in ischemic and toxic brain injury. Neurochem. Res. 25:745–753.

Schedin S., Sindelar P. J., Pentchev P., Brunk U., and Dallner G. (1997). Peroxisomal impairment in Niemann-Pick type C disease. J. Biol. Chem. 272:6245–6251.
Shoemaker W. J. and Bloom F. E. (1977). Effect of undernutrition on brain morphology. In: Wurtman R. J. and Wurtman J. J. (Eds.), *Nutrition and the Brain*. Raven, New York, pp. 147–192.
Sindelar P. J., Guan Z. Z., Dallner G., and Ernster L. (1999). The protective role of plasmalogens in iron-induced lipid peroxidation. Free Radic. Biol. Med. 26:318–324.
Singh I., Paintlia A. S., Khan M., Stanislaus R., Paintlia M. K., Haq E., Singh A. K., and Contreras M. A. (2004). Impaired peroxisomal function in the central nervous system with inflammatory disease of experimental autoimmune encephalomyelitis animals and protection by lovastatin treatment. Brain Res. 1022:1–11.
Stadelmann-Ingrand S., Pontcharraud R., and Fauconneau B. (2004). Evidence for the reactivity of fatty aldehydes released from oxidized plasmalogens with phosphatidylethanolamine to form Schiff base adducts in rat brain homogenates. Chem. Phys. Lipids 131:93–105.
van den Bosch H., Schrakamp G., Hardeman D., Zomer A. W. M., Wanders R. J. A., and Schutgens R. B. H. (1993). Ether lipid synthesis and its deficiency in peroxisomal disorders. Biochimie 75:183–189.
Viani P., Zini I., Cervato G., Biagini G., Agnati L. F., and Cestaro B. (1995). Effect of endothelin-1 induced ischemia on peroxidative damage and membrane properties in rat striatum synaptosomes. Neurochem. Res. 20:689–695.
Voelker D. R. (2003). New perspectives on the regulation of intermembrane glycerophospholipid traffic. J. Lipid Res. 44:441–449.
Vreken P., Valianpour F., Overmars H., Barth P. G., Selhorst J. J. M., Van Gennip A. H., and Wanders R. J. A. (2000). Analysis of plasmenylethanolamines using electrospray tandem mass spectrometry and its application in screening for peroxisomal disorders. J. Inherit. Metab. Dis. 23:429–433.
Wanders R. J. A., Purvis Y. R., Heymans H. S. A., Bakkeren J. A. J. M., Parmentier G. G., van Eldere J., Eyssen H., van den Bosch H., Tager J. M., and Schutgens R. B. H. (1986). Age-related differences in plasmalogen content of erythrocytes from patients with the cerebro-hepato-renal (Zellweger) syndrome: Implications for postnatal detection of the disease. J. Inherit. Metab. Dis. 9:335–342.
Wells K., Farooqui A. A., Liss L., and Horrocks L. A. (1995). Neural membrane phospholipids in Alzheimer disease. Neurochem. Res. 20:1329–1333.
Wing D. R., Harvey D. J., Hughes J., Dunbar P. G., McPherson K. A., and Paton W. D. (1982). Effects of chronic ethanol administration on the composition of membrane lipids in the mouse. Biochem. Pharmacol. 31:3431–3439.
Wissing D., Mouritzen H., Egeblad M., Poirier G. G., and Jäättelä M. (1997). Involvement of caspase-dependent activation of cytosolic phospholipase A_2 in tumor necrosis factor-induced apoptosis. Proc. Natl. Acad. Sci. USA 94:5073–5077.
Yanagihara T. and Cumings J. N. (1969). Alterations of phospholipids, particularly plasmalogens, in the demyelination of multiple sclerosis as compared with that of cerebral oedema. Brain 92:59–70.
Yehuda S., Rabinovitz S., and Mostofsky D. I. (1999). Essential fatty acids are mediators of brain biochemistry and cognitive functions. J. Neurosci. Res. 56:565–570.
Yehuda S., Rabinovitz S., Carasso R. L., and Mostofsky D. I. (2002). The role of polyunsaturated fatty acids in restoring the aging neuronal membrane. Neurobiol. Aging 23:843–853.
Zhang J. P. and Sun G. Y. (1995). Free fatty acids, neutral glycerides, and phosphoglycerides in transient focal cerebral ischemia. J. Neurochem. 64:1688–1695.
Zoeller R. A., Lake A. C., Nagan N., Gaposchkin D. P., Legner M. A., and Lieberthal W. (1999). Plasmalogens as endogenous antioxidants: Somatic cell mutants reveal the importance of the vinyl ether. Biochem. J. 338:769–776.
Zoeller R. A., Grazia T. J., LaCamera P., Park J., Gaposchkin D. P., and Farber H. W. (2002). Increasing plasmalogen levels protects human endothelial cells during hypoxia. Am. J. Physiol. Heart Circ. Physiol. 283:H671–H679.

Chapter 7
Synthesis of Platelet-Activating Factor in Brain

7.1 Introduction

Platelet-activating factor (PAF) is the trivial name for 1-*O*-alkyl-2-acetyl-*sn*-glycero-3-phosphocholine (Fig. 7.1). PAF is synthesized from a specific subclass of PtdCho, which contains an ether bond, rather than an ester bond at the *sn*-1 position of the glycerol backbone (Snyder, 1995; Bazan, 2003). PAF production is tightly regulated both at the synthetic and degradative levels. PAF is released by a wide variety of cells, including neural cells, macrophages, platelets, endothelial cells, mast cells, and neutrophils. It causes neutrophil adhesion, chemotaxis, increased vascular permeability, and vasodialation. Although the synthesizing enzymes have not been purified and fully characterized from brain tissue, reports on the synthesis of PAF in mammalian brain are beginning to emerge (Francescangeli et al., 2000). PAF is synthesized in neural cells either spontaneously or under appropriate stimulation. In neuronal and glial cell cultures, acetylcholine dramatically stimulates PAF synthesis, and addition of cholinergic receptor antagonist, atropine, produces the inhibition of PAF synthesis (Sogos et al., 1990). PAF induces a significant mobilization of intracellular free Ca^{2+}, which is inhibited by PAF antagonists. The increase in Ca^{2+} is not only caused by the release from intracellular stores, but also through calcium influx via calcium channels. In neurons, PAF receptors are linked through guanine nucleotide-binding proteins (G proteins) to phospholipase C (PLC), and receptor-operated Ca^{2+} channels that are modulated by protein kinase C (PKC). Both PTX-sensitive and insensitive G proteins appear to be coupled with the PAF receptor, producing the activation of PLC and the increase in intracellular Ca^{2+}. Thus, in neurons, PAF action is associated with PLC and PKC-mediated signal transduction processes (Yue et al., 1992; Farooqui et al., 2008). In astrocytes, PAF upregulates nerve growth factor mRNA in a time and concentration-dependent manner. This increase in nerve growth factor mRNA is suppressed by WEB 2086 and BN52021, potent PAF antagonists (Brodie, 1995; Yoshida et al., 2005). PAF may be involved in interactions between astroglial cells and neurons. Astrocytic PAF may provide a neurotrophic signal to injured neurons. It is suggested that interplay between PAF and the neurotrophic receptor may be involved in regenerative processes in the brain tissue. In brain, the physiological

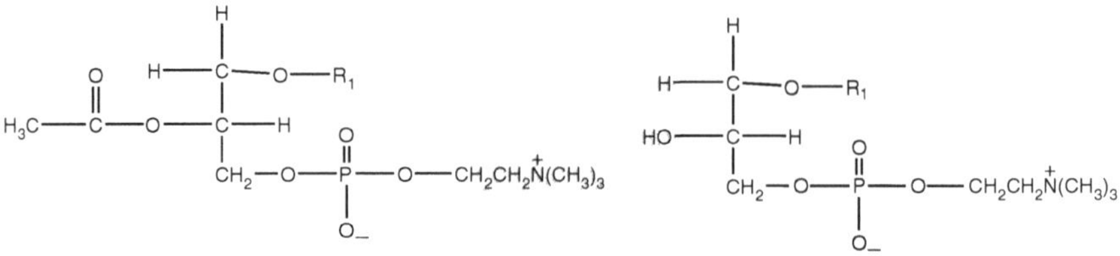

a b

Fig. 7.1 Structures of PAF (1-*O*-alkyl-2-acetyl-*sn*-glycero-3-phosphocholine) (**a**) and lyso-PAF (2-lyso-1-alkyl-*sn*-glycero-3-phosphocholine) (**b**). At the *sn*-1 position R1 is alkyl group and *sn*-2 position is occupied by an acetyl group.

activity of PAF is not limited to its proinflammatory function, neurotoxicity, apoptosis, and blood–brain barrier permeability, but also associated with neurotrophic effects. In nonneural tissues, PAF is also involved in a variety of other settings including reproduction, allergic reactions, and circulatory system disturbances such as atherosclerosis (Chao and Olson, 1993; Honda et al., 2002; Bazan, 2003).

7.2 Biosynthesis of PAF

Three different pathways of PAF synthesis are known to occur in mammalian tissues (Honda et al., 2002; Snyder, 1995). They include remodeling pathway, de novo synthesis, and oxidative fragmentation pathway. The remodeling pathway involves a structural modification of preexisting ether-linked phospholipids that serve as structural components of membranes (Snyder et al., 1996), and play a crucial role in inflammatory/hypersensitivity responses of cells, whereas the de novo reaction sequence appears to be of physiologically important for maintaining basal PAF levels in various tissues and blood (Stafforini et al., 1987). The balance between PAF biosynthesis and degradation determines its levels in various tissues. The degradation of PAF is catalyzed by platelet-activating factor hydrolase, which converts it to inactive lyso-PAF (see Chap. 8).

7.2.1 Remodeling Pathway (Deacylation/Reacylation Pathway)

The remodeling pathway occurs primarily in inflammatory cells (Fig. 7.2). In this pathway, the first step is the hydrolysis of arachidonate from 1-*O*-alkyl-2-arachidonyl-*sn*-glycero-3-phosphocholine by cytosolic phospholipase A_2 ($cPLA_2$). This reaction generates 1-*O*-alkyl-2-lyso-*sn*-glycero-3-phosphocholine (lyso-PAF) and arachidonic acid, simultaneously (Rubin et al., 2005). This reaction is the basis of

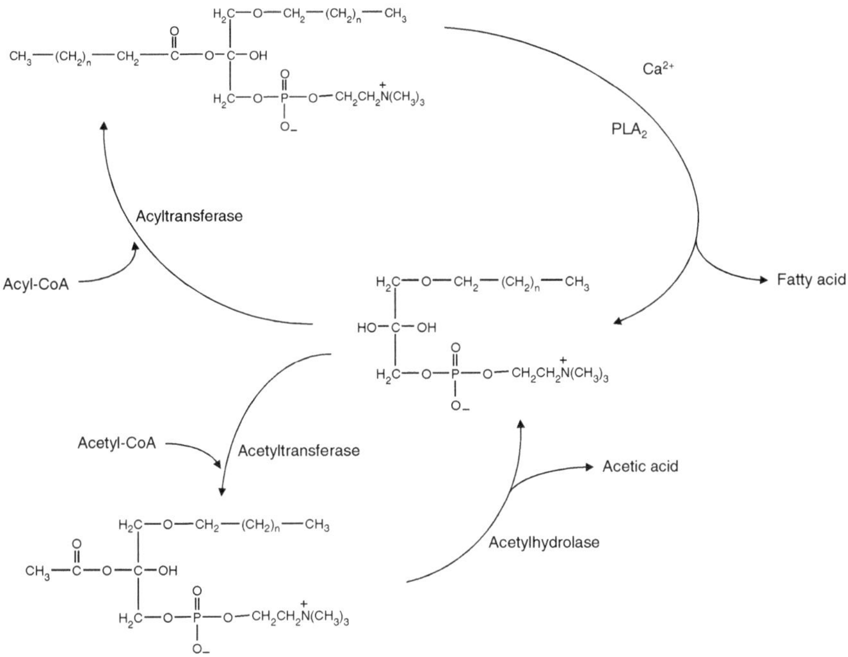

Fig. 7.2 Synthesis and degradation of PAF by the remodeling pathway.

the interrelationship between the synthesis of PAF and eicosanoids. Lyso-PAF is acetylated by acetyl-CoA/1-*O*-alkyl-2-lysophosphatidylcholine acetyltransferase to produce PAF. PAF can also be synthesized from plasmalogens. Thus, endothelin-mediated stimulation of brain PlsEtn-PLA_2 decreases PlsEtn levels and evokes the generation of PAF (Collado et al., 2003). The second step of PAF production requires the conversion of lyso-PAF to PAF by the enzyme acetyl-CoA: lyso-PAF acetyltransferase. Both the enzymes are activated by posttranslational phosphorylation (Prescott et al., 1990; Baker et al., 2002).

7.2.2 *Cytosolic Phospholipase A_2 ($cPLA_2$)*

$cPLA_2$ activity occurs in the brain tissue, but it has never been purified to homogeneity. $cPLA_2$ prefers arachidonic acid over other fatty acids, and does not use Ca^{2+} for catalysis, although submicromolar Ca^{2+} concentrations are needed for membrane binding (Clark et al., 1987; Farooqui et al., 2000). Owing to the presence of a Ca^{2+}-dependent phospholipid-binding domain at the N-terminal region, $cPLA_2$ is translocated in a Ca^{2+}-dependent manner from cytosol to the nuclear or other cellular membranes (Clark et al., 1987; Hirabayashi et al., 2004), where other downstream enzymes, including cyclooxygenases and lipoxygenases responsible for the metabolism of arachidonic acid to eicosanoids, are located. This gives $cPLA_2$ an access to its membrane-associated phospholipid substrate. The C-terminal region of $cPLA_2$ contains the phosphorylation site and the catalytic site. These sites may be involved in the regulation of the enzymic activity. The activation of $cPLA_2$ can be through serine residues, notably Ser^{505} and Ser^{727}, by mitogen-activated protein kinase (MAPK) and protein kinase C (PKC) (Hirabayashi and Shimizu, 2000; Hirabayashi et al., 2004). $cPLA_2$ also contains putative pleckstrin homology domains responsible for the ability of this enzyme to interact and bind with anionic phospholipids. Thus, $cPLA_2$ activity is also modulated through a cooperative binding mechanism with anionic phospholipids, such as phosphatidylinositol 4,5-bisphosphate (PtdIns-4,5-bis-P_2), phosphatidylinositol 3,4,5-trisphosphate (PtdIns-3,4,5-tris-P_3), and ceramide 1-phosphate (Hirabayashi et al., 2004; Pettus et al., 2004; Mosior et al., 1998).

In brain, several paralogs of $cPLA_2$ have been reported to occur (Table 7.1). They include $cPLA_2\alpha$, $cPLA_2$-β, $cPLA_2\gamma$, and $cPLA_2\delta$ (Diaz-Arrastia and Scott, 1999; Farooqui et al., 2000; Hirabayashi et al., 2004). $cPLA_2\alpha$, $cPLA_2\beta$, $cPLA_2\gamma$, and $cPLA_2\delta$ have molecular mass of 85, 114, 61, and 109 kDa, respectively. The basal expression of paralogs of $cPLA_2$ mRNA under normal conditions is very low in neuronal and glial cells of brain tissue (Owada et al., 1994; Stephenson et al., 1999). Recent electron microscopic, immunolabeling, and in situ hybridization studies have indicated that $cPLA_2\alpha$ is localized in somata and dendrites of Purkinje cells, whereas $cPLA_2\beta$ is present in granule cells of rat brain (Ong et al., 1999; Shirai and Ito, 2004). $cPLA_2\alpha$ is predominately found in astrocytes of gray matter (Farooqui et al., 2000; Pardue et al., 2003) as well as in hippocampal neurons (Sandhya et al., 1998; Kishimoto et al., 1999; Strokin et al., 2003), where under physiological conditions, $cPLA_2\alpha$ may be involved in second messenger generation and long-term potentiation (LTP), a mechanism involved in memory storage.

$cPLA_2\alpha$, $cPLA_2\beta$, and $cPLA_2\gamma$ contain two catalytic domains interspaced with paralog-specific sequences. The lipase motif, GXSGS, is located in the catalytic domain. $cPLA_2\beta$ is found mainly in the cerebellum and shares more similarities with $cPLA_2\alpha$ than with $cPLA_2\gamma$. $cPLA_2\gamma$ lacks the C2 domain, but contains a prenyl-group-binding motif, which behaves as a lipid anchor and allows binding of the enzyme to the membrane. Recombinantly expressed $cPLA_2\gamma$ liberates arachidonic acid from phosphatidylcholine. Unlike $cPLA_2\alpha$, $cPLA_2\gamma$ also acts on other

Table 7.1 Properties of $cPLA_2$ paralogs associated with remodeling pathway of PAF synthesis.

$cPLA_2$ paralog	Molecular mass (kDa)	CalB domain	Chromosomal localization	Reference
$cPLA_2\alpha$	85	Present	1	Shirai and Ito, 2004
$cPLA_2\beta$	114	Present	15	Ghosh et al., 2006; Shirai and Ito, 2004
$cPLA_2\gamma$	61	Absent	19	Shirai and Ito, 2004
$cPLA_2\delta$	93	–	–	Ohto et al., 2005
$cPLA_2\varepsilon$	100	–	–	Ohto et al., 2005
$cPLA_2$-ζ	95	–		Ohto et al., 2005

fatty acid residues at the *sn*-2 and *sn*-1 positions of glycerophospholipids. $cPLA_2\alpha$ has remarkable specificity for arachidonic acid at the *sn*-2 position. It also has *sn*-1 lysophospholipase activity, and a weak translocase activity associated with it. $cPLA_2\beta$ prefers to cleave fatty acids at the *sn*-1 position, and $cPLA_2\gamma$ efficiently hydrolyzes fatty acid at *sn*-1 as well as *sn*-2 positions of glycerol moiety (Song et al., 1999). The overexpression of $cPLA_2\gamma$ increases the proportions of polyunsaturated fatty acids in phosphatidylethanolamine, indicating that this paralog can modulate phospholipids composition (Asai et al., 2003). $cPLA_2\gamma$ is constitutively expressed in the endoplasmic reticulum, where it is involved in remodeling and maintaining membrane phospholipids composition under oxidative stress. $cPLA_2\beta$ displays much lower activity with [2-arachidonyl]PtdCho than do the other two paralogs. The genes for human $cPLA_2\alpha$, $cPLA_2\beta$, $cPLA_2\gamma$ have been mapped to chromosomes 1, 15, and 19, respectively. Mitogen-activated protein kinase phosphorylation sites are only present in $cPLA_2\alpha$ and not conserved in $cPLA_2\beta$ and $cPLA_2\gamma$. $cPLA_2\delta$ is mainly found in skin (Chiba et al., 2004). In contrast to other $cPLA_2$ paralogs, $cPLA_2\delta$ has a preference for linoleic acid release instead of arachidonic acid release. Roles of various paralogs of $cPLA_2$ in the brain tissue remain speculative (Farooqui et al., 2000; Hirabayashi et al., 2004). As $cPLA_2\delta$ mainly occurs in skin, it is proposed that this paralog plays a critical role in inflammation in psoriatic lesions (Chiba et al., 2004).

Besides $cPLA_2$, involvement of $sPLA_2$ and Ca^{2+}-independent $iPLA_2$ in PAF synthesis has also been reported (Bernatchez et al., 2001; McHowat et al., 2001). Treatment of human umbilical vein endothelial cells (HUVEC) cells with secretory PLA_2 ($sPLA_2$) inhibitors (SB203347 and LY311727) inhibits endothelial cell PAF synthesis by up to 90%. In contrast, $cPLA_2$ and $iPLA_2$ inhibitors have no effect on PAF synthesis in HUVEC cells. Collective evidence suggests that $sPLA_2$ provides substrate for endothelial cell PAF formation (Bernatchez et al., 2001). The treatment of endothelial cells with thrombin also results in PAF production (McHowat et al., 2001). On the basis of various pharmacological experiments, it is proposed that in HUVEC, PAF production requires Ca^{2+}-independent PLA_2 activation. This may occur through the CoA-independent transacylase remodeling pathway rather than as a direct result of the Ca^{2+}-independent PLA_2-mediated hydrolysis of alkylacylglycerophosphocholine (McHowat et al., 2001).

7.2.3 Acetyl-CoA/Lyso-PAF Acetyltransferase

This enzyme catalyzes the transfer of acetyl group from acetyl-CoA to the free hydroxyl at the *sn*-2 position of lyso-PAF. It is localized in microsomes. Acetyl-CoA/lyso-PAF acetyltransferase has been purified and cloned from nonneural tissues (Shindou et al., 2007). Acetyl-CoA/lyso-PAF acetyltransferase follows simple Michaelis–Menten kinetics with respect to acetyl-CoA (Table 7.2). In contrast, with respect to lyso-PAF, this enzyme follows simple saturation kinetics only up to 50 μM. Higher concentrations of lyso-PAF produce a sudden reduction in enzymic

Table 7.2 Properties of human kidney cortex acetyl-CoA/lyso-PAF acetyltransferase involved in remodeling pathway of PAF synthesis.

Parameter	Value	Reference
pH optimum	7.4	Nomikos et al., 2003
K_m (μM) for lyso-PAF	15.6	Nomikos et al., 2003
V_{max} (nmol min^{-1} mg^{-1}) for lyso-PAF	2.1	Nomikos et al., 2003
K_m (μM) for acetyl-CoA	90.4	Nomikos et al., 2003
V_{max} (nmol/min/mg) for acetyl-CoA	1.49	Nomikos et al., 2003
Detergents	Inhibition	Nomikos et al., 2003
Glycerol	Stabilization	Nomikos et al., 2003
Thermal stability	Very labile	Nomikos et al., 2003
DTNB	Inhibition	Nomikos et al., 2003
Subcellular localization	Microsomes	Nomikos et al., 2003; Prescott et al., 2000
Effect of Ca^{2+}	Stimulation	Prescott et al., 2000
Phosphorylation/dephosphorylation	Stimulation	Prescott et al., 2000

activity. This may be due to detergent-like effect of lyso-PAF. Acetyl-CoA/lyso-PAF acetyltransferase is a microsomal protein that has molecular mass of 60 kDa with three putative membrane-spanning domains. It is a very labile enzyme that is sensitive to detergents. Treatment with common detergents results in total loss of the enzymic activity. The addition of glycerol stabilizes the enzymic acitivity. DTT and mercaptoethanol significantly inhibit acetyl-CoA/lyso-PAF acetyltransferase, indicating the presence of disulfide bridges in the enzyme. Acetyl-CoA/lyso-PAF acetyltransferase can be induced by lipopolysaccharide (LPS), and suppressed by dexamethasone treatment. Acetyl-CoA/lyso-PAF acetyltransferase catalyzes not only biosynthesis of PAF from lyso-PAF, but also incorporation of arachidonoyl-CoA to produce PAF precursor membrane glycerophospholipids (lysophosphatidylcholine acyltransferase activity). Under physiological conditions, the enzyme prefers arachidonoyl-CoA and participates in membrane biogenesis. Following acute inflammatory stimulation with LPS, the activated enzyme utilizes acetyl-CoA more efficiently and produces PAF. Collective evidence suggests that a single enzyme catalyzes membrane biogenesis of inflammatory cells while producing a prophlogistic mediator in response to external stimuli (Shindou et al., 2007). In human umbilical vein endothelial cells (HUVEC), PAF synthesis is upregulated in response to hydrogen peroxide (H_2O_2) treatment and modulated by intracellular redox (Tosaki et al., 2007). Thus, treatment with antioxidants such as *N*-acetylcysteine, pyrrolidinecarbodithioic acid (PDTC), and Trolox reduces PAF production in H_2O_2-treated HUVEC. Addition of exogenous lyso-PAF has no effect on acetyl-CoA/lyso-PAF acetyltransferase activity. The acetyl-CoA/lyso-PAF acetyltransferase activity responds quickly to H_2O_2-treatment, but the activation seems to be transitory and caused by the overexpression of phospholipids hydroperoxide glutathione peroxidase, an antioxidant enzyme responsible for reducing peroxidized phospholipids produced in cell membranes and lipoproteins (Sakamoto et al., 2002).

Tyrosine kinase inhibitors and calmodulin antagonists inhibit acetyl-CoA/lyso-PAF acetyltransferase activity in H_2O_2-stimulated cells, suggesting that tyrosine kinase and calcium/calmodulin-dependent protein kinase are involved in regulating acetyltransferase activity (Tosaki et al., 2007). Similarly, in human neutrophils, acetyltransferase activity is modulated by mitogen-activated protein kinases, namely the p38 kinase (Nixon et al., 1999). Thus, nonneural acetyl-CoA/lyso-PAF acetyltransferase activity is modulated by phosphorylation/dephosphorylation process. In the endothelial cells, acetyl-CoA/lyso-PAF acetyltransferase is not only regulated by cytosolic Ca^{2+} and H^+, but through the activation of protein kinases C and A (Heller et al., 1991). Collectively, these studies suggest that acetyl-CoA/lyso-PAF acetyltransferase activity is regulated by phosphorylation/dephosphorylation processes.

7.2.4 CoA-Independent Transacetylase

Lyso-PAF formation can also be initiated by the selective transfer of arachidonate from 1-*O*-alkyl-2-arachidonyl-*sn*-glycero-3-phosphocholine to an acceptor lysophospholipid by a CoA-independent transacetylase activity (Prescott et al., 1990). The CoA-independent transacetylase has no requirement for CoA, Ca^{2+}, or Mg^{2+}, and exhibits a broad pH optimum (7.0–8.0). This enzyme has a K_m value of 12.0 μM (Lee et al., 1992). CoA-independent transacetylase participates in the biosynthesis of ethanolamine plasmalogen and acyl analogs of PAF, in vivo. A similar CoA-independent transacetylase activity that transfers acetyl group from PAF to sphingosine to form *N*-acetylsphingosine (C_2-ceramide) has been reported to occur in HL-60 cells (Lee et al., 1996). This enzyme is termed as PAF/sphingosine transacetylase. Kinetic studies have indicated that PAF/sphingosine transacetylase has a narrow substrate specificity, and strict stereochemical configuration requirements (Lee et al., 1996). Ceramide, sphingosylphosphocholine, stearylamine, sphingosine 1-phosphate, or sphingomyelin do not act as acetate acceptors, whereas sphinganine has only a limited capacity to accept the acetate from PAF. Tissue distribution studies have indicated that PAF transacetylase activity is widely distributed among several tissues. On the basis of the inhibitory effect of diethyl pyrocarbonate and *N*-ethylmaleimide, it is proposed that histidine and cysteine residues may be involved in the catalytic activity of this enzyme (Lee et al., 1996).

In rabbit cerebral nuclear preparations, PAF can be synthesized by the nuclear acetylation of alkylglycerophosphocholine (1-alkyl GPC), which is generated during nuclear transacylation reaction (Baker and Chang, 2000). Lysophospholipids promote the formation of 1-alkyl GPC from nuclear alkylacylglycerophosphocholine via transacylation. The formation of 1-alkyl GPC promotes PAF synthesis. The nuclear generation of PAF is of considerable interest because PAF plays regulatory roles in transcription events associated with inflammation and oxidative stress in the brain tissue (Baker and Chang, 2000).

C_2-Ceramide, generated via PAF/sphingosine transacetylase, may participate in modulation of physiological levels of C_2-ceramide in both undifferentiated and differentiated intact HL-60 cells (Lee et al., 1996). As C_2-ceramide has many biological activities that differ from that of PAF and sphingosine, the CoA-independent, PAF-dependent transacetylase serves as a modifier of PAF, and sphingosine functions by generating a unique of lipid mediator (C_2-ceramide) that modulates many cellular function including cell death (He et al., 2007). The breakdown of sphingomyelin and the release of ceramide are stimulated by PAF (Lang et al., 2005). In erythrocytes, this process is associated with apoptotic cell death. The molecular mechanism involved in apoptotic cell death in this system is not fully understood. However, PAF may activate an erythrocyte sphingomyelinase, and the generation of ceramide may lead to the activation of scramblase with subsequent phosphatidylserine exposure, a process associated with apoptotic cell death (Lang et al., 2005).

The transfer of the acetyl group from PAF to lysoplasmalogen in a CoA-independent manner is catalyzed by PAF/lysoplasmalogen transacetylase in HL-60 cells, endothelial cells, and a variety of rat tissues (Karasawa et al., 1999). Multiple column chromatographic procedures result in 13,700-fold purification of this enzyme from rat kidney mitochondrial plus microsomal membranes. On SDS-polyacrylamide gels, the purified enzyme migrates as a single band with an apparent molecular mass of 40 kDa. The purified enzyme catalyzes transacetylation of the acetyl group not only from PAF to lysoplasmalogen forming plasmalogen analogs of PAF, but also to sphingosine producing *N*-acetylsphingosine (C_2-ceramide) (Karasawa et al., 1999). In addition, this enzyme has PAF-acetylhydrolase activity in the absence of lipid acceptor molecules. Collective evidence suggests that PAF-dependent transacetylase is an enzyme that modifies the cellular functions of PAF through generation of other diverse lipid mediators (Karasawa et al., 1999).

The purified PAF/lysoplasmalogen transacetylase activity is stimulated by phosphatidyl-serine (PtdSer) with optimal activation occurring at 25 μM. Other acidic phospholipids, such as phosphatidylinositol (PtdIns) and phosphatidylinositol 4-phosphate (PtdIns-4-*P*), are partially effective, while diacylglycerol and free fatty acids have no effect on the PAF/lysoplasmalogen transacetylase activity (Lee et al., 2001). PtdSer exerts its effect on the PAF/lysoplasmalogen transacetylase activity through the increases of both K_m and V_{max}. In addition, *N*-ethylmalimide (NEM) and dithiobis-(2-nitro-5-thiobenzoic acid) (DTNB) not only inhibits the PAF/lysoplasmalogen transacetylase activity, but also partially blocks the PAF lysophospholipid transacetylase and PAF acetyl hydrolase activities of the purified enzyme in a concentration-dependent manner. The inhibition of PAF/lysoplasmalogen transacetylase by NEM and DTNB is partially protected by lysoplasmalogens. Furthermore, PAF fully protects the inhibition of PAF acetyl hydrolase, partially protects the inhibition of PAF lysophospholipid transacetylase, and does not protect the inhibition of PAF/lysoplasmalogen transacetylase by NEM. These results suggest that three individual catalytic activities of PAF-dependent transacetylase have different dependencies on the thiol-containing residues (cysteines) of the enzyme. Furthermore, the nonresponsiveness of the purified cytosolic PAF/lysoplasmalogen transacetylase activity to PtdSer activation is consistent with the view that

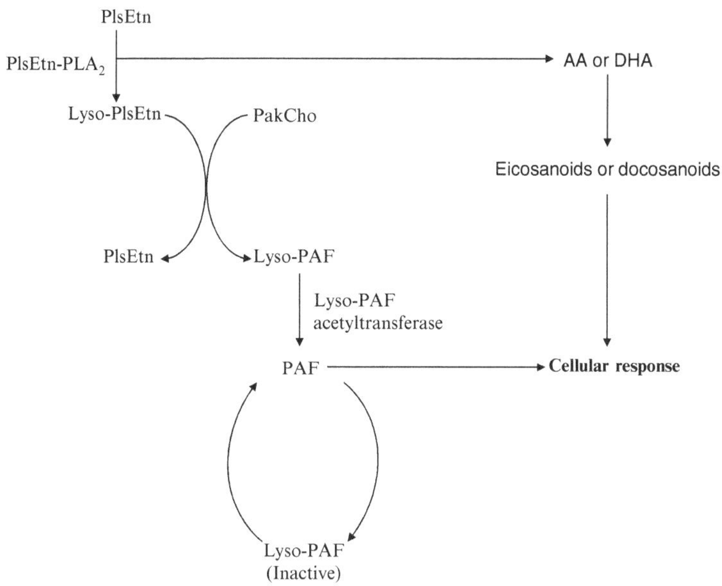

Fig. 7.3 Pathway showing relationship between plasmalogen degradation and PAF synthesis in neural and nonneural cells. Plasmalogen are hydrolyzed by PlsEtn-PLA_2 into lyso-PlsEtn. Transacylation of lysoPlsEtn (lysoplasmalogen) with PakCho (alkylacyl-glycerol-3-phosphocholine) results in the production of lyso-PAF. Acetylation of lyso-PAF by acetyl-CoA/1-alkyl-*sn*-glycero-3-phosphorylcholine 2-*O*-acetyltransferase results in synthesis of PAF.

membrane-bound PAF-dependent transacetylase and cytosolic PAF-dependent transacetylase may be distinct enzymes.

The addition of exogenous lysoplasmalogens or thrombin to endothelial cell cultures markedly stimulates the synthesis of PAF in a CoA-independent manner (McHowat and Creer, 2000). This suggests that plasmalogen-selective PLA_2, an enzyme that hydrolyzes plasmalogen into lysoplasmalogen, may modulate PAF synthesis (Fig. 7.3). On the basis of these observations, it can be proposed that cross-talk (interplay) occurs between plasmalogen and PAF metabolism in vivo. This interplay may be involved in fine-tuning not only PAF-receptor-induced biological responses, but also generation and maintenance of levels of other lipid mediators such as eicosanoids and docosanoids through the receptor-mediated plasmalogen degradation of plasmalogens (Farooqui and Horrocks, 2007).

7.3 De Novo Synthesis of PAF

De novo PAF synthesis in mammalian tissues requires the participation of three enzymes namely 1-alkyl-2-lyso-*sn*-glycero-3-phosphate (alkyllyso-GP)/acetyl-CoA acetyltransferase, 1-alkyl-2-acetyl-*sn*-glycero-3-phosphate phosphohydrolase,

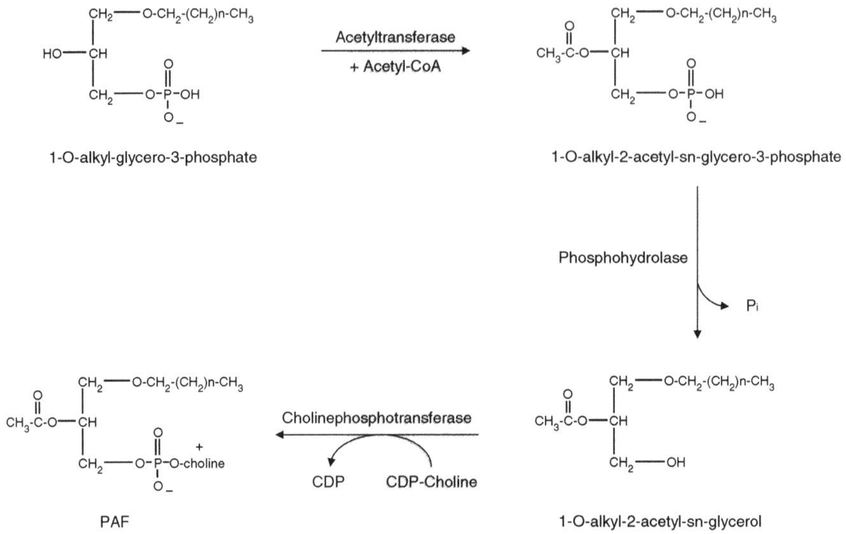

Fig. 7.4 Reactions showing de novo PAF synthesis in neural and non-neural cells.

and dithiothreitol (DTT)-insensitive 1-alkyl-2-acetyl-*sn*-glycerol/CDP-choline phosphotransferase (Heller et al., 1991; Snyder, 1995) (Fig. 7.4). In the de novo pathway, alkylglycerophosphate is converted to alkylacetylglycerophosphate by 1-alkyl-2-lyso-*sn*-glycero-3-phosphate (alkyllyso-GP)/acetyl-CoA acetyltransferase (Lee et al., 1986; Baker and Chang, 1993), and alkylacetylglycerophosphate is then dephosphorylated to 1-*O*-alkyl-2-acetyl-*sn*-glycerol by 1-alkyl-2-acetyl-*sn*-glycero-3-phosphate phosphohydrolase (Panwala et al., 1998). 1-*O*-alkyl-2-acetyl-*sn*-glycerol is transformed into PAF by 1-alkyl-2-acetyl-*sn*-glycerol/CDP-choline phosphotransferase. This pathway is analogous to the biosynthetic pathway of PtdCho except that the enzyme 1-alkyl-2-acetyl-*sn*-glycerol/CDP-choline phosphotransferase is insensitive to DTT. Less information is available on the de novo PAF synthesis than the remodeling pathway. De novo PAF synthesis has been reported to occur in a variety of cells and tissues, including renal medulla, umbilical vein endothelial cells, murine neuroblastoma cells, and rabbit cerebral cortices (Lee et al., 1989; Heller et al., 1991; Baker and Chang, 1993; Gimenez and Aguilar, 2001).

7.3.1 1-Alkyl-2-lyso-sn-glycero-3-phosphate (Alkyllyso-GP)/ Acetyl-CoA Acetyltransferase

The direct PAF precursors in de novo pathway are 1-alkyl-2-acetyl-*sn*-glycerol and 1-alkyl-2-acetyl-*sn*-glycerol-3-phosphate. 1-Alkyl-2-lyso-*sn*-glycero-3-phosphate (alkyllyso-GP)/acetyl-CoA acetyltransferase is a microsomal enzyme that has a broad pH optimum (8.0–8.8) at 23°C. To minimize the 1-alkyl-2-acetyl-*sn*-glycero-3-phosphate phosphohydrolase activity, it is important to determine the enzymic activity at 23°C. The apparent K_m for acetyl-CoA under these conditions is 226 µM, and the optimal concentration of alkyllyso-GP range between 16 and 25 µM (Lee et al., 1986). On the basis of pH optima, substrate inhibition studies, and sensitivities to preincubation temperatures of the microsomes, it appears that alkyllyso-GP/acetyl-CoA acetyltransferase differs from other acetyltransferases responsible for the transfer of acetate from acetyl-CoA to alkyllyso-GPC to form PAF. Both neural and nonneural tissues show high activities of alkyllyso-GP/acetyl-CoA acetyltransferase, which indicates that this pathway is operational in many neural and nonneural cells (Lee et al., 1986). The existence of a complete de novo biosynthetic pathway for the synthesis of PAF may be responsible for maintaining physiological levels of PAF for normal cell function.

In rabbit brain cerebral cortex, subcellular distribution studies have indicated that specific activity of alkyllyso-GP/acetyl-CoA acetyltransferase in nuclear fraction is three times higher than that of the microsomal fraction (Baker and Chang, 1996). Nuclear alkyllyso-glycerol phosphate/acetyl-CoA acetyltransferase shows optimal activity within the alkaline range (pH 8–9) (Table 7.3). Alkyllyso-glycerol phosphate/acetyl-CoA acetyltransferase is inhibited by MgATP or oleoyl CoA in a dose-dependent manner. However, the alkyllyso-glycerol phosphate/acetyl-CoA acetyltransferase can be distinguished from the nuclear lysoPAF acetyltransferase by a greater sensitivity to MgATP inhibition (Baker and Chang, 1996; Baker and Chang, 1998). Other nucleotides and the ATP nonhydrolyzable analog, AMP-PNP (5′-adenylylimido-diphosphate) show no comparable inhibition. Bovine serum albumin or the fatty acyl CoA synthetase inhibitor, Triacsin C, decreases MgATP inhibition. MgATP inhibition is increased when nuclei are preincubated in 50 mM Tris-HCl, pH 7.4/1 mM $MgCl_2$ at 37°C, and preincubations elevate levels of nuclear free fatty acid. Addition of exogenous free fatty acids promotes the MgATP-induced inhibition. Oleoyl CoA, in the absence of MgATP, also inhibits alkyllyso-GP/acetyl-CoA acetyltransferase. These results suggest that MgATP potentiates the conversion of nuclear free fatty acids to fatty acyl CoA. Fatty acyl CoA may

Table 7.3 Properties of 1-alkyl-2-lyso-*sn*-glycero-3-phosphate (alkyllyso-GP)/acetyl-CoA: acetyltransferase from nonneural tissues.

Parameter	Value
pH optimum	8.0
K_m for acetyl-CoA (µM)	22.6
V_{max} (nmol min^{-1} mg^{-1} protein)	73.5
Localization	Nucleus, microsomes
MgATP	Inhibition
Oleoyl-CoA	Inhibition

Summarized from (Baker and Chang, 1998) and Lee et al., (1986)

directly inhibit nuclear alkyllyso-GP/acetyl-CoA acetyltransferase. MgATP-induced inhibition is competitive with respect to the 1-alkyl-*sn*-glycero-3-phosphate substrate suggesting that an inhibitor resembling in structure to 1-alkyl-*sn*-glycero-3-phosphate. 1-Hexadecyl-2-arachidonoyl-*sn*-glycero-3-phosphate is identified as a competitive inhibitor for 1-alkyl-*sn*-glycero-3-phosphate in the acetylation reaction. Neuronal nuclei can convert 1-alkyl-*sn*-glycero-3-phosphate to 1-alkyl-2-acyl-*sn*-glycero-3-phosphate, a reaction dependent upon MgATP and the presence of acetyl-CoA or free CoA. This nuclear acylation is increased by free fatty acid addition and is seen using oleoyl CoA in the absence of MgATP. It is proposed that nuclear alkyllyso-GP/acetyl-CoA acetyltransferase may be regulated by the availability of MgATP (Baker and Chang, 2002).

7.3.2 1-Alkyl-2-acetyl-sn-glycero-3-phosphate Phosphohydrolase

This enzyme catalyzes the hydrolysis of phosphate moiety from 1-alkyl-2-acetyl-*sn*-glycero-3-phosphate and converts it to 1-alkyl-2-acetyl-*sn*-glycerol. It has similar role as phosphatidate phosphohydrolase in the biosynthesis of PtdCho. Subcellular distribution studies in rat spleen indicate that the bulk of the enzymic activity (53%) is located in the microsomal fraction, whereas 28% of the activity is present in mitochondrial fraction (Panwala et al., 1998). The microsomal enzyme has an optimal pH of 7.0–7.4, an "apparent" K_m of 31.8 μM for 1-alkyl-2-acetyl-*sn*-glycero-3-phosphate, and is widely distributed in various rat tissues. This enzyme is distinguished from phosphatidate phosphohydrolase activity from spleen on the basis of substrate specificity, pH profiles, temperature sensitivities, effect of detergents, and cations. 1-Alkyl-2-acetyl-*sn*-glycero-3-phosphate phosphohydrolase shows no notable substrate specificity with regard to variations in alkyl chain length (C16:0 vs. C18:0) at the *sn*-1 position or short chain acyl groups (C2:0–C6:0, with the exception of C3:0) at the *sn*-2 position of the glycerol moiety. The enzymic activity of alkyl-2-acetyl-*sn*-glycero-3-phosphate phosphohydrolase is 30- to 90-fold higher than alkyllyso-glycerol phosphate/acetyl-CoA acetyltransferase in most of the rat tissues (Panwala et al., 1998).

7.3.3 1-Alkyl-2-acetyl-sn-glycerol/CDP-choline Phosphotransferase

This microsomal enzyme transfers the phosphocholine moiety from CDP-choline to 1-alkyl-2-acetyl-*sn*-glycerol. The availability of CDP-choline is the rate-limiting step in the PAF synthesis (Snyder, 1997). Therefore, factors that regulate cytidylyltransferase activity play an important role in modulating the levels of PAF in various tissues. For example, fatty acids (Blank et al., 1988; Vallari et al., 1990) and

Table 7.4 Properties of brain PAF-synthesizing choline phosphotransferase.

Parameter	Effect
pH optimum	8.0
K_m (μM) for CDP-choline	42
V_{max} (nmol/min/mg protein)	3.0
Subcellular localization	Microsomal
Mg^{2+}	Stimulation
Ca^{2+}	Inhibition
Deoxycholate	Inhibition
Thermal stability	Inactivation
Dithiothreitol (DTT)	No effect

Summarized from Goracci and Francescangeli, (1991) and Snyder, (1997)

intracellular levels of CDP-choline (Lee et al., 1990) have been shown to upregulate PAF biosynthesis through the de novo synthesis. The two cholinephosphotransferases that catalyze the biosynthesis of phosphatidylcholine and PAF have been reported to occur in the rat kidney inner medulla. They are stimulated by Mg^{2+} or Mn^{2+} and are inhibited by Ca^{2+}. Topographic studies indicate that both the activities are located on the cytoplasmic face of microsomal vesicles (Woodard et al., 1987). PAF synthesis is slightly stimulated by 10mM DTT, whereas the enzymic synthesis of PtdCho is inhibited greater than 95% under the similar experimental conditions. These cholinephosphotransferases have different pH optima, substrate specificities, and are sensitive to temperature, deoxycholate, or ethanol (Table 7.4). Substrate specificities studies of the DTT-insensitive cholinephosphotransferase indicate that the enzyme prefers a lipid substrate with 16:0 or 18:1 *sn*-1-alkyl chains. Short chain esters at the *sn*-2 position (acetate or propionate) are utilized by the DTT-insensitive cholinephosphotransferase, but analogs with acetamide or methoxy substituents at the *sn*-2 position are not substrates. Also, CDP-choline is the preferred water-soluble substrate when compared with CDP-ethanolamine (Woodard et al., 1987).

Brain PAF-synthesizing 1-alkyl-2-acetyl-*sn*-glycerol/CDP-choline phosphotransferase shows similar subcellular localization, Mg^{2+} requirement, Ca^{2+} inhibitory properties, and thermal inactivation profiles as the kidney medulla enzyme (Goracci and Francescangeli, 1991; Francescangeli et al., 2000). Brain PAF-synthesizing phosphocholinetransferase shows optimal activity at pH 8.0. With 1-alkyl-2-acetylglycerol, the enzyme has K_m and V_{max} values of 42 μM and 3.0 nmol^{-1} min^{-1} mg protein, respectively. This is in contrast with PtdCho-synthesizing cholinephosphotransferases, which has a broad pH optimum (8.0–9.0) and K_m and V_{max} values of 55 μM and 2.2 nmol^{-1} min^{-1} mg protein (Goracci and Francescangeli, 1991). Subcellular distribution studies in 15-day-old rabbit cerebral cortices indicate that PAF-synthesizing cholinephosphotransferase shows highest specific activity in microsomal fraction. The microsomal fraction is subfractioned into rough and smooth microsomal fractions (Baker and Chang, 1993). Rough microsomal subfraction has higher specificity than smooth microsomal fraction. PAF-synthesizing cholinephosphotransferase is inhibited by Triton X-100. Detailed kinetic studies have indicated that the reversibility and direction of PAF-synthesizing cholinephosphotransferase-catalyzed reaction depends on the ratio [CDP-choline]/[CMP]. This ratio is related to the energy charge of the cell. It is proposed that the de novo pathway mainly contribute to PAF synthesis for maintaining its basal levels under physiological conditions (Francescangeli et al., 2000).

7.4 Oxidative Fragmentation Pathway for PAF Synthesis

The third pathway for the synthesis of PAF is the oxidative fragmentation of phosphatidylcholines. When exposed to oxidative conditions, 1-*O*-alkyl-2-arachidonyl-*sn*-glycerophosphocholine breaks down into a variety of species of 1-*O*-alkyl phospholipids containing different short chain substituents at the *sn*-2 position. These 1-*O*-alkyl phospholipids interact with PAF receptors and induce a variety of biological effects (Stafforini et al., 1996; Prescott et al., 2000). The fragmentation of 1-palmitoyl-2-arachidonoyl-*sn*-glycero-3-phosphocholine by ozonolysis results in 1-palmitoyl-2-(5-oxovaleroyl)-*sn*-glycero-3-phosphocholine (Smiley et al., 1991; Prescott et al., 2000). This phospholipid stimulates human neutrophils at submicromolar concentrations, and its effects are blocked by specific PAF-receptor antagonists (WEB2086, L659989, and CV3988) (Smiley et al., 1991). Similarly, fragmentation of 1-palmitoyl-2-arachidonoyl-*sn*-glycero-3-phosphocholine with soybean lipoxygenase produces 15-hydroperoxy derivatives, which do not activate neutrophils. The oxidation of 15-hydroperoxy derivatives under air generates numerous fragmented phospholipids, some of which activate polymorphonuclear leukocytes. The hydrolysis of *sn*-2 residues of these derivatives with PLA_2 or PAF acetylhydrolase retards their biologic activity. Furthermore, neutrophil activation is completely blocked by L659989, a specific PAF-receptor antagonist (Smiley et al., 1991). Similarly, oxidized PtdCho-mediated platelet aggregation is also retarded by FR-900452, an antagonist of PAF receptor (Tanaka et al., 1994). On the basis of these observations, oxidative fragmentation pathway of PAF analog production has been proposed.

The occurrence of plasmalogenic analogs of PAF (1-*O*-(1′-alkenyl)-2-acetyl-*sn*-glycero-3-phosphocholine (vinyl-PAF) or alkenyl-PAF) and 1-*O*-acyl-2-acetyl-*sn*-glycero-3-phosphocholine (acyl-PAF) and their involvement in inflammatory situations has also been reported (Fig. 7.5). These analogs stimulate the production of leukocyte superoxide radicals at higher concentration than PAF. Stimulation of superoxide radical production strongly depends on the structure of the polar heads of PAF analogs. Choline-containing plasmalogenic and acyl PAF analogs may act as specific lipid mediators of the neutrophil function (Lee et al., 1988). Both 1-acyl-PAF and 1-alkenyl-PAF stimulated chemotaxis of human leukocytes in agarose gel. PAF and 1-alkenyl-PAF promote rat paw edema in the range of doses 0.1–10 and 10–100 μg per paw, respectively. 1-Acyl-PAF-mediated paw edema (10–100 μg per paw) is more pronounced than PAF or 1-alkenyl-PAF induced paw edema (Kamal-Eldin and Yanishlieva, 2002; Phillis et al., 2006). Alkenyl-PAF also

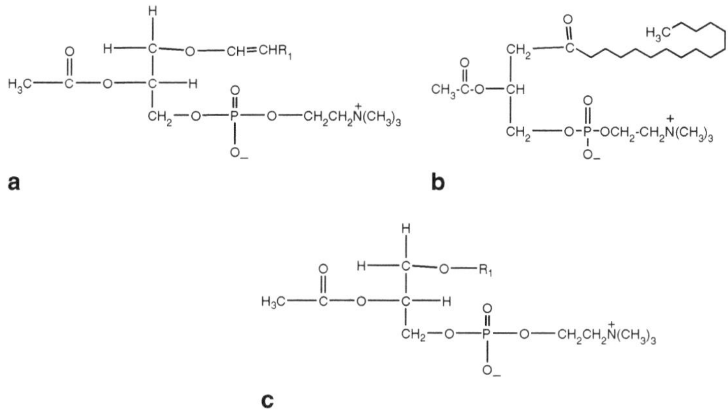

Fig. 7.5 Structures of three major molecular forms of platelet activating factor. Plasmalogenic analog of PAF or vinyl-PAF (1-alk-1′-enyl-2-acetyl-*sn*-glycero-3-phosphocholine) (**a**), 1-acyl-PAF (1-*O*-palmitoyl-2-*O*-acetyl-*sn*-glycero-3-phosphocholine (**b**), and alkyl-PAF (1-*O*-alkyl-2-acetyl-*sn*-glycero-3-phosphocholine) (**c**).

exhibits significant antiinflammatory effect by inhibiting PAF- or carrageenan-mediated rat paw edema, and this effect exceeds that of dexamethasone. In these models of inflammation, 1-acyl-PAF does not show any antiinflammatory activity. Thus, unregulated production of structural analogs of PAF by remodeling and fragmentation pathways has expanded the family of PAF-related signaling molecules. These molecules are called as "PAF-like ether lipids". The mode of "PAF-like ether lipids" family action may be similar to that of native PAF. It is quite likely that the action of plasmalogenic PAF analogs is also mediated by the PAF receptor, a G protein-coupled membrane-spanning molecule that is involved in multiple signaling pathways in various types of neural and nonneural cells.

7.5 Regulation of PAF Synthesis

As stated earlier that in nonneural cells, oxidative stress stimulates PAF synthesis (Sakamoto et al., 2002). Thus, treatment of HUVEC with H_2O_2 stimulates PAF synthesis in a dose-dependent manner, and antioxidants prevent PAF production in H_2O_2-treated HUVEC (Tosaki et al., 2007). H_2O_2 treatment stimulates acetyl-CoA/1-*O*-alkyl-2-lyso-*sn*-glycero-3-phosphocholine acetyltransferase, the enzyme, which catalyzes the last step of PAF synthesis. However, exogenous lyso-PAF addition has not effected to acetyltransferase activity. The acetyltransferase activity responds rapidly to H_2O_2-treatment, but in a transitory manner. Acetyl-CoA/1-*O*-alkyl-2-lyso-*sn*-glycero-3-phosphocholine acetyltransferase activity is regulated by tyrosine kinase and calcium/calmodulin-dependent protein kinase.

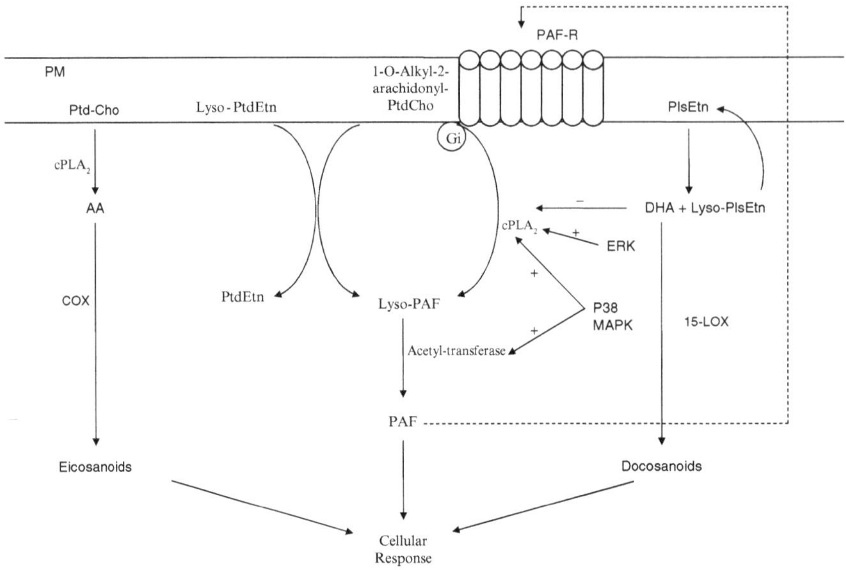

Fig. 7.6 Regulation of PAF synthesis by ERK and p38 in nonneural cells. *Lyso-PtdEtn* lysophosphatidyethanolamine, *PtdEtn* phosphatidylethanolamine; *lyso-PAF* lyso-platelet activating factor, *PAF* platelet-activating factor, *Lyso-PlsEtn* lysophosphatidyethanolamine, *cPLA*$_2$ cytosolic phospholipase A$_2$, *COX* cyclooxygenase, *15-LOX* 15-lipoxygenase, *AA* arachidonic acid, *DHA* docosahexaenoic acid, *ERK* extracellular-signal-regulated protein kinase, and *p38 MAPK* p38 mitogen-activated protein kinase, *Gi* G protein, *plus sign* indicates activation, and *minus sign* indicates inhibition.

Phospholipid hydroperoxide glutathione peroxidase, which requires selenium, negatively modulates PAF biosynthesis. Thus, overexpression of this enzyme in nonneural cells downregulates PAF synthesis through the involvement of cPLA$_2$ and acetyl-CoA/1-*O*-alkyl-2-lyso-*sn*-glycero-3-phosphocholine acetyltransferase and modulation of their activities via the extracellular signal-regulated kinases (ERKs) and p38 (Sakamoto et al., 2002) (Fig. 7.6). These studies are supported by the observation that selenium deficiency not only results in stimulation of cPLA$_2$ and accumulation of lipid peroxides, but also in downregulation of PAF synthesis.

Another mechanism that modulates PAF synthesis in endothelial cells is the incorporation of n-3 fatty acids (Mayer et al., 2002; Shikano et al., 1993). Supplementation of human eosinophilic leukemia (Eol-1) cells cultures with docosahexaenoic acid (DHA) and eicosapentaenoic acid (EPA) increases PAF synthesis with concurrent decrease in other unsaturated fatty acids. In DHA-treated cells, the liberation of arachidonic acid in response to an ionophore is siginificantly decreased. It is stated that DHA supplementation not only attenuates PLA$_2$ activity, but also decreases acetyl-CoA/1-alkylGPC acetyltransferase activity, suggesting that enrichment of DHA in glycerophospholipids attenuates PAF synthesis through the inhibition of arachidonic acid-specific cPLA$_2$ (Shikano et al., 1993).

Incorporation of docosahexaenoic acid (DHA) in cellular glycerophospholipids not only decreases PAF synthesis, but also reduces cytokine-induced expression of endothelial leukocyte adhesion molecules, secretion of inflammatory mediators, and leukocyte adhesion to cultured endothelial cells. DHA decreases the expression of vascular cell adhesion molecule 1 (VCAM-1), a process induced by interleukin (IL)-1, tumor necrosis factor-α (TNF-α), IL-4, or bacterial lipopolysaccharide (De Caterina et al., 1994). The reduction in expression of inflammatory mediators needs a prolonged exposure of endothelial cells to DHA, and correlates with the degree of DHA incorporation into cellular glycerophospholipids. DHA also limits cytokine-stimulated endothelial cell expression of E-selectin and intercellular adhesion molecule 1, and the secretion of IL-6 and IL-8 into the medium. Cyclooxygenase inhibitors do not block the effect of DHA on PAF synthesis and VCAM-1 expression. These properties of DHA may contribute to antiatherogenic and antiinflammatory effects of n-3 fatty acids (De Caterina et al., 1994; Mayer et al., 2002). Oxidants, overexpression of phospholipid hydroperoxide glutathione peroxidase, and incorporation of n-3 fatty acids in membranes may modulate of PAF synthesis in nonneural cells (De Caterina et al., 1994; Mayer et al., 2002; Sakamoto et al., 2002; Tosaki et al., 2007).

Adenosine, histamine, prostaglandin E_2, and CGS-21680, a selective agonist of the adenosine A2A receptor, potently inhibit PAF biosynthesis in agonist (formyl Met-Leu-Phe (fMLP))-dependent manner in thapsigargin-activated human polymorphonuclear leukocytes (PMN) (Flamand et al., 2006a). The observed inhibitions of PAF biosynthesis is reversed effectively by exogenous 1-*O*-alkyl-lyso-*sn*-glyceryl-3-phosphocholine (lyso-PAF), suggesting that these effects of CGS-21680 and histamine involve the blockade of cytosolic $PLA_2\alpha$ activity (Flamand et al., 2006a). This suggestion is supported by the observation that $cPLA_2\alpha$ inhibitor, pyrrophenone, completely blocks the PAF formation, and lyso-PAF similarly prevents this effect of pyrrophenone (Flamand et al., 2006b). The inhibitory effects of CGS-21680 and histamine on PAF biosynthesis can also be prevented by the protein kinase A inhibitor H-89, supporting roles for the Gs-coupled receptors A2A and H2, respectively. Collective evidence suggests that multiple and potent inhibitory effects of adenosine and histamine on leukocyte functions are mediated by PAF. Similarly, cytidine 5′-diphosphocholine (CDP-choline) decreases levels of PAF in the brain tissue. Thus, decrease of cerebral PAF levels may be caused not only by increase in PAF-acetyl hydrolase activity but also due to the inactivation of DTT-insensitive choline phosphotransferase activity (Gimenez and Aguilar, 2001).

7.6 Conclusion

PAF is a biologically active ether lipid, structurally identified as 1-*O*-alkyl-2-acetyl-*sn*-glycero-3-phosphocholine. It represents an important class of highly active lipid mediator that elicits a wide range of neurochemical responses associated with normal brain function and neurological disorders (Farooqui and Horrocks, 2004; Bazan, 2003). Although the occurrence and properties of PAF synthesizing enzymes in the brain have been reported (Goracci and Francescangeli, 1991; Francescangeli et al., 2000; Baker and Chang, 1993; Baker and Chang, 1998; Baker and Chang, 2000; Bazan, 2003), none of these enzymes have been purified and characterized from the brain tissue. Three distinct biosynthetic pathways have been reported to occur in brain. They include remodeling pathway, de novo synthesis pathway, and oxidative fragmentation pathway. Romodeling pathway may be responsible for the production of PAF in inflammatory cells (astrocytes, microglia, endothelial cells, and leukocytes) upon stimulation, while the de novo pathway may be associated with constitutive basal PAF generation in neurons. Unlike the remodeling pathway, the de novo synthesis pathway does not result in the simultaneous synthesis of eicosanoids. The significance of oxidative fragmentation pathway of PAF production in the brain is not known. However, actions of PAF are mimicked by fragmented-oxidized phospholipids because of their affinity for PAF receptors. PAF is not stored in the cell. It is synthesized and translocated to the cell exterior to interact with its receptors. At present, nothing is known about the translocation of PAF in neural cells. PAF-mediated signal transduction process is terminated by the action of PAF hydrolase (Prescott et al., 2000).

References

Asai K., Hirabayashi T., Houjou T., Uozumi N., Taguchi R., and Shimizu T. (2003). Human group IVC phospholipase A_2 ($cPLA_2\gamma$) – Roles in the membrane remodeling and activation induced by oxidative stress. J. Biol. Chem. 278:8809–8814.

Baker P. R. S., Owen J. S., Nixon A. B., Thomas L. N., Wooten R., Daniel L. W., O'Flaherty J. T., and Wykle R. L. (2002). Regulation of platelet-activating factor synthesis in human neutrophils by MAP kinases. Biochim. Biophys. Acta Mol. Cell Res. 1592:175–184.

Baker R. R. and Chang H. (1993). The potential for platelet-activating factor synthesis in brain: Properties of cholinephosphotransferase and 1-alkyl-*sn*-glycero-3-phosphate acetyltransferase in microsomal fractions of immature rabbit cerebral cortex. Biochim. Biophys. Acta Lipids Lipid Metab. 1170:157–164.

Baker R. R. and Chang H. (1998). MgATP has different inhibitory effects on the use of 1-acyl-lysophosphatidylcholine and lyso platelet-activating factor acceptors by neuronal nuclear acetyltransferase activities. Biochim. Biophys. Acta 1392:351–360.

Baker R. R. and Chang H. Y. (1996). Alkylglycerophosphate acetyltransferase and lyso platelet activating factor acetyltransferase, two key enzymes in the synthesis of platelet activating factor, are found in neuronal nuclei isolated from cerebral cortex. Biochim. Biophys. Acta 1302:257–263.

Baker R. R. and Chang H. Y. (2000). The regulation of CoA-independent transacylation reactions in neuronal nuclei by lysophospholipid, free fatty acid, and lysophospholipase: the control of nuclear lyso-platelet-activating factor metabolism. Mol. Cell. Biochem. 215:135–144.

Baker R. R. and Chang H. Y. (2002). MgATP may depress de novo neuronal nuclear PAF generation by promoting the formation of alkylacylglycerophosphate, an inhibitor of alkylglycerophosphate acetyltransferase. Biochim. Biophys. Acta Mol. Cell Biol. Lipids 1585:44–49.

Bazan N. G. (2003). Synaptic lipid signaling: Significance of polyunsaturated fatty acids and platelet-activating factor. J. Lipid Res. 44:2221–2233.

Bernatchez P. N., Winstead M. V., Dennis E. A., and Sirois M. G. (2001). VEGF stimulation of endothelial cell PAF synthesis is mediated by group V 14 kDa secretory phospholipase A_2. Brit. J. Pharmacol. 134:197–205.

Blank M. L., Lee Y. J., Cress E. A., and Snyder F. (1988). Stimulation of the de novo pathway for the biosynthesis of platelet-activating factor (PAF) via cytidylyltransferase activation in cells with minimal endogenous PAF production. J. Biol. Chem. 263:5656–5661.

Brodie C. (1995). Platelet activating factor induces nerve growth factor production by rat astrocytes. Neurosci. Lett. 186:5–8.

Chao W. and Olson M. S. (1993). Platelet-activating factor: Receptors and signal transduction. Biochem. J. 292:617–629.

Chiba H., Michibata H., Wakimoto K., Seishima M., Kawasaki S., Okubo K., Mitsui H., Torii H., and Imai Y. (2004). Cloning of a gene for a novel epithelium-specific cytosolic phospholipase A_2, $cPLA_2\delta$, induced in psoriatic skin. J. Biol. Chem. 279:12890–12897.

Clark M. A., Conway T. M., Shorr R. G. L., and Crooke S. T. (1987). Identification and isolation of a mammalian protein which is antigenically and functionally related to the phospholipase A_2 stimulatory peptide melittin. J. Biol. Chem. 262:4402–4406.

Collado A. P., Latorre E., Fernández I., Aragonés A. D., and Catalán R. E. (2003). Endothelin-1 decreases ethanolamine plasmalogen levels and evokes PAF production in brain microvessels. Microvasc. Res. 66:197–203.

De Caterina, R., Cybulsky, M.I., Clinton, S.K., Gimbrone, M.A., and Libby, P. (1994) The omega-3 fatty acid docosahexaenoate reduces cytokine-induced expression of proatherogenic and proinflammatory proteins in human endothelial cells. Atherioscler. Thromb. 14:1829–1836.

Diaz-Arrastia R. and Scott K. S. (1999). Expression of cPLA2-β and cPLA2-γ, novel paralogs of group IV cytosolic phospholipase A_2 in mammalian brain. Soc. Neurosci. Abs. 25:2206.

Farooqui A. A. and Horrocks L. A. (2004). Plasmalogens, platelet-activating factor, and other ether lipids. In: Nicolaou A. and Kokotos G. (eds.), *Bioactive Lipids*. Oily Press, Bridgwater, England, pp. 107–134.

Farooqui A. A. and Horrocks L. A. (2007). *Glycerophospholipids in the Brain: Phospholipases A_2 in Neurological Disorders*, pp. 1–394. Springer, New York.

Farooqui A. A., Ong W. Y., and Horrocks (2008). *Neurochemical Aspects of Excitotoxicity*, pp.1–290. Springer, New York.

Farooqui A. A., Ong W. Y., Horrocks L. A., and Farooqui T. (2000). Brain cytosolic phospholipase A_2: Localization, role, and involvement in neurological diseases. Neuroscientist 6:169–180.

Flamand N., Lefebvre J., Lapointe G., Picard S., Lemieux L., Bourgoin S. G., and Borgeat P. (2006a). Inhibition of platelet-activating factor biosynthesis by adenosine and histamine in human neutrophils: involvement of $cPLA_2\alpha$ and reversal by lyso-PAF. J. Leukocyte Biol. 79:1043–1051.

Flamand N., Picard S., Lemieux L., Pouliot M., Bourgoin S. G., and Borgeat P. (2006b). Effects of pyrrophenone, an inhibitor of group IVA phospholipase A_2, on eicosanoid and PAF biosynthesis in human neutrophils. Brit. J. Pharmacol. 149:385–392.

Francescangeli E., Boila A., and Goracci G. (2000). Properties and regulation of microsomal PAF-synthesizing enzymes in rat brain cortex. Neurochem. Res. 25:705–713.

Ghosh M., Loper R., Gelb M. H., and Leslie C. C. (2006). Identification of the expressed form of human cytosolic phospholipase $A_2\beta$ ($cPLA_2\beta$) – $cPLA_2\beta3$ is a novel variant localized to mitochondria and early endosomes. J. Biol. Chem. 281:16615–16624.

Gimenez R. and Aguilar J. (2001). Cytidine (5′) diphosphocholine-induced decrease in cerebral platelet activating factor is due to inactivation of its synthesizing enzyme cholinephosphotransferase in aged rats. Neurosci. Lett. 299:209–212.

Goracci G. and Francescangeli E. (1991). Properties of PAF-synthesizing phosphocholinetransferase and evidence for lysoPAF acetyltransferase activity in rat brain. Lipids 26:986–991.

He X., Guan X. L., Ong W. Y., Farooqui A. A., and Wenk M. R. (2007). Expression, activity, and role of serine palmitoyltransferase in the rat hippocampus after kainate injury. J. Neurosci. Res. 85:423–432.

Heller R., Bussolino F., Ghigo D., Garbarino G., Pescarmona G., Till U., and Bosia A. (1991). Stimulation of platelet-activating factor synthesis in human endothelial cells by activation of the *de novo* pathway. Phorbol 12-myristate 13-acetate activates 1-alkyl-2-lyso-*sn*-glycero-3-phosphate:acetyl-CoA acetyltransferase and dithiothreitol-insensitive 1-alkyl-2-acetyl-*sn*-glycerol:CDP-choline cholinephosphotransferase. J. Biol. Chem. 266:21358–21361.

Hirabayashi T., Murayama T., and Shimizu T. (2004). Regulatory mechanism and physiological role of cytosolic phospholipase A_2. Biol. Pharm. Bull. 27:1168–1173.

Hirabayashi T. and Shimizu T. (2000). Localization and regulation of cytosolic phospholipase A_2. Biochim. Biophys. Acta 1488:124–138.

Honda Z., Ishii S., and Shimizu T. (2002). Platelet-activating factor receptor. J. Biochem. 131:773–779.

Kamal-Eldin A. and Yanishlieva N. V. (2002). N-3 fatty acids for human nutrition: Stability considerations. Eur. J. Lipid Sci. Technol. 104:825–836.

Karasawa K., Qiu X., and Lee T. (1999). Purification and characterization from rat kidney membranes of a novel platelet-activating factor (PAF)-dependent transacetylase that catalyzes the hydrolysis of PAF, formation of PAF analogs, and C_2-ceramide. J. Biol. Chem. 274:8655–8661.

Kishimoto K., Matsumura K., Kataoka Y., Morii H., and Watanabe Y. (1999). Localization of cytosolic phospholipase A_2 messenger RNA mainly in neurons in the rat brain. Neuroscience 92:1061–1077.

Lang P. A., Kempe D. S., Tanneur V., Eisele K., Klarl B. A., Myssina S., Jendrossek V., Ishii S., Shimizu T., Waidmann M., Hessler G., Huber S. M., Lang F., and Wieder T. (2005). Stimulation of erythrocyte ceramide formation by platelet-activating factor. J. Cell Sci. 118:1233–1243.

Lee T. C., Malone B., Blank M. L., Fitzgerald V., and Snyder F. (1990). Regulation of the synthesis of platelet-activating factor and its inactive storage precursor (1-alkyl-2-acyl-sn-glycero-3-phosphocholine) from 1-alkyl-2-acetyl-sn-glycerol by rabbit platelets. J. Biol. Chem. 265:9181–9187.

Lee T. C., Malone B., Longobardi L., and Balestrieri M. L. (2001). Differential regulation of three catalytic activities of platelet-activating factor (PAF)-dependent transacetylase. Arch. Biochem. Biophys. 387:41–46.

Lee T. C., Malone B., and Snyder F. (1986). A new de novo pathway for the formation of 1-alkyl-2-acetyl-sn-glycerols, precursors of platelet activating factor. Biochemical characterization of 1-alkyl-2-lyso-sn-glycero-3-P/acetyl-CoA acetyltransferase in rat spleen. J. Biol. Chem. 261: 5373–5377.

Lee T. C., Malone B., and Snyder F. (1988). Formation of 1-alkyl-2-acetyl-sn-glycerols via the de novo biosynthetic pathway for platelet-activating factor. Characterization of 1-alkyl-2-acetyl-sn-glycero-3-phosphate phosphohydrolase in rat spleens. J. Biol. Chem. 263:1755–1760.

Lee T. C., Malone B., Woodard D., and Snyder F. (1989). Renal necrosis and the involvement of a single enzyme of the de novo pathway for the biosynthesis of platelet-activating factor in the rat kidney inner medulla. Biochem. Biophys. Res. Commun. 163:1002–1005.

Lee T. C., Ou M. C., Shinozaki K., Malone B., and Snyder F. (1996). Biosynthesis of *N*-acetylsphingosine by platelet-activating factor: Sphingosine CoA-independent transacetylase in HL-60 cels. J. Biol. Chem. 271:209–217.

Lee T. C., Uemura Y., and Snyder F. (1992). A novel CoA-independent transacetylase produces the ethanolamine plasmalogen and acyl analogs of platelet-activating factor (PAF) with PAF as the acetate donor in HL-60 cells. J. Biol. Chem. 267:19992–20001.

Mayer K., Merfels M., Muhly-Reinholz M., Gokorsch S., Rosseau S., Lohmeyer J., Schwarzer N., Krull M., Suttorp N., Grimminger F., and Seeger W. (2002). Omega-3 Fatty acids suppress monocyte adhesion to human endothelial cells: role of endothelial PAF generation. Am. J. Physiol. Heart Circ. Physiol. 283:H811–H818.

McHowat J., and Creer M. H. (2001). Selective plasmalogen substrate utilization by thrombin-stimulated Ca^{2+}-independent PLA_2 in cardiomyocytes. Am. J. Physiol. Heart Circ. Physiol. 278:H1933–H1940.

McHowat J., Kell P. J., O'Neill H. B., and Creer M. H. (2001). Endothelial cell PAF synthesis following thrombin stimulation utilizes Ca^{2+}-independent phospholipase A_2. Biochemistry 40:14921–14931.

Mosior M., Six D. A., and Dennis E. A. (1998). Group IV cytosolic phospholipase A_2 binds with high affinity and specificity to phosphatidylinositol 4,5-bisphosphate resulting in dramatic increases in activity. J. Biol. Chem. 273:2184–2191.

Nixon A. B., O'Flaherty J. T., Salyer J. K., and Wykle R. L. (1999). Acetyl-CoA:1-*O*-alkyl-2-lyso-*sn*-glycero-3-phosphocholine acetyltransferase is directly activated by p38 kinase. J. Biol. Chem. 274:5469–5473.

Nomikos T. N., Iatrou C., and Demopoulos C. A. (2003). Acetyl-CoA/1-*O*-alkyl-*sn*-glycero-3-phosphocholine acetyltransferase (lyso-PAF AT) activity in cortical and medullary human renal tissue. Eur. J. Biochem. 270:2992–3000.

Ohto T., Uozumi N., Hirabayashi T., and Shimizu T. (2005). Identification of novel cytosolic phospholipase A_2s, murine $cPLA_2\delta$, ε, and ζ, which form a gene cluster with $cPLA_2\beta$. J. Biol. Chem. 280:24576–24583.

Ong W. Y., Sandhya T. L., Horrocks L. A., and Farooqui A. A. (1999). Distribution of cytoplasmic phospholipase A_2 in the normal rat brain. J. Hirnforsch. 39:391–400.

Owada Y., Tominaga T., Yoshimoto T., and Kondo H. (1994). Molecular cloning of rat cDNA for cytosolic phospholipase A_2 and the increased gene expression in the dentate gyrus following transient forebrain ischemia. Mol. Brain Res. 25:364–368.

Panwala C. M., Jones J. C., and Viney J. L. (1998). A novel model of inflammatory bowel disease: Mice deficient for the multiple drug resistance gene, mdr1a, spontaneously develop colitis. J. Immunol. 161:5733–5744.

Pardue S., Rapoport S. I., and Bosetti F. (2003). Co-localization of cytosolic phospholipase A_2 and cyclooxygenase-2 in Rhesus monkey cerebellum. Molec. Brain Res. 116:106–114.

Pettus B. J., Bielawska A., Subramanian P., Wijesinghe D. S., Maceyka M., Leslie C. C., Evans J. H., Freiberg J., Roddy P., Hannun Y. A., and Chalfant C. E. (2004). Ceramide 1-phosphate is a direct activator of cytosolic phospholipase A_2. J. Biol. Chem. 279:11320–11326.

Phillis J. W., Horrocks L. A., and Farooqui A. A. (2006). Cyclooxygenases, lipoxygenases, and epoxygenases in CNS: Their role and involvement in neurological disorders. Brain Res. Rev. 52:201–243.

Prescott S. M., Zimmerman G. A., and McIntyre T. M. (1990). Platelet-activating factor. J. Biol. Chem. 265:17381–17384.

Prescott S. M., Zimmerman G. A., Stafforini D. M., and McIntyre T. M. (2000). Platelet-activating factor and related lipid mediators. Annu. Rev. Biochem. 69:419–445.

Rubin B. B., Downey G. P., Koh A., Degousee N., Ghomashchi F., Nallan L., Stefanski E., Harkin D. W., Sun C. X., Smart B. P., Lindsay T. F., Cherepanov V., Vachon E., Kelvin D., Sadilek M., Brown G. E., Yaffe M. B., Plumb J., Grinstein S., Glogauer M., and Gelb M. H. (2005). Cytosolic phospholipase A_2-α is necessary for platelet-activating factor biosynthesis, efficient neutrophil-mediated bacterial killing, and the innate immune response to pulmonary infection – $cPLA_2$-α does not regulate neutrophil NADPH oxidase activity. J. Biol. Chem. 280:7519–7529.

Sakamoto H., Tosaki T., and Nakagawa Y. (2002). Overexpression of phospholipid hydroperoxide glutathione peroxidase modulates acetyl-CoA, 1-*O*-alkyl-2-lyso-sn-glycero-3-phosphocholine acetyltransferase activity. J. Biol. Chem. 277:50431–50438.

Sandhya T. L., Ong W. Y., Horrocks L. A., and Farooqui A. A. (1998). A light and electron microscopic study of cytoplasmic phospholipase A_2 and cyclooxygenase-2 in the hippocampus after kainate lesions. Brain Res. 788:223–231.

Shikano M., Masuzawa Y., and Yazawa K. (1993). Effect of docosahexaenoic acid on the generation of platelet-activating factor by eosinophilic leukemia cells, Eol-1. J. Immunol. 150:3525–3533.

Shindou H., Hishikawa D., Nakanishiu H., Harayama T., Ishii S., Taguchi R., and Shimizu T. (2007). A single enzyme catalyzes both platelet-activating factor production and membrane biogenesis of inflammatory cells – Cloning and characterization of acetyl-CoA/lyso-PAF acetyltransferase. J. Biol. Chem. 282:6532–6539.

Shirai Y. and Ito M. (2004). Specific differential expression of phospholipase A_2 subtypes in rat cerebellum. J. Neurocytol. 33:297–307.

Smiley P. L., Stremler K. E., Prescott S. M., Zimmerman G. A., and McIntyre T. M. (1991). Oxidatively fragmented phosphatidylcholines activate human neutrophils through the receptor for platelet-activating factor. J. Biol. Chem. 266:11104–11110.

Snyder F. (1995). Platelet-activating factor: The biosynthetic and catabolic enzymes. Biochem. J. 305:689–705.

Snyder F. (1997). CDP-choline/alkylacetylglycerol cholinephosphotransferase catalyzes the final step in the de novo synthesis of platelet-activating factor. Biochim. Biophys. Acta 1348:111–116.

Snyder F., Fitzgerald V., and Blank M. L. (1996). Biosynthesis of platelet-activating factor and enzyme inhibitors. Adv. Exp. Med. Biol. 416:5–10.

Sogos V., Bussolino F., Pilia E., Torelli S., and Gremo F. (1990). Acetylcholine-induced production of platelet-activating factor by human fetal brain cells in culture. J. Neurosci. Res. 27:706–711.

Song C., Chang X. J., Bean K. M., Proia M. S., Knopf J. L., and Kriz R. W. (1999). Molecular characterization of cytosolic phospholipase $A_{2\text{-}\beta}$. J. Biol. Chem. 274:17063–17067.

Stafforini D. M., McIntyre T. M., Carter M. E., and Prescott S. M. (1987). Human plasma platelet-activating factor acetylhydrolase association with lipoprotein particles and role in the degradation of platelet-activating factor. J. Biol. Chem. 262:4215–4222.

Stafforini D. M., Prescott S. M., Zimmerman G. A., and McIntyre T. M. (1996). Mammalian platelet-activating factor acetylhydrolases. Biochim. Biophys. Acta 1301:161–173.

Stephenson D., Rash K., Smalstig B., Roberts E., Johnstone E., Sharp J., Panetta J., Little S., Kramer R., and Clemens J. (1999). Cytosolic phospholipase A_2 is induced in reactive glia following different forms of neurodegeneration. Glia 27:110–128.

Strokin M., Sergeeva M., and Reiser G. (2003). Docosahexaenoic acid and arachidonic acid release in rat brain astrocytes is mediated by two separate isoforms of phospholipase A_2 and is differently regulated by cyclic AMP and Ca^{2+}. Brit. J. Pharmacol. 139:1014–1022.

Tanaka T., Iimori M., Tsukatani H., and Tokumura A. (1994). Platelet-aggregating effects of platelet-activating factor-like phospholipids formed by oxidation of phosphatidylcholines containing an *sn*-2-polyunsaturated fatty acyl group. Biochim. Biophys. Acta Lipids Lipid Metab. 1210:202–208.

Tosaki T., Sakamoto H., Kitahara J., Imai H., and Nakagawa Y. (2007). Enhancement of acetyl-CoA: 1-*O*-alkyl-2-lyso-*sn*-glycero-3-phosphocholine acetyltransferase activity by hydrogen peroxide. Biol. Pharm. Bull. 30:272–278.

Vallari D. S., Record M., and Snyder F. (1990). Conversion of alkylacetylglycerol to platelet-activating factor in HL-60 cells and subcellular localization of the mediator. Arch. Biochem. Biophys. 276:538–545.

Woodard D. S., Lee T. C., and Snyder F. (1987). The final step in the de novo biosynthesis of platelet-activating factor. Properties of a unique CDP-choline/1-alkyl-2-acetyl-sn-glycerol choline-phosphotransferase in microsomes from the renal inner medulla of rats. J. Biol. Chem. 262:2520–2527.

Yoshida H., Imaizumi T., Tanji K., Sakaki H., Metoki N., Hatakeyama M., Yamashita K., Ishikawa A., Taima K., Sato Y., Kimura H., and Satoh K. (2005). Platelet-activating factor enhances the expression of nerve growth factor in normal human astrocytes under hypoxia. Molec. Brain Res. 133:95–101.

Yue T. L., Stadel J. M., Sarau H. M., Friedman E., Gu J. L., Powers D. A., Gleason M. M., Feuerstein G., and Wang H. Y. (1992). Platelet-activating factor stimulates phosphoinositide turnover in neurohybrid NCB-20 cells: Involvement of pertussis toxin-sensitive guanine nucleotide-binding proteins and inhibition by protein kinase C. Molec. Pharmacol. 41:281–289.

Chapter 8
Degradation of Platelet-Activating Factor in Brain

8.1 Introduction

Platelet-activating factor (PAF), a phospholipid mediator of inflammation, is rapidly synthesized by neural (neurons and glial cells) and nonneural cells in response to neurotransmitters (glutamate and its analogs), cytokines (tumor necrosis factor-α, interferon-γ, interleukin-1), and pathological situations (Sogos et al., 1990; Prescott et al., 2000; Karasawa et al., 2003; Kunievsky and Yavin, 1994). PAF is normally present in the mammalian tissues in picomolar concentrations, and is found both in the cytosol and body fluids including blood plasma, cerebrospinal fluid, urine, and amniotic fluid (Lynch and Hensen, 1986; Cox et al., 1981; Billah et al., 1983). PAF is not stored in cells. Its levels in brain, other body tissues, and plasma are modulated by PAF-acetyl hydrolases. This enzyme inactivates PAF by removing the acetyl group from the *sn*-2 position of glycerol moiety and generating lyso-PAF, which is biologically inactive (Fig. 8.1). The lyso-PAF can be reacylated by an acyl-CoA/1-radyl-*sn*-glycero-3-phosphocholine acyltransferase. Alkyl-PAF is less potent than PAF. The alkyl moiety of lyso-PAF is degraded to an aldehyde by a tetrahydropiridine-dependent alkyl monooxygenase (Lee et al., 1981). Alternatively, a lysophospholipase D (lyso-PLD) can hydrolyze phosphocholine moiety to generate an analog of phosphatidic acid, or catalyze a phosphate transfer by a transphosphatidylation reaction (Wykle and Schremmer, 1974).

PAF-acetyl hydrolase selectively hydrolyzes short acyl chains (C_2 to C_9) at the *sn*-2 position (Fig. 8.1). This enzyme shows no activity with acyl chains longer than C_9, but unusual *sn*-2 acyl group containing a carbonyl group at the ω-end of the acyl chain act as substrate for this enzyme. PAF structural analogs are hydrolyzed by PAF-acetyl hydrolase. These analogs competitively inhibit PAF-acetyl hydrolase activity (Stafforini et al., 1997). PAF receptor antagonists also inhibit PAF-acetyl hydrolase activity. This is in contrast to $cPLA_2$ and $sPLA_2$ that require Ca^{2+}. $cPLA_2$ uses Ca^{2+} for binding to phospholipid substrate in membrane whereas $sPLA_2$ utilizes Ca^{2+} for the catalysis of phospholipid substrate (Farooqui et al., 2006). Besides short chain phospholipids, PAF-acetyl hydrolase also hydrolyzes short-chain diacylglycerols, triacylglycerols, and acetylated alkanols (Tselepis and Chapman, 2002; Min et al., 2001). Unlike $cPLA_2$ and $sPLA_2$, PAF-acetyl hydrolase is not an

A. A. Farooqui et al., *Metabolism and Functions of Bioactive Ether Lipids in the Brain*

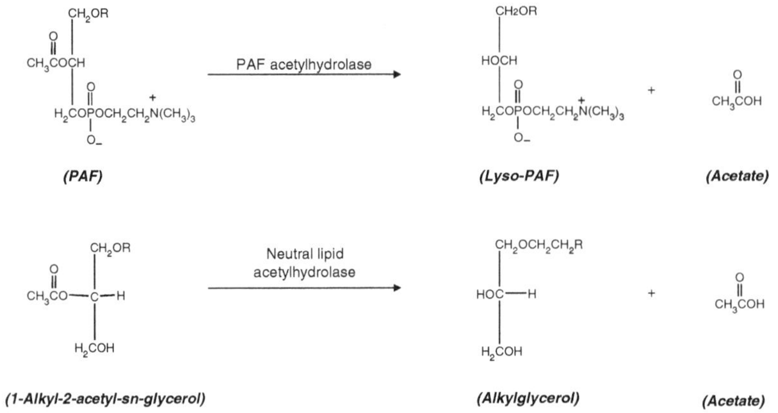

Fig. 8.1 Reaction catalyzed by PAF-acetyl hydrolase and alkylacetylglycerol acetyl hydrolase.

interfacial enzyme, and has broad substrate specificity as an esterase (Gelb et al., 2000). PAF-acetyl hydrolase hydrolyzes PAF, its analogs, and short chain oxidized phospholipids in a calcium-independent manner. These substrates are more water soluble than two long fatty acyl chains of native phospholipids that are interfacially hydrolyzed by $cPLA_2$ and $sPLA_2$.

8.2 PAF-Acetyl Hydrolases in Brain and Plasma

PAF-acetyl hydrolases are a family of distinct enzymes with the common property of hydrolyzing and inactivating PAF (Snyder, 1995; Tjoelker and Stafforini, 2000; Arai, 2002; Arai et al., 2002). It has been shown that the structure and the biochemical behavior of these enzymes depend on their cellular origin. At least three PAF-acetyl hydrolases are known to occur in mammalian tissues. These enzymes show different biochemical and molecular properties, and are encoded by different genes (Tables 8.1, 8.2, and 8.3). Two PAF-acetyl hydrolases, type I and type II, are intracellularly found in brain and other visceral organs. The third PAF-acetyl hydrolase is found in the plasma (Arai et al., 2002; Tjoelker and Stafforini, 2000; Arai, 2002). Although PAF-acetyl hydrolases are serine-dependent enzymes, but their ability to hydrolyze PAF is quite different. Brain type I PAF-acetyl hydrolases are classified into Ia and Ib isoforms. Ia is a heterodimeric isoform, which is composed of α_1 and α_2 subunits. In contrast, Ib isoform is a G protein-like complex, which is composed of α_1, α_2, and β subunits. The majority of adult tissues, except the brain, express only α_2 subunit. Type I PAF-acetyl hydrolase is also expressed in macrophages and tissues containing a high content of inflammatory cells (Derewenda and Derewenda,

Table 8.1 Properties of brain type I PAF-acetyl hydrolase.

Parameter	Value	Reference
pH optimum	7.0	Arai, 2002; Karasawa et al. 2003
K_m value (μM)	4.9	Arai, 2002; Karasawa et al. 2003
V_{max} (pmol^{-1} min^{-1} mg^{-1})	238.0	Arai, 2002; Karasawa et al. 2003
Localization	Cytosol	Arai, 2002; Karasawa et al. 2003
Effect of PMSF	Inhibition	Arai, 2002; Karasawa et al. 2003
Effect of DTNB	Inhibition	Arai, 2002; Karasawa et al. 2003
Effect of DTT	Stimulation	Arai, 2002; Karasawa et al. 2003

Table 8.2 Properties of kidney type II PAF-acetyl hydrolase.

Parameter	Value	Reference
pH optimum	7.4	Hattori et al. 1995b; Arai, 2002
K_m value (μM)	9.1	Hattori et al. 1995b; Arai, 2002
V_{max} (nmol^{-1} min^{-1} mg^{-1})	50,000	Hattori et al. 1995b; Arai, 2002
Diisopropylfluorophosphate	Inhibition	Hattori et al. 1995b; Arai, 2002
Molecular mass (kDa)	40	Hattori et al. 1995b; Arai, 2002

Type IIPAF acetyl hydrolases from other intracellular sources have similar physicochemical and kinetic properties.

Table 8.3 Properties of mammalian plasma PAF-acetyl hydrolase.

Parameter	Value	Reference
pH optimum	7.4	Karasawa et al., 2003; Rice et al., 1998
K_m value (μM)	13.7	Karasawa et al., 2003; Rice et al., 1998
V_{max} (μmol^{-1} min^{-1} mg^{-1})	170	Karasawa et al., 2003; Rice et al.,1998
Nature	Glycoprotein	Karasawa et al., 2003; Rice et al., 1998
Diisopropylfluorophosphate	Inhibition	Karasawa et al., 2003; Rice et al., 1998
Molecular mass (kDa)	43–67	Karasawa et al., 2003; Rice et al., 1998

1998; Arai, 2002; McMullen, 2000). Of the three intracellular isoforms, Ib is the most abundant in brain tissue (Hattori et al., 1996). Type II PAF-acetyl hydrolase and plasma PAF-acetyl hydrolase have been purified and characterized from various mammalian tissues (Karasawa et al., 2003; Hattori et al., 1996).

8.3 Purification and Properties of PAF-Acetyl Hydrolases

Bovine brain cytosolic PAF-acetyl hydrolase activity can be separated into three distinct peaks, all of which have pH optima in the neutral to mild alkaline region and are unaffected by EDTA (Hattori et al., 1996). Major PAF-acetyl hydrolase activity has been purified to homogeneity, using multiple column chromatographic procedures (Hattori et al., 1996). The purified enzyme has a molecular mass of about 100 kDa, as estimated by gel filtration chromatography. SDS-gel electrophoresis indicates the presence of three distinct bands of 45, 30, and 29 kDa, respectively (Hattori et al., 1996). These polypeptides exclusively comigrate with the enzymic activity, during the course of purification. Diisopropyl fluorophosphate completely inhibits the enzymic activity at 0.1 mM. [^{3}H]Diisopropyl fluorophosphate labeled only the 29-kDa polypeptide, suggesting that this polypeptide possesses an active serine residue(s). cDNA for the 29-kDa catalytic subunit has been cloned. The predicted sequence of 232 amino acids is unique, and is not homologous to any other PAF-acetyl hydrolase. cDNA encodes for PAF-acetyl hydrolase activity can be expressed in *E. coli* and COS7 cells (Hattori et al., 1994). These results indicate that this subunit is responsible for the catalytic activity. The amino acid sequence of active site is different from the consensus sequence of the serine esterase family. The sequence of about 30 amino acids located six residues downstream from the active serine site exhibits significant homology to the first transmembrane region of the PAF receptor. These data demonstrate that the catalytic subunit of brain PAF-acetyl hydrolase is a novel type of serine esterase (Hattori et al., 1994).

8.3.1 Types I PAF-Acetyl Hydrolases in Mammalian Tissues

Brain intracellular PAF-acetyl hydrolase type I is composed of mutually homologous α_1 and α_2 subunits, both of which account for catalytic activity along interactions with β subunit. Thus, this PAF-acetyl hydrolase is a heterotrimeric enzyme. Studies on determination of biochemical differences in three possible catalytic dimers, α_1/α_1, α_1/α_2, and α_2/α_2 indicate that the α_2/α_2 homodimer exhibits different substrate specificity from the α_1/α_1 homodimer and the α_1/α_2 heterodimer, both of which display similar substrate specificity (Manya et al 1999). Among 1-*O*-alkyl-2-acetyl-phospholipids, α_2/α_2 homodimer hydrolyzes PAF and 1-*O*-alkyl-2-acetyl-*sn*-glycero-3-phosphorylethanolamine most efficiently. In contrast, both α_1/α_1 and α_1/α_2 degrades 1-*O*-alkyl-2-acetyl-sn-glycero-3-phosphoric acid more efficiently than PAF. This metabolite is an intermediate of de novo PAF synthesis, where it is generated through the acetylation of alkylglycerophosphate (Manya et al 1999; Baker, 1995; Baker, 2002). 1-*O*-alkyl-2-acetyl-*sn*-glycero-3-phosphorylethanolamine is the poorest substrate for these enzymes. Collective evidence suggests that substrate specificity of α_1/α_2 heterodimer is similar to the α_1/α_1 homomer but not to the α_2/α_2 homodimer (Manya et al 1999). β subunit interacts with all three catalytic dimmers, but modulates the enzymic activity in a catalytic dimer composition-dependent manner. The β subunit strongly accelerates the enzymic activity of the α_2/α_2 homodimer, but suppresses the activity of the α_1/α_1 homodimer and has little effect on that of the α_1/α_2 heterodimer. Thus, the enzyme activity of type I PAF-acetyl hydrolase may be regulated not only by switching the composition of the catalytic subunit, but also by manipulating the β subunit (Manya et al., 1999; Arai, 2002).

Compositional studies have shown that α_1 subunit of type I acetyl hydrolase shares 63% sequence identity with α_2 subunit. Both α_1 and α_2 have a catalytic center (Hattori et al., 1994; Hattori et al., 1995a; Hattori et al 1995b; Manya et al

1999). The β subunit does not have catalytic center, but has an unique domain structure called a WD repeat. The WD repeats are typically found in multimeric protein complexes. They mediate protein–protein interactions (Neer et al., 1994). The α subunit of type I acetyl hydrolase strongly interacts with β-spectrin (Wang et al., 1995), a cytoskeletal protein associated with stability of neural membranes. The regulatory β subunit of type I PAF-acetyl hydrolase interacts with mNUDC, a product of nudC-like gene. NUDC increases the catalytic activity of type I PAF-acetyl hydrolase (Riera et al., 2007). The regulatory activity of NUDC is located in the carboxyl terminal half of NUDE protein and is highly conserved. This suggests that mammalian nudC-like gene products modulate type I PAF-acetyl hydrolase, and play an antiinflammatory role during the inflammatory process (Riera et al., 2007).

The gene for β subunit is identical to the human *L1S1* gene. *L1S1* gene is the causative gene for type I lissencephaly, ("smooth brain," from "lissos," meaning smooth, and "encephalos," meaning brain), which is a severe developmental disorder in which neuronal migration is impaired, leading to a thickened cerebral cortex whose normally folded contour is simplified and smooth (see Chap. 10) (Garg et al., 2007; Fleck et al., 2000; McMullen, 2000).

The expression of α_1/α_2 subunits in brain is developmentally regulated. The fetal brain contains α_1/α_2 heterodimers, while adult brain predominantly contains the α_2 homomer (Manya et al., 1998). Studies on primary cultured cells isolated from neonatal rat brain indicate that α_1 protein is expressed only in neurons. In contrast, α_2 and β transcripts and proteins have been detected both in neural and nonneural tissues, and their expression level remains constant from fetal stages through adulthood. These results indicate that α_1 expression is restricted to actively migrating neurons in rats and that switching of catalytic subunits from the α_1/α_2 heterodimer to the α_2/α_2 homodimer occurs during brain development. On the basis of these studies, it is suggested that PAF-acetyl hydrolase plays a key role(s) in neuronal migration (Manya et al., 1998; Arai, 2002). Most of the above-mentioned information has been obtained from in vitro investigations. It is not known whether such alterations in subunit composition occur in vivo and play some role in modulation of PAF-mediated processes or not. To gain more information on in vivo role of type I PAF-acetyl hydrolase, α_1 and α_2-deficient mice have been generated by the targeted disruption (Koizumi et al., 2003). α_1(–/–) Mice are indistinguishable from wild-type mice, whereas, α_2(–/–) male mice show a significant reduction in testis size. Both catalytic subunits are expressed at high levels in testis as well as in brain in mutant mice. Levels of LIS1 protein are significantly reduced in α_2(–/–) and double-mutant mice, indicating that the catalytic subunits, especially α_2, are a determinant of LIS1 expression level (Koizumi et al., 2003).

8.3.2 Types II PAF-Acetyl Hydrolases in Mammalian Tissues

Mammalian tissue cytosol contains types II PAF-acetyl hydrolase. Purified bovine liver Type II PAF-acetyl hydrolase is a 40 kDa monomer. It shares 43% homology with extracellular, plasma PAF-acetyl hydrolase. cDNA clones of bovine and human type II PAF-acetyl hydrolase have been isolated and characterized (Hattori et al 1996). From the longest open reading frame of the cloned cDNAs, both bovine and human PAF-acetyl hydrolases II contain 392 amino acid residues and exhibit 88% identity with each other at the amino acid level. Both bovine and human enzymes contain a Gly-X-Ser-X-Gly motif that is characteristic of lipases and serine esterases. A region surrounding the active site of type II PAF-acetyl hydrolase shares homology with other lipases. Expression of isoform II cDNA in COS7 cells results in a marked increase in PAF-acetyl hydrolase activity. Immunoblotting studies using an established monoclonal antibody against the bovine enzyme indicate that the recombinant protein is located in the membranes as well as cytosol. Type II acetyl hydrolase is expressed most abundantly in bovine liver and kidney, but low levels are detected in other tissues. Type II PAF-acetyl hydrolase is myristoylated at the N-terminus, and like other N-myristoylated proteins, is distributed in both the cytosol and membranes. The amino acid sequence deduced from the cDNA of isoform II shows no homology with any subunit of isoform Ib (Hattori et al., 1996).

Collectively, these studies suggest that type I and type II PAF-acetyl hydrolases are trimeric enzymes consisting of two catalytic subunits and a regulatory subunit with catalytic serine residues (Arai, 2002). The purified enzymes display similar activity against PAF and oxidatively modified phosphatidylcholine, but does not hydrolyze phosphatidylcholine or phosphatidylethanolamine with two long chain acyl groups (Hattori et al., 1993; Hattori et al., 1996; Manya et al., 1998; Manya et al., 1999). The intracellular PAF-acetyl hydrolase isozymes are distinguished from each other by differences in primary sequence, tissue localization, subunit composition, and substrate preferences. The most thoroughly characterized intracellular isoform, Ib, is a G-protein-like complex with two catalytic subunits (α_1 and α_2) and a regulatory β subunit (Manya et al., 1999).

The intracellular type II PAF-acetyl hydrolase shares homology with plasma acetyl hydrolase and catalyzes the hydrolysis of oxidized arachidonic acid from oxidized glycerophospholipids. Both the intracellular type II and plasma PAF-acetyl hydrolases have high affinity for esterified F_2-isoprostanes, a series of prostaglandin-like compounds formed in vivo from the free radical catalyzed peroxidation of arachidonic acid independent of cyclooxygenase. However, the rate of esterified F_2-isoprostane hydrolysis is slower than other substrates (Stafforini et al., 2006). Studies on PAF-acetyl hydrolase knockout mice demonstrate that these animals have a higher capacity to release F_2-isoprostanes compared with nontransgenic littermates. These studies suggest that PAF-acetyl hydrolases play key roles in the hydrolysis of F_2-isoprostanes esterified on phospholipids in vivo (Stafforini et al., 2006).

8.3.3 PAF-Acetyl Hydrolases in Mammalian Plasma

Homogeneous preparations of guinea pig plasma PAF-acetyl hydrolase have been obtained using multiple column chromatographic procedures. The guinea pig plasma

PAF-acetyl hydrolase migrates as a broad band on SDS-PAGE with a molecular mass of 58–63 kDa, which is larger than the human enzyme (43–67 kDa) (Karasawa et al., 1996; 1999; 2003). The binding of the enzyme to Concanavalin A indicates the glycoprotein nature of this enzyme. The purified enzyme contains N-linked heterogenous sugar chains (9 kDa) containing sialic acid. Treatment of human plasma PAF-acetyl hydrolase with peptide N-glycosidase F shifts the mobility of protein band to a lower molecular mass position on the SDS-PAGE, indicating that human plasma PAF-acetyl hydrolase is extensively N-glycosylated. Enzyme shows optimal activity at pH 7.4, and has K_m and V_{max} values of 13.7 μM and 170 μmol^{-1} min^{-1} mg^{-1} protein. Inhibition of purified enzyme by diisopropyl fluorophosphates indicates that PAF-acetyl hydrolase is a serine-dependent hydrolase (Karasawa et al., 1996; 1999; 2003). Substrate specificity studies indicate that plasma PAF-acetyl hydrolase show a broad substrate specificity. When compared with acetyl group at the *sn*-2 position, with five-carbon *sn*-2 acyl group, enzyme shows 50% of hydrolytic activity. Nine-carbon long acyl chains are cleaved but with lower rates.

To elucidate the molecular structure of plasma PAF-acetyl hydrolase and clarify its relationships with plasma PAF-acetyl hydrolases of other species, cDNA sequencing studies have been performed. Guinea pig plasma PAF-acetyl hydrolase cDNA contains an open reading frame encoding 436 amino acids. Its predicted molecular mass (49 kDa) is lower than that of the native enzyme, suggesting that guinea pig plasma PAF-acetyl hydrolase, unlike the human enzyme, is modified posttranslationally, perhaps by glycosylation (Karasawa et al., 2003). The primary structure of human plasma PAF-acetyl hydrolase contains a GXSXG motif. This motif is found in most serine esterases and lipases. The active site of these esterases contains serine, aspartate, and histidine. Site-directed mutagenesis studies indicate the presence of Ser^{273}, Asp^{296}, and His^{351} in a putative catalytic triad (Stafforini et al., 1997). The human plasma PAF-acetyl hydrolase gene is located on chromosome 6p12–21.1, and is composed of 12 exons (Stafforini et al., 1996a, b). In addition, a G^{994}→T missense mutation has been identified in exon 9 of this gene. This mutation that is located within the catalytic domain of human PAF-acetyl hydrolase has been detected in members of 14 Japanese families with PAF-acetyl hydrolase deficiency (Stafforini et al., 1996a,b).

Site-directed mutagenesis demonstrates that Ser^{273} (of the GXSXG motif), Asp^{296}, and His^{351} are essential for PAF hydrolysis. These residues are conserved in PAF-acetyl hydrolase sequences isolated from bovine, dog, mouse, and chicken (Tjoelker et al., 1995). The linear orientation and spacing of these catalytic residues are consistent with the α/β hydrolase conformation of other lipases and esterases. This model is supported by analysis of systematic truncations of PAF-acetyl hydrolase and reveals that deletions beyond 54 amino acids from the NH_2 terminus and 21 from the COOH terminus result in the loss of enzymic activity. These observations demonstrate that although plasma PAF-acetyl hydrolase is a phospholipase A_2, but has structural properties characteristic of the neutral lipases and esterases (Tjoelker et al., 1995). This enzyme is regulated by proinflammatory mediators such as LPS, TNF-α, IL-1, IL-8, and IFN-α. PAF-acetyl hydrolase residues (Tyr^{205}, Tryp^{115}, and Leu^{116}) bind directly to COOH terminal of LDL-apo β 100. Up to 70%

of plasma PAF-acetyl hydrolase is associated with low density lipoproteins, and remainder with high density lipoproteins (Prescott et al., 2000).

In vivo, the synthesis and secretion of the plasma PAF-acetyl hydrolase are hormonally regulated. Thus, 17α-ethynylestradiol administration to female and male rats results in a fivefold decrease in plasma PAF-acetyl hydrolase activity (Miyaura et al., 1991; Yasuda et al., 1996; Contador et al., 1997). A similar decrease is also observed when estrogen is administered at low doses. Dexamethasone injections to male and female rats also produce a threefold increase in the plasma PAF-acetyl hydrolase activity (Miyaura et al., 1991). The activity returns to the values prior to hormone treatment four days after treatment cessation. Testosterone and progesterone have no effect on plasma acetylhydrolase activity. It is stated that estrogen-mediated alterations in PAF-acetyl hydrolase activity are associated with activity changes in the HDL fraction and not due to the presence of an inhibitor or activator in the plasma of the hormone-treated animals. Human serum obtained from a group of women, in which the 17β-estradiol concentration is elevated in preparation for an in vitro fertilization procedure, shows an inverse relationship between the plasma estrogen concentration and the PAF-acetyl hydrolase activity. Collectively, these studies indicate that in vivo plasma PAF-acetyl hydrolase activity is hormonally regulated (Miyaura et al., 1991; Yasuda et al., 1996; Contador et al., 1997).

8.4 Other PAF-Acetyl Hydrolases

Besides above-mentioned PAF-acetyl hydrolases, other PAF-acetyl hydrolases have been reported to occur in mammalian tissues (Karasawa et al., 1996; Karasawa et al., 1999; Sheffield et al., 2001). An enzyme from erythrocytes is a 25-kDa homomer, whereas PAF-acetyl hydrolase from mammalian kidney and liver is a 40–44 kDa polypeptide (Rice et al., 1998). This enzyme is inactivated by diisopropyl fluorophosphate and 5,5′-dithiobis(2-nitrobenzoic acid), suggesting that both serine and cysteine residues are required for the enzyme activity. Labeling of 40 kDa polypeptide with [^{3}H]Diisopropyl fluorophosphate confirms the identity of this enzyme. This enzyme protects biomembrane glycerophospholipids from oxidative damage. Human erythrocytes contain a PAF-acetyl hydrolase. This enzyme has been purified using ion exchange and hydrophobic chromatographies (Stafforini et al., 1993). The enzyme has a molecular weight of 25 kDa, and behaves as a dimer during gel filtration. Like other PAF-acetyl hydrolases, the erythrocyte PAF-acetyl hydrolase requires sulfhydryl agents for maximal activity. Enzymic activity is inhibited by 5,5′-dithiobis(2-nitrobenzoic acid), NaF, diisopropyl fluorophosphate, diethylpyrocarbonate, and p-bromophenacylbromide. It hydrolyzes phospholipid containing short and/or oxidized acyl groups at the *sn*-2 position (Stafforini et al., 1993). PAF-acetyl hydrolase activity of bovine seminal plasma has been purified. This enzyme has a molecular mass of 60 kDa (Soubeyrand et al., 1998). It also hydrolyzes PtdCho. The kinetic properties of this enzyme are similar to those of the human serum PAF-acetyl hydrolase. Although capable of hydrolyzing long-chain phosphatidylcholine, it shows a highly preferential activity toward PAF. The enzyme activity toward phosphatidylcholine, but not PAF, was Ca^{2+}-dependent. Biochemical characterization of bovine seminal plasma indicates that the enzyme is extensively N-glycosylated. It exists predominantly as a dimer in solution. Western blotting reveals that the enzyme is highly heterogeneous in charge and has an isoelectric point of 5.7. It is expressed exclusively in the seminal vesicles and the ampulla. No association of the enzyme with either epididymal or ejaculated spermatozoa can be detected (Soubeyrand., et al 1998). Collective evidence suggests that PAF-acetyl hydrolases are widely distributed in the intracellular as well as the extracellular compartments. They are not only involved in reproduction but are also associated with a defense mechanism that protects the cells against the toxic effects of PAF and other biologically active oxidized phospholipids (Farooqui and Horrocks, 2004).

8.5 Regulation and Roles of PAF-Acetyl Hydrolases in Brain

PAF-acetyl hydrolases are unique serine esterases that are involved in the hydrolysis of PAF and its analogs in mammalian tissues. These enzymes convert PAF into its inactive metabolite, the lyso-PAF (Blank et al., 1981). The levels of PAF-acetyl hydrolase are critical for the regulation of the circulatory PAF and its physiological and pathological activities (Tjoelker and Stafforini, 2000). Some forms of PAF-acetyl hydrolases are specific for PAF hydrolysis, while other PAF-acetyl hydrolases act on a variety of PAF analogs. Thus, PAF-acetyl hydrolases act as general scavengers of glycerophospholipids species, which may accumulate inappropriately (Tjoelker and Stafforini, 2000). In brain tissue, PAF-acetyl hydrolases are associated with a variety of functions including neuroprotection, inhibition of inflammation, modulation of long-term potentiation, modulation of gene expression, and neuronal migration (Umemura et al., 2007; Tjoelker and Stafforini, 2000; Bate et al., 2004; Bazan, 2003; Grassi et al., 1998) (Fig. 8.2). PAF enhances presynaptic glutamate release, and act as a retrograde messenger in long-term potentiation (LTP), and enhances memory formation. PAF receptor antagonist, BN-52021, blocks the induction of LTP (Arai and Lynch, 1992), and the application of carbamyl-PAF, a nonhydrolyzable PAF analog, induces LTP formation (Kato et al., 1994). The upregulation of PAF-acetyl hydrolase may inhibit LTP and memory formation in brain tissue. Similarly, expression of gene associated with FOS/JUN/AP-1 transcriptional signaling system as well as transcription of COX-2 (inducible prostaglandin synthase) is also modulated by PAF-acetyl hydrolase activity (Bazan, 2003). PAF analogs, which show PAF receptor-antagonistic activity (CV-6209 and CV-3988) and PAF receptor-agonists such as carbamyl PAF block granule cell migration (Tokuoka et al., 2003; Adachi et al., 1997). These analogs also inhibit the bovine brain PAF-acetyl hydrolase activity. Granule cell migration is restored when inhibitors are removed by washing the treated cells with buffer suggesting that the inhibitory effect of PAF analogs is reversible. On the basis of these findings, it is

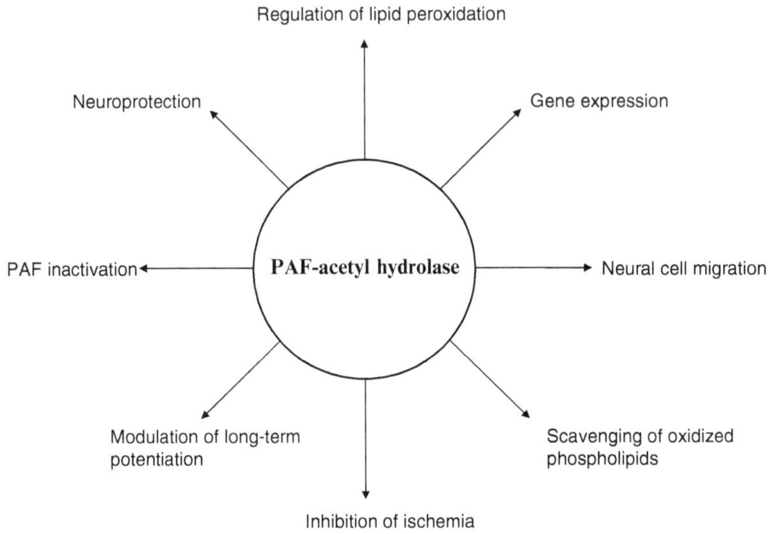

Fig. 8.2 Roles of PAF-acetyl hydrolase in brain.

suggested that the catalytic activity of PAF-acetyl hydrolase may play a crucial role in neuroprotection and neural cell migration.

As stated earlier that brain intracellular type I PAF-acetyl hydrolase is a complex enzyme composed of homologous α_1 and α_2 subunits, both of which account for catalytic activity, and the β subunit that modulates type I PAF-acetyl hydrolase activity. During brain development, Type I PAF-acetyl hydrolase activity is regulated not only by the combination of the catalytic subunits but also through the manipulation the β-subunit (Manya et al., 1998; Manya et al., 1999). The β subunit modulates activity of catalytic dimer in a composition-dependent manner. The β subunit accelerates the enzymic activity of α_2/α_2 homomer but has no effect on the activity of α_1/α_2 heterodimers. Thus, temporal modulation of expression of the α_1 subunit acts as an effective switch of subunit composition. Its coupling with β subunit provides a mechanism by which PAF levels are controlled at various stages of brain development (Manya et al., 1999). The exact composition of enzyme complex depends on the expression patterns of the α_1 and α_2 genes, exhibiting tissue specificity and developmental control. As stated earlier, all three possible dimers (α_1/α_1, α_1/α_2, and α_2/α_2) have been identified in various tissues. The α_1/α_2 heterodimer is thought to play an important role in fetal brain, while adult brain predominantly contains the α_2 homomer (Manya et al., 1998; Arai, 2002). The occurrence and expression of heteromeric subunit complexes of type I PAF-acetyl hydrolase at various stages of brain development suggest the existence of a versatile mechanism that may not only allow the accumulation of PAF and modulation of PAF-mediated signal transduction processes at various stages of developing brain but also stabilization of PAF-acetyl hydrolase activity in the adult brain (Arai, 2002).

PAF-acetyl hydrolases are associated with apoptotic cell death (Bonin et al., 2004). Using pharmacological inhibition of type I and II PAF-acetyl hydrolases and downregulation of type I PAF-acetyl hydrolase catalytic subunits by RNA interference (RNAi), it is demonstrated that the PAF-receptor-independent death pathway is regulated mainly by type I PAF-acetyl hydrolase and to a lesser extent by type II PAF-acetyl hydrolase. Moreover, the antiapoptotic actions of type I PAF-acetyl hydrolase are subunit-specific (Bonin et al., 2004). Type I PAF-acetyl hydrolase α_1 regulates intracellular PAF concentrations under normal physiological conditions, but expression is not sufficient to downregulate an acute rise in intracellular PAF levels. Type I PAF-acetyl hydrolase α_2 expression is induced when cells are deprived of serum or exposed to apoptogenic PAF concentrations limiting the duration of pathological cytosolic PAF accumulation. To inhibit PAF receptor-independent death pathway, several PAF antagonists (Fig. 8.3) have been screened. FR 49175 upregulate PAF hydrolysis, and prevent PAF-mediated caspase 3 activation. Both antagonists act indirectly to stimulate type I PAF-acetyl hydrolase α_2 homodimer activity by down-regulating type I PAF-acetyl hydrolase α_1 expression. Collective evidence suggest that type I PAF-acetyl hydrolase α_2 acts as a potent antiapoptotic protein, and provides a new means of pharmacologically targeting type I PAF-acetyl hydrolase to block PAF-mediated cell death (Bonin et al., 2004).

It is well known that type II PAF-acetyl hydrolase is N-myristoylated. Oxidative stress favors the myristoylation, which allows reversible translocation of this

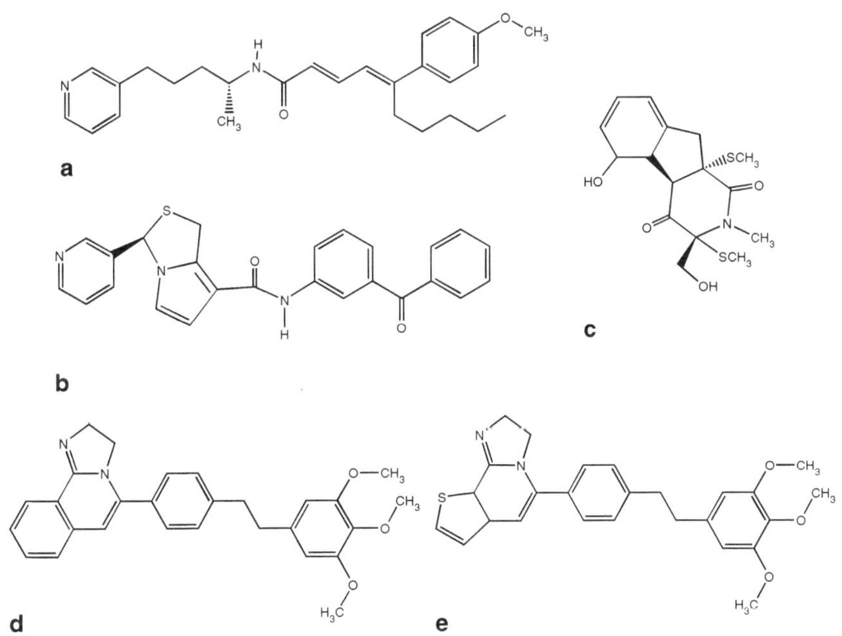

Fig. 8.3 Chemical structures of PAF receptor antagonists: Ro 24–0238 (**a**), RP 59227 (**b**), FR 49175 (**c**), SDZ 64–412 (**d**), and SDZ 65–123 (**e**).

enzyme from cytosol to membranes. Thus, in Chinese hamster ovary-K1cells, type II PAF-acetyl hydrolase is translocated from cytosol to membranes within 20 min. However, in cells treated with antioxidants, this enzyme is translocated from membranes to cytosol (Matsuzawa et al., 1997). Overexpression of PAF-acetyl hydrolase in Chinese hamster ovary-K1 cells suppresses oxidative stress-induced cell death, which occurs by apoptosis. These findings suggest that intracellular type II PAF-acetyl hydrolase translocates between cytosol and membranes in response to a redox state of the cell and protects the cell against oxidative stress most probably by hydrolyzing oxidized phospholipids (Matsuzawa et al., 1997).

PAF-acetyl hydrolase catalytic activity is sensitive to oxidants. Thus, peroxynitrite, an oxidant species generated after cellular activation, mediates oxidative inactivation of PAF-acetyl hydrolase (MacRitchie et al., 2007). Peroxynitrite-mediated inactivation of recombinant protein involves methionine and two tyrosine residues. Oxidation of LDL-associated PAF-acetyl hydrolase partially dissociates the enzyme from the LDL particles. Similarly, oxidation of the purified enzyme in the absence of lipoproteins inhibits subsequent association with LDL. Oxidant-mediated inactivation of PAF-acetyl hydrolase involves the dissociation of the enzyme from LDL particle (MacRitchie et al., 2007). The transfection of the plasma PAF-acetyl hydrolase gene has been shown to attenuate glutamate-mediated apoptotic cell death in cortical cultured neurons (Hirashima et al., 2000), and in transgenic mice, the overexpression of type II human PAF-acetyl hydrolase protects neurons from ischemic injury (Umemura et al., 2007). These studies are consistent with a potential role of PAF-acetyl hydrolases in protecting membrane glycerophospholipids against oxidative stress.

Plasma PAF-acetyl hydrolase acts exclusively in circulation and PAF associated with various blood cells (endothelial cell, leukocyte, and platelet) surfaces. As stated earlier that PAF-acetyl hydrolase not only hydrolyzes PAF, but also hydrolyzes oxidized derivatives of phosphatidylcholine that have a short-chain acyl residue at the *sn*-2 position (Fig. 8.4). Enzymic activity progressively falls as the length of the substrate *sn*-2 chain increases. The activity falls to zero with *sn*-2 chains of approximately eight carbons or longer (Stafforini et al., 1997). This enzyme also hydrolyzes oxidized phospholipids and plays a role in maintaining erythrocyte deformability (Yoshida et al., 1993). Plasma PAF-acetyl hydrolase and type II PAF-acetyl hydrolase liberate F_2-isoprostanes from *sn*-2 position of oxidized PtdCho with high affinity, but at a rate, which is much lower than other substrate (Stafforini et al., 2006). This suggests that PAF-acetyl hydrolases participate in cellular homeostasis because of their ability to initiate in vivo metabolic steps associated with repairing oxidized cellular membrane damage. Thus PAF-acetyl hydrolases plays a key role in the degradation of proinflammatory oxidized phospholipids and in the generation of lyso-PtdCho and oxidized fatty acids and their metabolites such as F_2-isoprostanes (Stafforini et al., 1997; Yoshida et al., 1993; Stafforini et al., 2006).

In human plasma, PAF-acetyl hydrolase circulates in a complex with low density lipoproteins (LDL) and high density lipoproteins (HDL). To characterize PAF-acetyl hydrolase-lipoprotein interactions, site-directed mutagenesis has been performed. Two domains within the primary sequence of human PAF-acetyl hydrolase,

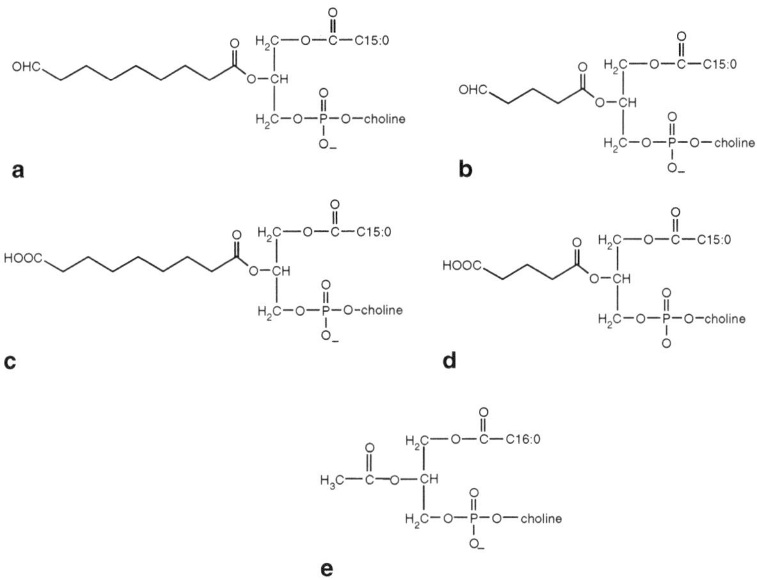

Fig. 8.4 Structures of PAF and PAF-like oxy-PtdCho: 9-Oxovaleroylphosphatidylcholine, *9-CHO-PtdCho* (**a**), 5-oxo-valeroylphosphatidylcholine, *5-CHO-PtdCho* (**b**), Azelaoylphosphatid ylcholine, *9-COOH-PtdCho* (**c**), Glutaroylphosphatidylcholine, *5-COOH-PtdCho* (**d**), and platelet-activating factor, *PAF* (**e**).

tyrosine[205] and residues 115 and 116, are important for its binding to LDL (Stafforini et al., 1999). Mutation or deletion of those sequences prevents the association of the enzyme with lipoproteins (Stafforini et al., 1999). These studies indicate that the carboxyl terminus of apoB plays a key role in the binding of PAF-acetyl hydrolase with LDL. This binding defines the physical state of PAF-acetyl hydrolase, confers a long half-life, and is a major determinant of its catalytic efficiency in vivo. Collective evidence suggests that plasma PAF-acetyl hydrolase have a dual role in metabolism. It not only inactivates PAF released from inflammatory cells, but is also involved in the elimination of oxidized fatty acid residues in neural membranes and high density plasma lipoproteins (Tjoelker and Stafforini, 2000; Arai, 2002). In vivo these actions may be critical in preventing inflammation and oxidative stress.

PAF-acetyl hydrolases are moderately inhibited by organophosphorus pesticides and related toxicants. Recently, new organophosphorus compounds have been designed for maximum in vitro potency and selectivity for mouse brain PAF-acetyl hydrolase. These compounds include a series of benzodioxaphosphorin 2-oxides (Quistad et al., 2005). Ultrahigh potency and selectivity has been achieved with *n*-alkyl methylphosphonofluoridates (long-chain sarin analogs). The increase in carbon chain length increases the potency of *n*-alkyl methylphosphonofluoridates for brain PAF-acetyl hydrolase. It is quite likely that the discovery of new PAF-AH inhibitors may facilitate investigations on other aspects of PAF metabolism and action in brain tissue (Quistad et al., 2005).

8.6 PAF Hydrolyzing Phospholipase C

Rat liver contains a phospholipase C (PLC) that hydrolyzes PAF into alkylacetylglycerol and phosphocholine (Fig. 8.5). Subcellular studies indicate that the enzyme is of lysosomal origin (Nishihira and Ishibashi, 1986). This PLC can be solubilized with 2% Triton X–100 from rabbit liver light mitochondria. PAF hydrolyzing PLC is purified 600- to 700-fold, using multiple column chromatographic procedures. The enzyme occurs in two forms. Low molecular mass form has a PI of 4.7, whereas high molecular mass form has a PI of 5.8. Each form has been purified to homogeneity. The purified enzymes migrate as single bands with molecular mass of 33 kDa and 75 kDa, respectively. The purified enzymes of low molecular mass and high molecular mass form have pH optima of 8.2 and 8.5 and apparent K_m values of 55.6 and 45.5 μM for PAF, respectively. PAF hydrolyzing PLC activity is inhibited by 1 mM EDTA. Addition of Ca^{2+} results in the complete restoration of enzymic activity. *p*-Chloromercuribenzoate markedly inhibits the enzyme activity,

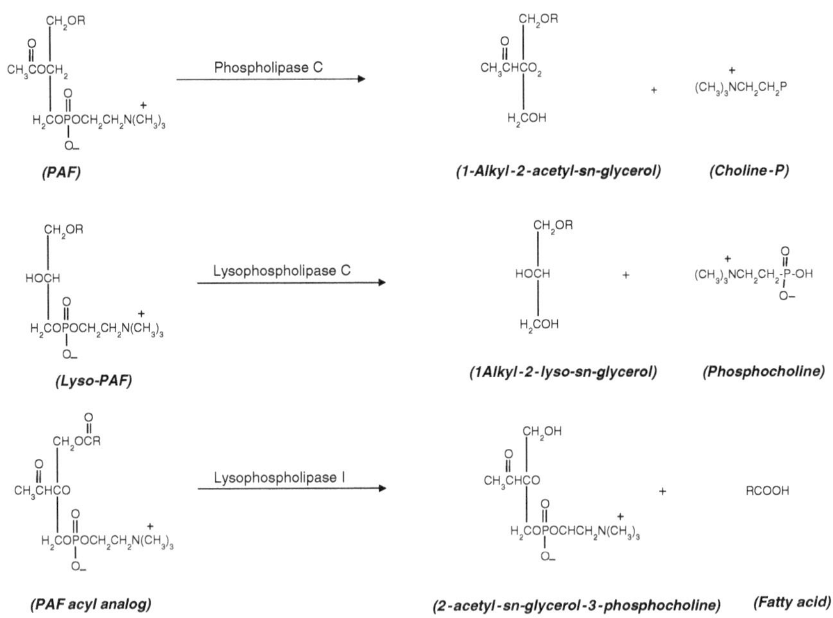

Fig. 8.5 Hydrolysis of PAF by PLC and lysophospholipases.

suggesting that PLC contains –SH groups (Nishihira and Ishibashi, 1986). Similar PAF hydrolyzing PLC is also found in the intestine (Wu et al., 2006). This enzyme shows optimal at pH 7.5. It is inhibited by EDTA, and stimulated by 0.1–0.25 mM Zn^{2+}. The activity is abolished by site mutation of the predicted metal-binding sites that are conserved in all nucleotide pyrophosphatase/phosphodiesterase (NPP) members. PAF hydrolyzing PLC is stimulated by bile salt, particularly taurocholate and taurochenodeoxycholate. The occurrence of an acto-lysophospholipase C in plasma membranes of porcine kidney epithelial cell line LLC-PK1 has also been reported (Tsutsumi et al., 2007). This enzyme hydrolyzes lysoPAF analog, Bodipy-lysoPAF, and participates in catabolism of lyso-PAF in kidney tissue.

8.7 Other PAF Hydrolyzing Lipases

These enzymes act on *sn*-1 position of acyl PAF analogs and hydrolyze them into 2-acetyl-*sn*-glycero-3-phosphocholine and fatty acid. Two lysophospholipases, namely lysophospholipase I and lysophospholipase II, have been purified from bovine liver (van den Bosch et al., 1991). Lysophospholipase II deacetylates PAF to lysoPAF at a faster rate than lysophospholipase I (Fig. 8.5). Detailed investigations on lysophospholipase-mediated degradation have not been performed. However, it is proposed that lysophospholipases may modulate the relative proportion of the acyl vs. ether-linked forms of PAF, and this may be associated with the intensity of PAF receptor-mediated signal transduction processes. A lyso-PAF hydrolyzing lysophospholipase D (lysoPLD) (Fig. 8.6) has been recently purified to apparent homogeneity from rat brain nuclear fraction (Sugimoto et al., 2006). The purified enzyme preparation migrates as a single band with a molecular weight of 35 kDa. It hydrolyzes 1-palmitoyl-glycerophosphorylcholine, lyso-PtdCho, PAF, and Lyso-PAF. It shows highest activity at pH 7.0–7.5 and requires Mg^{2+}. The K_m and V_{max} values for 1-palmitoyl-glycerophosphorylcholine were 176 μM and 0.3 μmol^{-1} min^{-1} mg^{-1}, respectively. It is proposed that lysoPLD purified from rat brain nuclear fractions hydrolyzes lysoPAF, PAF, and LPC to liberate choline (Sugimoto et al., 2006).

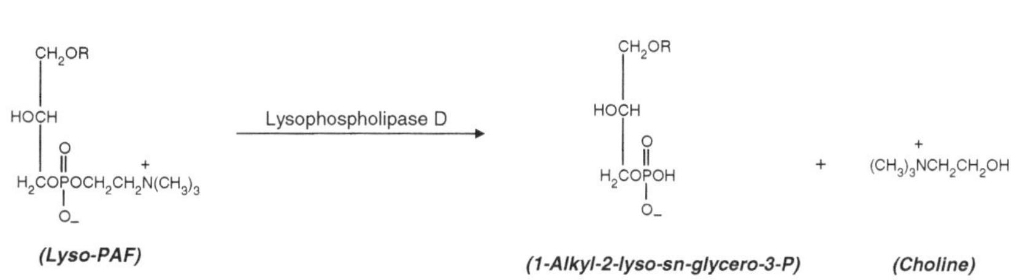

Fig. 8.6 Reaction catalyzed by lyso-PAF hydrolyzing lysophospholipase D.

8.8 Conclusion

PAF-acetyl hydrolases inactivate PAF by removing the acetyl group from the *sn*-2 position of glycerol moiety. These enzymes are found intracellularly in brain tissue and extracellularly in biological fluids such as plasma. Unlike PLA_2, which hydrolyze long acyl chains from the *sn*-2 position of glycerol moiety interfacially in the presence of Ca^{2+}, PAF-acetyl hydrolases are Ca^{2+}-independent enzymes that do not act in interfacially. They have substrate specificity for acyl chain of C_2 to C_9 length and hydrolyze a number of glycerophospholipids with oxidatively truncated fatty acyl chains. Three PAF-acetyl hydrolases are known to occur in mammalian tissues. They include type I, type II, and plasma PAF-acetyl hydrolase. Some intracellular type I PAF-acetyl hydrolases contain catalytic subunits, which organize themselves into homodimeric and heterodimeric forms. Type I PAF-acetyl hydrolase, Ib, is a G-protein-like complex of two catalytic subunits (α_1 and α_2) and a regulatory β subunit. The β subunit is a product of the LIS1 gene, mutations of which cause type I lissencephaly. Recent studies indicate that LIS1/β is important in cellular functions such as induction of nuclear movement and control of microtubule organization. Type II PAF-acetyl hydrolase is a 40 kDa monomer. Human PAF-acetyl hydrolase isoform II has been isolated and characterized. Type II PAF-acetyl hydrolase is myristoylated at the N-terminus, and like other N-myristoylated proteins is distributed in both the cytosol and the membranes. The amino acid sequence deduced from the cDNA of isoform II shows no homology with any subunit of isoform Ib. Plasma PAF-acetyl hydrolase is a glycoprotein that has a molecular mass of 58–63 kDa. Plasma PAF-acetyl hydrolase is specific for short acyl groups at the *sn*-2 position of the phospholipid substrate and with the exception of PAF. It hydrolyzes oxidized phospholipids to generate lyso-PtdCho and oxidized fatty acids. Thus, PAF-acetyl hydrolases may play a key role in the degradation of proinflammatory oxidized phospholipids and act as general scavenger of oxidized phospholipids species, which may accumulate inappropriately during neural trauma and neurodegenerative diseases. PAF-acetyl hydrolases may also be involved in modulation of long term potentiation, neural cell migration, gene expression, and neuroprotection.

References

Adachi T., Aoki J., Manya H., Asai H., Arai H. and Inoue K. (1997). PAF analogues capable of inhibiting PAF acetylhydrolase activity suppress migration of isolated rat cerebellar granule cells. Neurosci. Lett. 235:133–136.

Arai A. and Lynch G. (1992). Antagonists of the platelet-activating factor receptor block long-term potentiation in hippocampal slices. Eur. J. Neurosci. 4:411–419.

Arai H. (2002). Platelet-activating factor acetylhydrolase. Prostaglandins Other Lipid Mediat. 6869:83–94.

Arai H., Koizumi H., Aoki J., and Inoue K. (2002). Platelet-activating factor acetylhydrolase (PAF-AH). J. Biochem. 131:635–640.

Baker R. R. (1995). Enzymes of platelet activating factor synthesis in brain. Neurochem. Res. 20:1345–1351.

Baker R. R. (2002). Lipid acetylation reactions and the metabolism of platelet-activating factor. Neurochem. Res. 25:677–683.

Bate C., Salmona M. and Williams A. (2004). The role of platelet activating factor in prion and amyloid-beta neurotoxicity. Neuroreport. 15:509–513.

Bazan N. G. (2003). Synaptic lipid signaling: significance of polyunsaturated fatty acids and platelet-activating factor. J. Lipid Res. 44:2221–2233.

Billah M. M. and Johnston J. M. (1983). Identification of phospholipid platelet-activating factor (1-*O*-alkyl-2-acetyl-*sn*-glycero-3-phosphocholine) in human amniotic fluid and urine. Biochem. Biophys. Res. Commun. 113:51–58.

Blank M. L., Lee T., Fitzgerald V., and Snyder F. (1981). A specific acetylhydrolase for 1-alkyl-2-acetyl-*sn*-glycero-3-phosphocholine (a hypotensive and platelet-activating lipid). J. Biol. Chem. 256:175–178.

Bonin F., Ryan S. D., Migahed L., Mo F., Lallier J., Franks D. J., Arai H., and Bennett S. A. L. (2004). Anti-apoptotic actions of the platelet-activating factor acetylhydrolase I α_2 catalytic subunit. J. Biol. Chem. 279:52425–52436.

Contador M., Moya F. R., Zhao B., Furukawa M., Frenkel R. A., Guzzetta P., and Johnston J. M. (1997). Effect of dexamethasone on rat plasma platelet activating factor acetylhydrolase during the perinatal period. Early Hum. Dev. 47:167–176.

Cox C. P., Wardlow M. C., Jorgensen R., and Farr R. S. (1981). The presence of platelet-activating factor (PAF) in normal human mixed saliva. J. Immunol. 127:46–50.

Derewenda Z. S. and Derewenda U. (1998). The structure and function of platelet-activating factor acetylhydrolases. Cell Mol. Life Sci. 54:446–455.

Farooqui A. A. and Horrocks L. A. (2004). Plasmalogens, platelet-activating factor, and other ether lipids. In: Nicolaou A. and Kokotos G. (eds.), *Bioactive Lipids*. Oily Press, Bridgwater, England, pp. 107–134.

Farooqui A. A., Ong W. Y., and Horrocks L. A. (2006). Inhibitors of brain phospholipase A_2 activity: Their neuropharmacological effects and therapeutic importance for the treatment of neurologic disorders. Pharmacol. Rev. 58:591–620.

Fleck M. W., Hirotsune S., Gambello M. J., Phillips-Tansey E., Suares G., Mervis R. F., Wynshaw-Boris A., and McBain C. J. (2000). Hippocampal abnormalities and enhanced excitability in a murine model of human lissencephaly. J. Neurosci. 20:2439–2450.

Garg A., Sridhar M. R., and Gulati S. (2007). Autosomal recessive type I lissencephaly. Indian J. Pediatr. 74:199–201.

Gelb M. H., Min J. H., Jain M. K. (2000). Do membrane-bound enzymes access their substrates from the membrane or aqueous phase: Interfacial versus non-interfacial enzymes. Biochim. Biophy. Acta. 1488:20–27.

Grassi S., Francescangeli E., Goracci G., and Pettorossi V. E. (1998). Role of platelet-activating factor in long-term potentiation of the rat medial vestibular nuclei. J. Neurophysiol. 79:3266–3271.

Hattori M., Arai H., and Inoue K. (1993). Purification and characterization of bovine brain platelet-activating factor acetylhydrolase. J. Biol. Chem. 268:18748–18753.

Hattori M., Adachi H., Tsujumoto M., Arai H., and Inoue K. (1994). The catalytic subunit of bovine brain platelet-activating factor acetylhydrolase is a novel type of serine esterase. J. Biol. Chem. 269:23150–23155.

Hattori M., Adachi H., Aoki J., Tsujimoto M., Arai H., and Inoue K. (1995a). Cloning and expression of a cDNA encoding the beta-subunit (30-kDa subunit) of bovine brain platelet-activating factor acetylhydrolase. J Biol. Chem. 270:31345–31352.

Hattori K., Hattori M., Adachi H., Tsujimoto M., Arai H., and Inoue K. (1995b). Purification and characterization of platelet-activating factor acetylhydrolase II from bovine liver cytosol. J. Biol. Chem. 270:22308–22313.

Hattori K., Adachi H., Matsuzawa A., Yamamoto K., Tsujimoto M., Aoki J., Hattori M., Arai H., and Inoue K. (1996). cDNA cloning and expression of intracellular platelet-activating factor

(PAF) acetylhydrolase II. Its homology with plasma PAF acetylhydrolase. J. Biol. Chem. 271:33032–33038.

Hirashima Y., Ueno H., Karasawa K., Yokoyama K., Setaka M., and Takaku A. (2000). Transfection of the plasma-type platelet-activating factor acetylhydrolase gene attenuates glutamate-induced apoptosis in cultured rat cortical neurons. Brain Res. 885:128–132.

Karasawa K., Kuge O., Kawasaki K., Nishijima M., Nakano Y., Tomita M., Yokoyama K., Setaka M., ans Nojima S. (1996). Cloning, expression and characterization of plasma platelet-activating factor-acetylhydrolase from guinea pig. J. Biochem. (Tokyo). 120:838–844.

Karasawa K., Harada A., Satoh N., Inoue K., and Setaka M. (2003). Plasma platelet activating factor-acetylhydrolase (PAF-AH). Prog. Lipid Res. 42:93–114.

Karasawa K., Qiu X., and Lee T. (1999). Purification and characterization from rat kidney membranes of a novel platelet-activating factor (PAF)-dependent transacetylase that catalyzes the hydrolysis of PAF, formation of PAF analogs, and C_2-ceramide. J. Biol. Chem. 274:8655–8661.

Kato K., Clark G. D., Bazan N. G., and Zorumski C. F. (1994). Platelet-activating factor as a potential retrograde messenger in CA1 hippocampal long-term potentiation. Nature 367:175–179.

Koizumi H., Yamaguchi N., Hattori M., Ishikawa T. O., Aoki J., Taketo M. M., Inoue K., and Arai H. (2003). Targeted disruption of intracellular type I platelet activating factor-acetylhydrolase catalytic subunits cause severe impairment in spermatogenesis. J. Biol. Chem. 278:12489–12494.

Kunievsky B. and Yavin E. (1994). Production and metabolism of platelet-activating factor in the normal and ischemic fetal rat brain. J. Neurochem. 63:2144–2151.

Lee T. C., Blank M. L., Fitzgerald V., and Snyder F. (1981). Substrate specificity in the biocleavage of the *O*-alkyl bond: 1-alkyl-2-acetyl-*sn*-glycero-3-phosphocholine (a hypotensive and platelet-activating lipid) and its metabolites. Arch. Biochem. Biophys. 208:353–357.

Lynch J. M. and Hensen P. M. (1986). The intracellular retention of newly synthesized platelet-activating factor. J. Immunol. 137:2653–2661.

MacRitchie A. N., Gardner A. A., Prescott S. M., and Stafforini D. M. (2007). Molecular basis for susceptibility of plasma platelet-activating factor acetylhydrolase to oxidative inactivation. FASEB J. 21:1164–1176.

Manya H., Aoki J., Watanabe M., Adachi T., Asou H., Inoue Y., Arai H., and Inoue K. (1998). Switching of platelet-activating factor acetylhydrolase catalytic subunits in developing rat brain. J. Biol. Chem. 273:18567–18572.

Manya H., Aoki J., kato H., Ishii J., Hino S., Arai H., and Inoue K. (1999). Biochemical characterization of various catalytic complexes of the brain platelet-activating factor acetylhydrolase. J.Biol. Chem. 274:31827–31832.

Matsuzawa A., Hattori K., Aoki J., Arai H., and Inoue K. (1997). Protection against oxidative stress-induced cell death by intracellular platelet-activating factor-acetylhydrolase II. J Biol. Chem. 272:32315–32320.

McMullen T. W., Li J., Sheffield P. J., Aoki J., Martin T. W., Arai H., Inoue K., and Derewenda Z. S. (2000). The functional implications of the dimerization of the catalytic subunits of the mammalian brain platelet-activating factor acetylhydrolase (Ib). Protein Eng. 13:865–871.

Min J. H., Wilder C., Aoki J., Arai K., Inoue K., Paul L., and Gelb M. H. (2001). Platelet-activating factor acetylhydrolases: Broad substrate specificity and lipoprotein binding does not modulate the catalytic properties of the plasma enzyme. Biochemistry 40:4539–4549.

Miyaura S., Maki N., Byrd W., and Johnston J. M. (1991). The hormonal regulation of platelet-activating factor acetylhydrolase activity in plasma. Lipids 26:1015–1020.

Neer E. J., Schmidt C. J., Nambudripad R., and Smith T. F. (1994). The ancient regulatory-protein family of WD-repeat proteins. Nature 371:297–300.

Nishihira J. and Ishibashi T. (1986). A phospholipase C with a high specificity for platelet-activating factor in rabbit liver light mitochondria. Lipids 21:780–785.

Prescott S. M., Zimmerman G. A., Stafforini D. M., and McIntyre T. M. (2000). Platelet-activating factor and related lipid mediators. Annu. Rev. Biochem. 69:419–445.

Quistad G. B., Fisher K. J., Owen S. C., Klintenberg R., and Casida J. E. (2005). Platelet-activating factor acetylhydrolase: Selective inhibition by potent *n*-alkyl methylphosphonofluoridates. Toxicol. Appl. Pharmacol. 205:149–156.

Rice S. Q. J., Southan C., Boyd H. F., Terrett J. A., Macphee C. H., Moores K., Gloger I. S., and Tew D. G. (1998). Expression, purification and characterization of a human serine-dependent phospholipase A_2 with high specificity for oxidized phospholipids and platelet activating factor. Biochem. J. 330:1309–1315.

Riera J., Rodriguez R., Carcedo M. T., Campa V. M., Ramos S., and Lazo P. S. (2007). Isolation and characterization of nudC from mouse macrophages, a gene implicated in the inflammatory response through the regulation of PAF-AH(I) activity. FEBS Lett. 581:3057–3062.

Sheffield R. J., McMullen T. W., Li J., Ho Y. S., Garrard S. M., Derewenda U., and Derewenda Z. S. (2001). Characterization of a platelet activating factor acetylhydrolase from rat adipocyte. Life Sci. 67:2807–2825.

Snyder F. (1995). Platelet-activating factor: The biosynthetic and catabolic enzymes. Biochem. J. 305:689–705.

Sogos V., Bussolino F., Pilia E., Torelli S., and Gremo F. (1990). Acetylcholine-induced production of platelet-activating factor by human fetal brain cells in culture. J. Neurosci. Res. 27:706–711.

Soubeyrand S., Lazure C., and Manjunath P. (1998). Phospholipase A_2 from bovine seminal plasma is a platelet-activating factor acetylhydrolase. Biochem. J. 29:41–47.

Stafforini D. M., Rollins E. N., Prescott S. M., and McIntyre T. M. (1993). The platelet-activating factor acetylhydrolase from human erythrocytes. Purification and properties. J. Biol. Chem. 268:3857–3865.

Stafforini D. M., Satoh K., Atkinson D. L., Tjoelker L. W., Eberhard C., Yoshida H., Imaizumi T., Takamatsu S., Zimmerman G. A., McIntyre T. M., Gray P. W., and Prescott S. M. (1996a). Platelet-activating factor acetylhydrolase deficiency A missense mutation near the active site of an anti-inflammatory phospholipase. J. Clin. Invest. 97:2784–2791.

Stafforini D. M., Prescott S. M., Zimmerman G. A., and McIntyre T. M. (1996b). Mammalian platelet-activating factor acetylhydrolases. Biochim. Biophys. Acta. 1301:161–173.

Stafforini D. M., McIntyre T. M., Zimmerman G. A., and Prescott S. M. (1997). Platelet-activating factor acetylhydrolases. J. Biol. Chem. 272:17895–17898.

Stafforini D. M., Tjoelker L. W., McCormick S. P., Vaitkus D., McIntyre T. M., Gray P. M., Young S. G., and Prescott S. M. (1999). Molecular basis of the interaction between plasma platelet-activating factor acetylhydrolase and low density lipoprotein. J. Biol. Chem. 274:7018–7024.

Stafforini D. M., Sheller J. R., Blackwell T. S., Sapirstein A., Yull F. E., McIntyre T. M., Bonventre J. V., Prescott S. M., and Roberts L. J. (2006). Release of free F_2-isoprostanes from esterified phospholipids is catalyzed by intracellular and plasma platelet-activating factor acetylhydrolases. J. Biol. Chem. 281:4616–4623.

Sugimoto S., Sugimoto H., Aoyama C., Aso C., Mori M., and Izumi T. (2006). Purification and characterization of lysophospholipase D from rat brain. Biochim. Biophys. Acta Mol. Cell Biol. Lipids. 1761:1410–1418.

Tjoelker L. W., Eberhardt C., Unger J., Le Trong H., Zimmerman G. A., McIntyre T. M., Stafforini D. M., Prescott S. M., and Gray P. W. (1995). Plasma platelet-activating factor acetylhydrolase is a secreted phospholipase A_2 with a catalytic triad. J. Biol. Chem. 270:25481–25487.

Tjoelker L. W. and Stafforini D. M. (2000). Platelet-activating factor acetylhydrolases in health and disease. Biochim. Biophys. Acta. 1488:102–123.

Tokuoka S. M., Ishii S., Kawamura N., Satoh M., Shimada A., Sasaki S., Hirotsune S., Wynshaw-Boris A., and Shimizu T. (2003). Involvement of platelet-activating factor and LIS1 in neuronal migration. Eur. J. Neurosci. 18:563–570.

Tselepis A. D. and Chapman M. J. (2002). Inflammation, bioactive lipids and atherosclerosis: potential roles of a lipoprotein-associated phospholipase A_2, platelet activating factor-acetylhydrolase. Atherosclerosis Suppl. 3:57–68.

Tsutsumi T., Morishiqe J., Fukuzawa K., and Tokumura A. (2007). Movement of monoglyceride derived from hydrolysis of fluorescence-labeled lyso platelet-activating factor by lysophospholipase C through plasma membranes of porcine kidney epithelial cell line LLC-PK1. Prostaglandin Other Lipid Mediat. 83:33–41.

Umemura K., Kato I., Hirashima Y., Ishii Y., Inoue T., Aoki J., Kono N., Oya T., Hayashi N., Hamada H., Endo S., Oda M., Arai H., Kinouchi H., and Hiraga K. (2007). Neuroprotective role of transgenic PAF-acetylhydrolase II in mouse models of focal cerebral ischemia. Stroke 38:1063–1068.

van den Bosch H., Slurk A., ten Cate J. W., and Aarsman A. J. (1991). Studies on the selectivity of enzymes involved in platelet-activating factor formation in stimulated cells. Lipids 26:967–973.

Wang D. S., Shaw R., Hattori M., Arai H., Inoue K., and Shaw G. (1995). Binding of pleckstrin homology domains to WD40/beta-transducin repeat containing segments of the protein product of the Lis-1 gene. Biochem. Biophys. Res. Commun. 209:622–629.

Wu J., Nilsson A., Jonsson B. A. G., Stenstad H., Agace W., Cheng Y. J., and Duan R. D. (2006). Intestinal alkaline sphingomyelinase hydrolyses and inactivates platelet-activating factor by a phospholipase C activity. Biochem. J. 394:299–308.

Wykle R. L. and Schremmer J. M. (1974). A lysophospholipase D pathway in the metabolism of ether-linked lipids in brain microsomes. J. Biol. Chem. 249:1742–1746.

Yoshida H., Satoh K., and Takamatsu S. (1993). Platelet-activating factor acetylhydrolase in red cell membranes. Does decreased activity impair erythrocyte deformability in ischemic stroke patients? Stroke 24:14–18.

Yasuda K., Furukawa M., and Johnston J. M. (1996). Effect of estrogens on plasma platelet-activating factor acetylhydrolase and the timing of parturition in the rat. Biol. Reprod. 54:224–229.

Chapter 9
Roles of Platelet-Activating Factor in Brain

9.1 Introduction

Platelet-activating factor (PAF) is a bioactive phospholipid that activates a number of cells including neural cells (neurons, astrocytes, oligodendrocytes, and microglia), platelets, leukocytes, monocytes, macrophages, endothelial cells, and smooth muscle cells (Aihara et al., 2000) (Montrucchio et al., 2000). A variety of stimuli, including those producing inflammation, promote the synthesis and release of PAF from neural and nonneural cells. As PAF interacts with many types of nonneural cells, it mediates processes as diverse as wound healing, physiological inflammation, angiogenesis, apoptosis, and reproduction (Montrucchio et al., 2000). Physiological concentrations (1–100 nM) of PAF promote differentiation in developing neurons and increase the strength of synaptic transmission in the mature brain. Higher concentrations of PAF (μM) that occur in pathological conditions such as head and spinal cord trauma and ischemia trigger neuronal cell death (Bazan et al., 1997; Kornecki et al., 1996). In brain tissue, PAF may be associated with neural cell migration, gene expression, calcium mobilization, noniception, and long-term potentiation (Fig. 9.1). PAF interacts with neural and nonneural cells by binding to specific receptors called as PAF receptors (PAF-Rs). These receptors have been cloned and characterized from nonneural tissues (Honda et al., 1991). Like G protein-coupled receptors, PAF-Rs possess seven transmembrane helices and signals through several G proteins such as Gαo, Gαi, Gβγ, and Gαq. PAF-Rs are associated with multiple intracellular signaling pathways (Honda et al., 1991; Clark et al., 2000). Because neural and most nonneural cells both synthesize and release PAF and express PAF receptors, PAF has potent biological actions on brain and visceral tissues. Many PAF analogs with varying PAF activity have been synthesized. These analogs include oxidized phospholipids and glucose-containing PAF (Glc-PAF). The most potent of the nonenzymically generated PAF analogs are the butanoyl (C4-PAF) and butenoyl (C4:1-PAF), which are one-tenth as potent as PAF (Fig. 9.2). These PAF analogs initiate and modulate many of the cellular events in inflammatory process and oxidative stress. Like PAF, oxidized glycerophospholipids induce and propagate chronic inflammatory processes (Leitinger, 2003). Collective evidence suggests that complex mixtures of oxidized phospholipids or

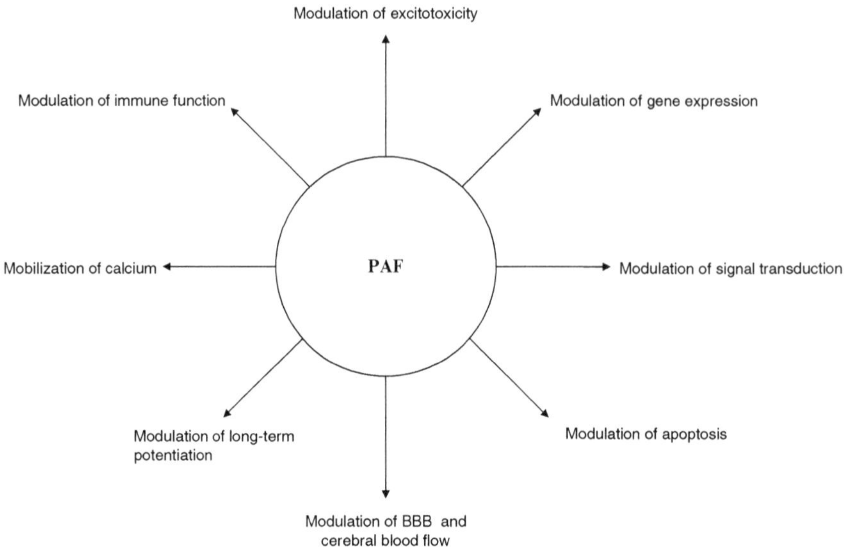

Fig. 9.1 Roles of PAF in brain tissue.

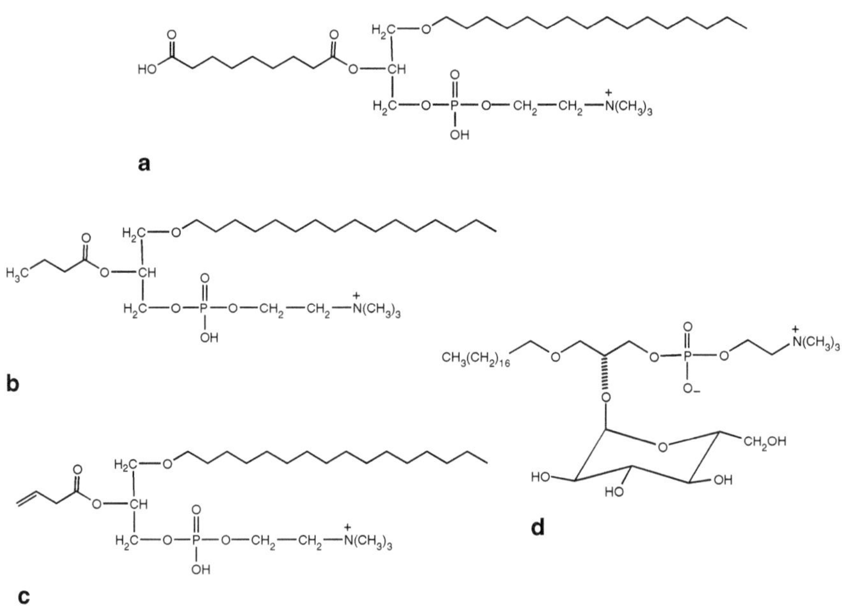

Fig. 9.2 Chemical structures of synthetic PAF analogs: Azelaoyl-PtdCho (**a**), butanoyl-PAF (**b**), butenoyl-PAF (**c**), Glc-PAF (**d**).

specific synthetic oxidized glycerophospholipids produce an inflammatory reaction and response in nonneural cells through the induction of proinflammatory genes (such as *COX-2*, *MCP-1*, *IL-8*, *VEGF*, and *TNF-α*). In human aortic endothelial cells, oxidized glycerophospholipids modulate the expression of a number of genes related to angiogenesis, atherosclerosis, inflammation, and wound healing (Furnkranz and Leitinger, 2004). In addition, PAF analogs activate platelets, induce differentiation of monocytes, and induce dedifferentiation of smooth muscle cells – processes related to plaque formation (Montrucchio et al., 2000). Oxidized glycerophospholipids act via transcription factors such as peroxisome proliferator-activated receptors (PPARs) α and γ, and via nuclear factor of activated T cells (NFAT) and Egr-1 (Furnkranz and Leitinger, 2004). They also modulate the fate of an inflammatory response by intervening into such processes as removal of apoptotic cells and by blocking bacterial-induced inflammation (Montrucchio et al., 2000). Inappropriate activation of PAF-mediated signaling pathway is associated with many diseases in which inflammation is thought to be one of the underlying features (Farooqui et al., 2007a). PAF has potent actions on cerebral vessels and cerebral metabolism when administered in vivo. Chemical structures of some PAF agonists and antagonists are shown in Figs. 9.3 and 9.4. These agonists and antagonists

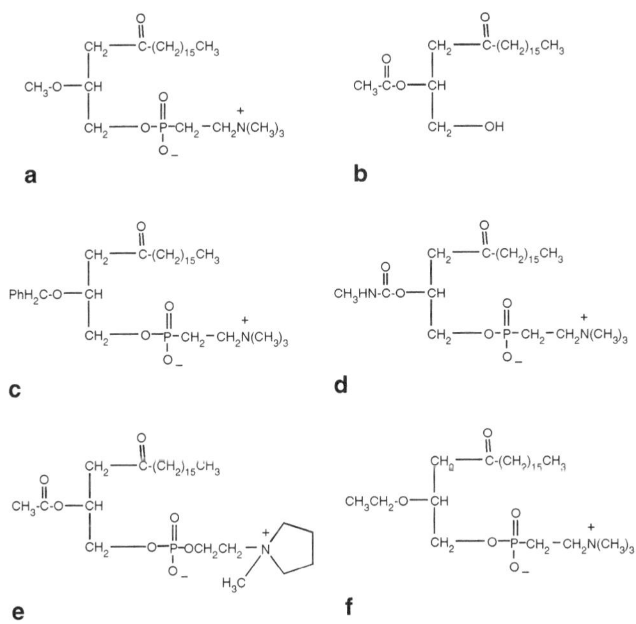

Fig. 9.3 Chemical structures of PAF analogs and PAF receptor agonists. PAF analogs include 2-*O*-methyl-PAF (1-hexadecyl-2-*O*-methyl-*sn*-glycero-3-phosphocholine) (**a**); 1-*O*-hexadecyl-2-methyl-*sn*-glycerol (**b**); 1-*O*-2-*O*-benzyl-PAF (1-*O*-hexadecyl-2-*O*-benzyl-*sn*-glycero-3-phosphocholine (**c**); and PAF agonists include C-PAF (carbamyl-PAF) (1-*O*-hexadecyl-2-*N*-methylcarbamyl-*sn*-glycero-3-phosphocholine (**d**); pyrrolidino-PAF (1-*O*-hexadecyl-2-*O*-acetyl-*sn*-glycero-3-phospho-(-*N*-methylpyrrolidino)-ethanolmine (*e*), and E-PAF (1-O-hexadecyl-2-O-ethyl-*sn*-glycero-3-phosphocholine).

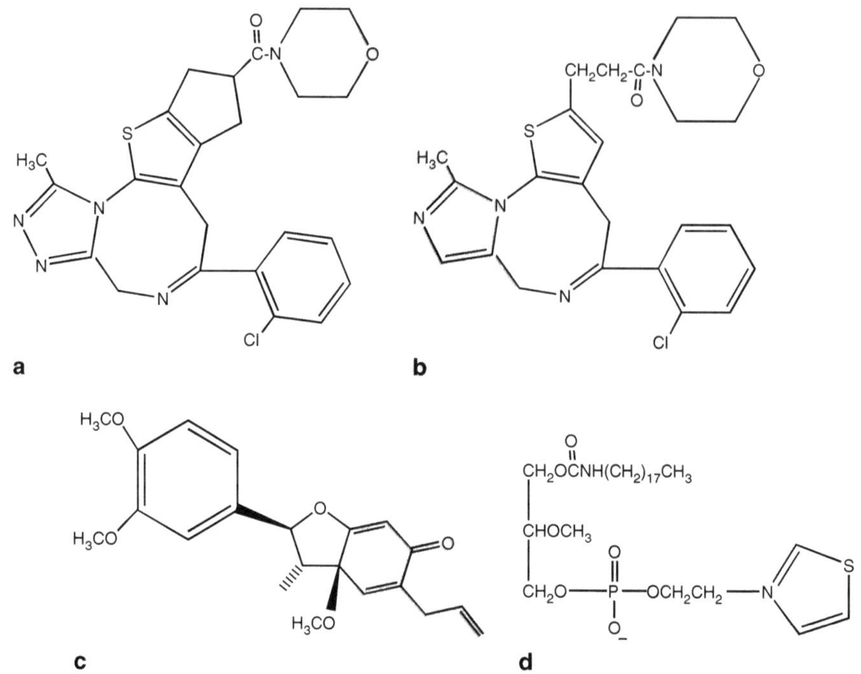

Fig. 9.4 Structures of PAF receptors of PAF antagonists: WEB-2170 (**a**), WEB-2086 (**b**), kadsurenone (**c**), and CV-3988 (**d**).

have been used not only to study properties of PAF receptors, but also to discover the role of PAF in neural and nonneural tissues.

In the cardiovascular system, PAF plays a role in embryogenesis because it regulates endothelial cell migration and angiogenesis, and may modulate cardiac function because it exhibits mechanical and electrophysiological actions on cardiomyocytes (Montrucchio et al., 2000). Moreover, PAF may contribute to the modulation of blood pressure mainly by affecting the renal vascular circulation. In pathological conditions, PAF has been involved in the hypotension and cardiac dysfunctions occurring in various cardiovascular stress situations such as cardiac anaphylaxis and hemorrhagic, traumatic, and septic shock syndromes (Montrucchio et al., 2000).

9.2 PAF Receptors in Brain

PAF receptor (PAF-R) consists of 341–342 amino acids with a putative seven transmembrane spanning domain. The receptor is expressed ubiquitously with the highest expression in leukocytes. Brain contains a small but significant amount of

PAF-R mRNA (Bito et al., 1992). Expression of PAF-Rs in brain has been established by radioligand-binding assay and Northern blotting and in situ hybridization studies in rats and mice (Marcheselli et al., 1990; Ishii et al., 1996). PAF receptor mRNA has been detected in the hypothalamus, medulla-pons, olfactory bulb, hippocampus, cerebral cortex, spinal cord, thalamus, and cerebellum of the rat brain.

In situ hybridization and Northern blotting studies have indicated that mRNA for PAF-R is expressed in both neurons and glial cells (Mori et al., 1996). Two populations (high affinity and low affinity) of PAF receptors have been reported to occur in rat brain hypothalamic membranes (Junier et al., 1988). Studies on PAF binding site in gerbil brain have indicated that maximal number of PAF receptors occur in the midbrain and hippocampus, and olfactory bulb, frontal cortex and cerebellum contain less PAF-binding sites (Hosford et al 1990). PAF receptors have also been reported in neuroblastoma x glioma hybrid NG 108–15 cells. Subcellular distribution studies have indicated the presence of three distinct PAF-binding sites in the rat cerebral cortex, two high affinity intracellular sites on microsomal membranes, and one low affinity-binding site on synaptosomal plasma membrane (Marcheselli et al., 1990). The synthesis of PAF in neurons and glial cells is upregulated following acetylcholine stimulation and downregulated by atropine, a muscarinic cholinergic receptor antagonist.

9.3 Translocation of PAF from Synthetic Site to Cell Surface Receptors

In response to various stimuli, PAF is rapidly synthesized within neural and non-neural cells. Although most of it is secreted into the extracellular milieu but some PAF remains associated with membranes (Prescott et al., 2002). The ability of the PAF to transduce the signal depends on its interaction with PAF-Rs on the cell surface. The translocation of PAF from its synthesizing site to the outer plasma membrane requires its transport. Very little information is available on mechanisms associated with PAF transport across cellular membranes. The human multidrug-resistance (MDR1) P-glycoprotein (Pgp), an ATP-binding-cassette transporter (ABCB1 with molecular mass of 170 kDa), which is ubiquitously expressed in non-neural and neural cells, plays an important role in PAF transport. Thus, [^{14}C]PAF synthesized intracellularly from exogenous alkylacetylglycerol and [^{14}C]choline become accessible to albumin in the extracellular medium of pig kidney epithelial LLC-PK1 cells in the absence of vesicular transport (Raggers et al., 2001). PAF translocation across the apical membrane is greatly stimulated by the expression of MDR1 Pgp, and blocked by the MDR1 inhibitors, PSC833, and cyclosporin A. (Raggers et al., 2001). In addition to PAF, MDR1 Pgp also transports other membrane lipids with shortened acyl chains across the plasma membrane. MDR1 Pgp transporter is insensitive to the MRP1 inhibitors, indomethacin (COX1 and COX2 inhibitor), and to depletion of GSH, which is required for MRP1 activity (Raggers et al., 2001). Mice deficient in MDR1 (MDR1a–/–) are susceptible to develop a severe and spontaneous intestinal inflammation. These knockouts mice develop a form of colitis similar to ulcerative colitis, which can be treated by antibiotics. This indicates a barrier function for P-glycoprotein against the invasion of bacteria or toxins (Panwala et al., 1998). It is stated that these processes at least in part are mediated by PAF synthesis and transport in response to bacterial toxin (Raggers et al., 2001; Panwala et al., 1998).

9.4 PAF-Receptor-Mediated Signal Transduction

PAF-Rs are linked through G proteins to several intracellular signal transduction pathways, including activation of PLA_2, PLC, and PLD, activation of GTPase activity, turnover of phosphatidylinositol, calcium mobilization, and activation of kinases (Table 9.1) (Maclennan et al., 1996; Mori et al., 1996; Honda et al., 2002; Kornecki and Ehrlich, 1991; Clark et al., 2000). Activation of PLA_2, PLC, and PLD results in arachidonic acid release, synthesis of eicosanoids, enhancement of polyphosphoinositide turnover, and generation of diacylglycerol, and inositol 1,4,5-trisphosphate ($InsP_3$), causing elevation in intracellular calcium concentration (Fig. 9.5). Activation of guanylate cyclase results in generation of cGMP, inhibition of the adenylate cyclase causing a decrease in cAMP levels, and phosphorylation of proteins through the activation of various kinases, including MAP-kinases, phosphatidylinositol 3-kinase, and tyrosine kinases (Fig. 9.6) (Chao and Olson, 1993; Honda et al., 2002; Ishii and Shimizu, 2000). Arachidonic acid is metabolized to eicosanoids, diacylglycerols activate protein kinase C, and $InsP_3$ mobilizes calcium from intracellular stores (Izumi and Shimizu, 1995; Ishii and Shimizu, 2000). PAF analogs and PAF receptor antagonists inhibit most of these processes. Thus, PAF acts as an intracellular lipid mediator (Marcheselli and Bazan, 1994). The binding of PAF to intracellular sites also elicits gene expression in neuronal and glial cell

Table 9.1 Enzymic activities responding to PAF receptor activation in neural and nonneural cells.

Enzyme	Mechanism of coupling	Effect	Reference
Phospholipase A_2	G protein	Stimulation	Izumi and Shimizu, 1995
Phospholipase C	G protein	Stimulation	Izumi and Shimizu, 1995
Phospholipase D	G protein	Stimulation	Izumi and Shimizu, 1995
Adenylate cyclase	G protein	Inhibition	Izumi and Shimizu, 1995
PtdIns-3-kinase	G protein	Stimulation	Izumi and Shimizu, 1995
MAP-kinase	–	Stimulation	DeCoster et al., 1998
Protein tyrosine kinase	–	Stimulation	Izumi and Shimizu, 1995
Cyclooxygenase-2	–	Stimulation	Bazan et al., 1991
Metalloproteinase-2	–	Stimulation	Ottino et al., 2005
Metalloproteinase-9	–	Stimulation	Taheri and Bazan, 2007
Caspase-3	–	Stimulation	Hostettler and Carlson, 2002

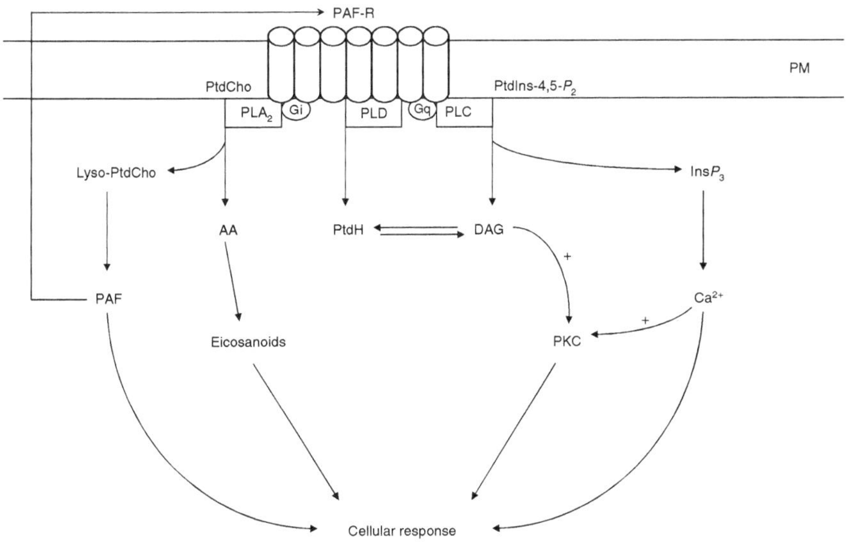

Fig. 9.5 Coupling of PAF receptor with phospholipases A_2, C, and D. *PtdCho* Phosphatidylcholine, *lyso-PtdCho* lyso-phosphatidylcholine, *AA* arachidonic acid, *PAF* platelet-activating factor; *PAF-R* platelet-activating receptor, *PtdH* phosphotidic acid, *DAG* diacylglycerol; *PtdIns-4,5-P_2* phosphatidylinositol 4,5-bisphosphate, *InsP_3* inositol 2,4,5-trisphosphate, *PKC* protein kinase C, and G_i and G_q are G proteins.

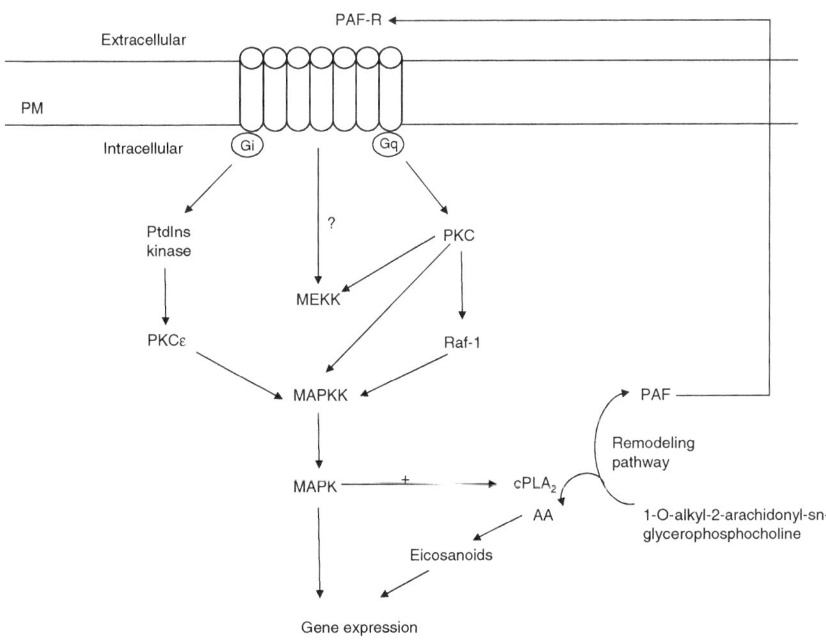

Fig. 9.6 Coupling of various kinases with PAF receptor. *PtdIns kinase* Phosphatidylinositol 3-kinase, *PKCζ* protein kinase Cζ, *MEKK* mitogen-activated protein kinase kinase kinase, *PKC* protein kinase C, *raf-1* product of oncogene c-raf. This protein functions in the MARK/ERK signal transduction, mitogen-activated protein kinase kinase (*MARKK*), mitogen-activated protein kinase (*MARK*), platelet-activating factor (*PAF*), and platelet-activating factor receptor (*PAF-R*).

lines (Bazan et al., 1994). Furthermore, PAF receptors are also involved in the release of prostaglandin E_2 from astrocytes. PAF also exerts its biological effects by activating leukocytes, stimulating platelet aggregation, and inducing the release of cytokines, and expression of cell adhesion molecules (Maclennan et al., 1996; Honda et al., 2002). PAF-R activation results in elevation of calcium through opening of calcium channels as well as from intracellular stores. This elevation in calcium is known to cause neuronal growth cone collapse (Izumi and Shimizu, 1995; Ishii and Shimizu, 2000).

PAF upregulates the expressions of NGF mRNA and NGF protein in astrocytes, in a time and dose-dependent manner (Yoshida et al., 2005). After 48 h of stimulation, PAF increases the levels of NGF protein in astrocyte-conditioned medium by 1.4-fold. The PAF-mediated stimulation of NGF expression can be further enhanced (2.1-fold of the control) in the cells under hypoxic culture condition. BN52021 (Ginkgolide B), an antagonist for PAF-binding sites (Fig. 9.7), prevents the effect

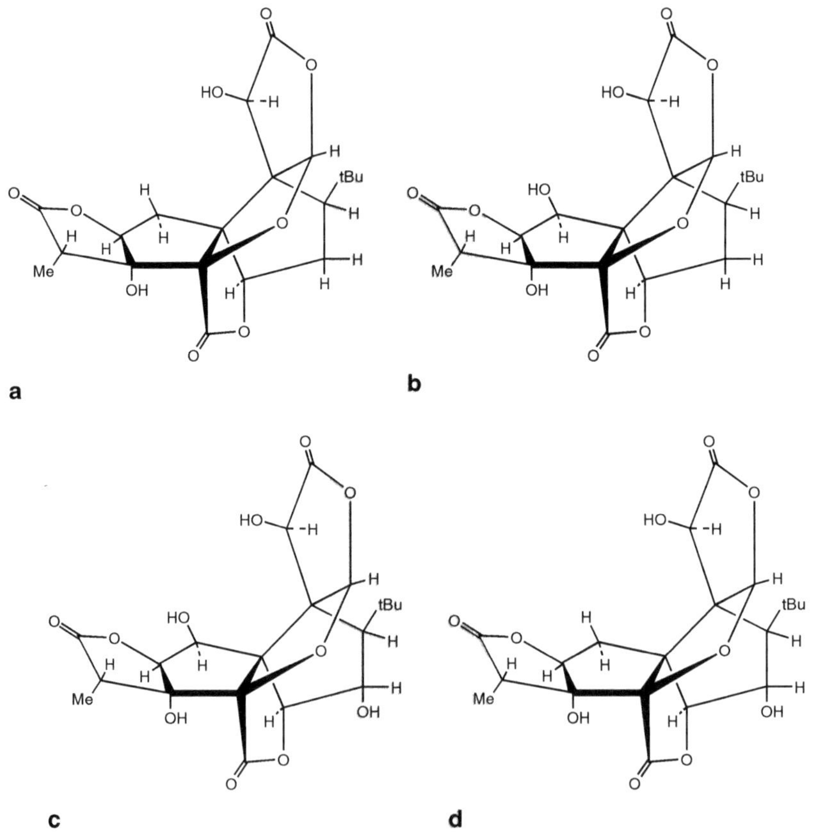

Fig. 9.7 Structures of PAF receptor antagonists: BN 52020 (**a**), BN 52021 (**b**), BN 52022 (**c**), and BN 52024 (**d**).

Table 9.2 Effect of PAF on gene expression of enzymes, cytokines, and growth factors in neural and nonneural tissues.

Enzyme/cytokine/growth factor	Effect	Reference
Cyclooxygenase	Upregulation	Bazan et al., 1991
Metalloproteinase-2	Upregulation	Han et al., 2004
Metalloproteinase-9	Upregulation	Taheri and Bazan, 2007
Nitric oxide synthase	Downregulation	Qu et al., 1999
Cytokines	Upregulation	Wang and Sun, 2000
NF-κB	Upregulation	Ko et al., 2002
Vascular endothelial growth factor	Upregulation	Ko et al., 2006; Yoshida et al., 2002
Heparin-binding growth factor	Upregulation	Pan et al., 1995
Nerve growth factor	Upregulation	Yoshida et al., 2005; Brodie, 1995
Immediate early genes	Upregulation	Squinto et al., 1989; Bazan et al., 1991

of PAF. PAF enhances *NGF* gene expression in human astrocytes, and the PAF-mediated increase in the expression of NGF under hypoxia may benefit the protection of the brain tissue by promoting neuronal survival (Yoshida et al., 2005). In nonneural cells, PAF upregulates the expression of vascular endothelial growth factor (VEGF) and heparin-binding growth factor (Table 9.2). These growth factors play an important role in angiogenesis. Collective evidence suggests that the binding of PAF to PAF receptors activates diverse intracellular signal transduction pathways (Figs. 9.5 and 9.6) associated with calcium mobilization, the stimulation of phospholipases A_2, C, and D activities, and activation of various kinases that ultimately results in the stimulation of transcription factors and induction of gene expression. In brain tissue, PAF also upregulates the expression of nerve growth factor, and generation of several lipid mediators. These processes modulate neurodegeneration as well as neuroprotection in brain tissue.

9.5 Roles of PAF in brain

9.5.1 PAF in Gene Expression

PAF promotes transcriptional activation of a number of genes, including c-fos, c-jun, and krox-24, genes for cytokines, enzymes, and growth factors (Table 9.2). In neuroblastoma cells, PAF treatment results in a 7- and 12-fold increase in c-fos mRNA in 15 and 30 min, respectively (Squinto et al., 1989; Bazan et al., 1991). The activation of these genes by PAF can be blocked by PAF antagonist, BN 52021. Similarly, PAF also stimulates c-jun transcription. Interestingly, c-jun response is slower and lower than c-fos response. It reaches maximal level (eightfold) in 60 min (Squinto et al., 1989). Pretreatment with PAF antagonist inhibits c-jun transcription (Bazan et al., 1991). PAF induces a rapid and transient elevation of c-fos and krox-24 in rat astroglial cell cultures. This effect is completely blocked by pretreatment with a PAF receptor antagonist. The induction of the expression of immediate early genes indicates that PAF possesses proliferative and differentiating effects in neural cells.

PAF also stimulates the inducible isoform of prostaglandin synthase or cyclooxygenase (COX-2) in a dose-dependent manner (Fig. 9.8). This enzyme catalyzes the rate-limiting step in the oxidation of arachidonic acid to prostaglandins. An immediate early gene encodes COX-2, which is responsible for prostaglandin synthesis in neuropathological processes. Preincubation of cells with the PAF antagonist, BN 50730, blocks the stimulation of the immediate early gene responsible for COX-2 (Bazan et al., 1997). Thus, PAF is associated with short-and long-term responses of cells to neural cell stimulation or neural trauma. Similarly, PAF upregulates the expression of COX-2 and production of PGE_2, as well as secretion of the inflammatory cytokine, interleukin-8 (IL-8) in the immortalized sebaceous gland cell line

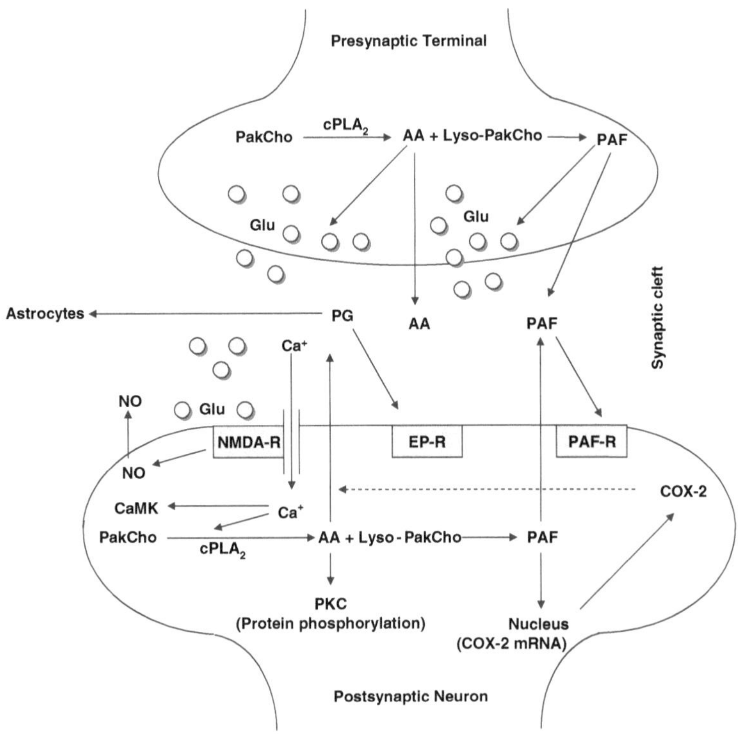

Fig. 9.8 Hypothetical model showing the generation of involvement of PAF and its interaction with glutamate, calcium, and arachidonic acid metabolism. *PakCho* Alkylacyl-glycerophosphocholine, *lyso-PakCho* Alkyl-lyso-glycerophosphocholine, *cPLA*$_2$ cytosolic phospholipase A_2, *Glu* glutamate, *AA* arachidonic acid, *NO* nitrix oxide, *NMDA-R* NMDA receptor, *PG* prostaglandin, *EP-R* prostaglandin receptor, *PAF* platelet-activating factor, *PAF-R* platelet-activating factor receptor, *CaMK* calmodulin-dependent protein kinase, *PKC* protein kinase C, and *COX-2* cyclooxygenase-2.

SZ95 (Zhang et al., 2006). This upregulation can be prevented by PAF receptor antagonists. Collective evidence suggests that PAF, under these pathological conditions, behaves as a neuronal injury messenger by at least two mechanisms: (a) enhancing glutamate release (see later) and (b) by sustained augmentation of COX-2 transcription. These events link PAF with neurodegeneration.

PAF augments angiogenesis by promoting the synthesis of various angiogenic factors, via the activation of NF-κB (Ko et al., 2005). Matrix metalloproteinase-2 and 9 (MMP)-2 and 9 play an important role in PAF-induced angiogenesis. In ECV304 cells, PAF increases mRNA expression, protein synthesis, and MMP-9 activity in a NF-κB-dependent manner. PAF-mediated increase in MMP-9 promoter activity in ECV304 is inhibited by WEB2107 (Fig. 9.4) and NF-κB inhibitors. Transfected NF-κB subunits, p65 or p50, increases luciferase activity in the reporter plasmid MMP-9, resulting in an increase not only of MMP-9 luciferase activity, but also of mRNA expression in MMP-9. MMP-9 or NF-κB inhibitors significantly block PAF-mediated angiogenesis, in a dose-dependent manner, in an in vivo mouse Matrigel implantation model (Ko et al., 2005). In parallel to the Matrigel implantation study, MMP-9 or NF-κB inhibitors prevent PAF-mediated sprouting of porcine pulmonary arterial endothelial cells. These data indicate that NF-κB-dependent MMP-9 plays a key role in PAF-induced angiogenesis.

In human corneal epithelial cells (HCECs), PAF modulates the expression of *MMP-9* gene. In HCECs, DNA-binding activity of NF-κB, Sp1, and AP-1 are upregulated by PAF. Mutation of the -79AP-1 or -600 NF-κB motif reduces the activity of MMP-9 promoter and the induction of gene expression by PAF. In untreated HCECs, mutation of the -558Sp1 motif upregulates gene expression, but it produces a significant decrease in the promoter activity mediated by PAF (Taheri and Bazan, 2007). Inhibition of MEK activity blocks the PAF-mediated phosphorylation and activation of Sp1 and abolishes the upregulation of MMP-9 expression and activity. Interactions between several regulatory elements are required for the induction of MMP-9 promoter activity by PAF and that PAF overturns the repressor effect of Sp1 through activation of the MEK/ERK signaling cascade (Taheri and Bazan, 2007). Similarly, in A375SM cells PAF-mediated stimulation of CREB and ATF-1 phosphorylation requires pertussis toxin-insensitive Gαq protein and adenylate cyclase activity (Melnikova et al., 2006). This stimulation is antagonized by a cAMP-dependent protein kinase A and p38 MAPK inhibitors. Furthermore, PAF also mediates the stimulation of gelatinase activity of MMP-2 by activating transcription and MMP-2 expression. MMP-2 activation correlates with the PAF-mediated increase in the expression of an MMP-2 activator, membrane type 1 MMP. PAF-mediated expression of pro-MMP-2 is causally related to PAF-mediated CREB and ATF-1 phosphorylation. This process is inhibited by PAF-R antagonist and inhibitors of p38 MAPK and protein kinase A, and is abrogated upon quenching of CREB and ATF-1 activities by forced overexpression of a dominant-negative form of CREB. PAF-mediated MMP-2 activation is also downregulated by p38 MAPK and protein kinase A inhibitors. Finally, PAFR antagonist PCA4248 blocks the development of A375SM lung metastasis in nude mice. Collective evidence suggests that PAF acts as a promoter of melanoma metastasis in vivo (Melnikova et al., 2006). PAF also modulates the expression of immediate early genes and enzymes associated with neuroinflammation and angiogenesis in brain (Farooqui et al., 2007b).

9.5.2 PAF in Neural Cell Migration

Neuronal migration is a critical event in brain development. Intracellular type I PAF acetyl hydrolase is a heterotrimeric enzyme composed of α_1, α_2, and β subunits. This trimeric enzyme may be involved in neural cell migration. Noncatalytic β subunit of this enzyme is a product of the *LIS1* gene. This gene is a causative gene for Miller-Dieker syndrome (Lo Nigro et al., 1997; Reiner et al., 1993). This syndrome is a form of lissencephaly that is characterized by widespread agyria of the brain and defects of neuronal cell migration. The deficiency of this PAF-acetyl hydrolase in Miller-Dieker syndrome may lead to an overabundance of PAF, during cerebral cortical development. Overabundance of PAF not only initiates premature growth cone collapse, but also blocks radial migration of neurons to the cortical plate and results in disorientation of neurons, which migrate abnormally (Clark et al., 2000). Thus, *L1S1* gene plays as a pivotal molecule that links PAF action and neuronal cell migration in studies, both in vivo and in vitro (Reiner et al., 1993).

In granule cell prepared from rat cerebellum at postnatal day 0, neural cell migration is observed on a layer coated with laminin. This granule cell migration is effectively prevented by PAF analogs, which display PAF-R-antagonistic activity (CV-6209 and CV-3988) (Fig. 9.4) and PAF-R-agonistic activity (carbamoyl PAF) (Adachi et al., 1997). These PAF analogs also inhibit the activity of bovine brain PAF acetyl hydrolase. Cell migration is restored when the inhibitors are removed by washing the treated cells with buffer. This observation suggests that the inhibitory effect of PAF analogs is reversible. Structurally-unrelated PAF antagonists (SM-12502, TCV-309, and YM-264), none of which show any appreciable inhibitory activity against PAF acetyl hydrolase, do not inhibit granule cell migration under the same conditions. It is proposed that the catalytic activity of PAF acetyl hydrolase plays a crucial role in neural cell migration (Adachi et al., 1997).

LIS1 mutant mice display defects in neuronal migration and layering in vivo, and also in cerebellar granule cell migration in vitro. This observation is supported by the studies on PAF-R-deficient mice. PAF-R-deficient mice display histological abnormalities in the embryonic cerebellum. Cerebellar granule neurons from PAF-R deficient mice migrate more slowly in vitro than wild-type neurons. Furthermore, PAF-R antagonists reduce the migration of wild-type neurons in vitro. Synergistic reduction of neuronal migration is observed in a double mutant of PAF-R and LIS1 (Tokuoka et al., 2003). Unexpectedly, PAF also modulates the migration of PAF-R-deficient neurons, suggesting a PAF-R-independent pathway for PAF action. The PAF-R-independent response to PAF is abolished in granule neurons derived from the double mutant mice. Thus, the migration of cerebellar granule cells is regulated by PAF through receptor-dependent and receptor-independent pathways, and that

LIS1 is a pivotal molecule that links PAF action and neuronal cell migration both in vivo and in vitro (Tokuoka et al., 2003).

9.5.3 PAF in Long-Term Potentiation

Long-term potentiation (LTP) is a long-lasting enhancement of synaptic efficacy because of repeated stimulation of postsynaptic NMDA receptors. Hippocampal LTP can be induced in the dentate gyrus, CA1, and CA3 subfields. LTP formation in the dentate gyrus and CA1 subfields involves the NMDA type of glutamate receptor. In contrast, LTP formation in CA3 subfield occurs via a NMDA receptor-independent mechanism (Harris and Cotman, 1986). Postsynaptic NMDA receptor stimulation, influx of Ca^{2+}, activation of $cPLA_2$, and the release of arachidonic acid accompany LTP. LTP also depends on gene expression, protein synthesis, and the establishment of new neuronal connections. PAF has been proposed to act as a potential retrograde messenger in hippocampal LTP (Kornecki et al., 1996; Bazan et al., 1997). The application of PAF to the CA1 subfield of rat hippocampal slices enhances LTP in a presynaptic manner (Kato et al., 1994). PAF antagonist, BN 52021, blocks the development of LTP (Kato et al., 1994), indicating that PAF modulates LTP. Similarly NO donor, sodium nitroprusside, also mediates the induction of LTP and an increase in glutamate release (Pettorossi and Grassi, 2001). The mc-PAF-mediated LTP can be blocked by the NO scavenger, while NO-mediated LTP is only reduced by BN-52021. This observation supports the view that both PAF and NO are associated with LTP formation. Their retrograde messenger action may involve two different phases of vestibular LTP: NO in the induction phase and PAF in the full expression phase (Pettorossi and Grassi, 2001). These results are supported by the memory enhancement effect of PAF in rats performing inhibitory avoidance task and water maze task (Izquierdo et al., 1995). In the inhibitory avoidance task, PAF mediates memory enhancement when injected at defined times after learning specific regions of the limbic system. In contrast, in water maze task, PAF is effective when infused into the striatum. PAF antagonists impair spatial learning and inhibitory avoidance, whereas treatment with a synthetic nonhydrolyzable analog of PAF (mc PAF, 1-*O*-hexadecyl-2-methylcarbamoyl-*sn*-glycerol-3-phosphocholine) enhances memory (Packard et al., 1996). The molecular mechanism associated with generation of PAF and its role in memory enhancement is not fully understood. However, it is known that PAF modulates the release of glutamate, which stimulates postsynaptic NMDA receptor. This process results in the elevation in intracellular Ca^{2+}, which stimulates Ca^{2+}-dependent calmodulin kinase. This enzyme activates acetyl-CoA/lyso-PAF acetyltransferase (Snyder, 1995). The activation of this enzyme is required for the synthesis of PAF via remodeling pathway. PAF synthesis in the postsynaptic terminal at a sufficiently high level may result in its diffusion in synaptic cleft, where PAF interacts with presynaptic PAF receptors and promote PAF-mediated glutamate release. Generation of PAF also promotes the release of arachidonic acid in brain (Kunievsky et al., 1992) through potentiation of

NMDA receptor and activation of PLA_2 during LTP formation (Miller et al., 1992). It is known for quite some time that arachidonic acid acts as a potential retrograde messenger in hippocampal LTP formation (Williams et al., 1989; Miller et al., 1992). Thus, it is quite likely that PAF modulates memory-related processes through arachidonic acid release (Bazan et al., 1997; Kornecki et al., 1996). As stated earlier, the target of PAF action at the presynaptic level is the stimulation of glutamate release via Ca^{2+}-dependent vesicular exocytosis (Kornecki and Ehrlich, 1991). Glutamate through the stimulation of postsynaptic NMDA receptor initiates a cascade of signal transduction processes that result in arachidonic acid, eicosanoids, and PAF release. The molecular mechanism by which PAF and its receptors mediate and modulate glutamate exocytosis is not fully understood, and additional studies are required on this important topic.

9.5.4 *PAF in Glutamate-Mediated Neurotoxicity*

PAF is known to augment the presynaptic release of glutamate. Excessive activation of postsynaptic glutamate receptors and subsequent downstream signals leads to excitotoxicity (Fig. 9.8) (Olney et al., 1979; Farooqui and Horrocks, 1994). Several studies have indicated that the NMDA receptor signal pathway may be involved in PAF-induced neurotoxicity (Nogami et al., 1997; Xu et al., 2004). Thus, the exposure of primary neuronal to PAF for 24h enhances neuronal death in a dose-dependent manner. The PAF-mediated neuronal death is significantly prevented not only by BN52021, a PAF antagonist, but also by MK-801, a NMDA antagonist (Fig. 9.9). PAF-mediated neuronal cell death also involves nitric oxide and L-NAME, an NO synthase (NOS) inhibitor also blocks PAF-mediated neuronal death. Moreover, the increases in NOS activity and neuronal NOS expression induced by chronic exposure of the cultured neurons to PAF are dramatically blocked by BN52021 and MK-801, respectively. The NMDA receptor/NO signaling pathway contributes to the pathological mechanism of cell death triggered via PAF receptor activation (Xu et al., 2004). PAF induces apoptotic cell death in cultured astroglial and oligodendroglial cells, and PAF receptor antagonists, WEB 2170 and BN 52021, block this PAF-mediated apoptotic cell death. To determine whether PAF-R expression is altered during excitotoxicity, changes in PAF-R mRNA localization have been compared with markers of neuronal apoptosis and reactive gliosis following systemic injection of kainic acid (Bennett et al., 1998). PAF-R mRNA is normally present in neurons and microglia of rat hippocampus. In kainic acid neurotoxicity expression of PAF-R mRNA becomes restricted to apoptotic neurons and to glia involved in phagocytosing apoptotic debris. PAF-R mRNA is rarely detected in surviving neurons. These results support the view that PAF-R-expressing neurons may be preferentially susceptible to excitotoxic challenge. On the basis of the immunocytochemical findings and knockout mice (–/–) deficient in the caspase-3 gene, it is suggested that caspase-3 activation may also be involved in PAF-mediated cell death (Hostettler and Carlson, 2002).

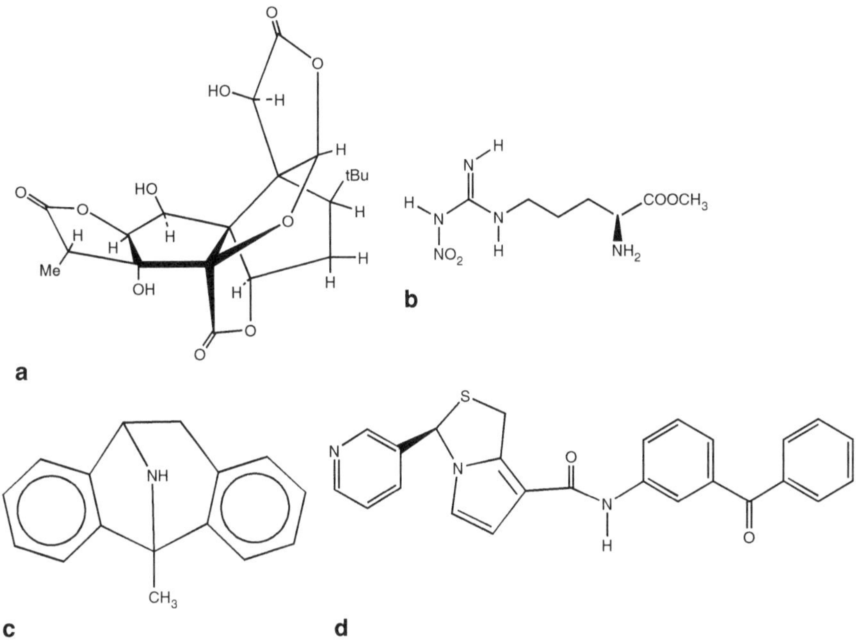

Fig. 9.9 Chemical structures of PAF antagonists, nitric oxide synthase inhibitor, and NMDA receptor antagonist. BN 52021 (**a**), L-NAME (**b**), MK 801 (**c**), and RP59227, tulopafant (**d**).

Postsynaptic density protein-93 (PSD-93), a guanylate kinase, binds to and clusters the *N*-methyl-D-aspartate (NMDA) receptor and assembles a specific set of signaling proteins such as neuronal nitric oxide synthase, nNOS, around the NMDA receptor at synapses in the central nervous system. Targeted disruption of *PSD-93* gene significantly blocks NMDA receptor-nitric oxide signaling-dependent neurotoxicity triggered via PAF receptor activation (Xu et al., 2004). In addition, the deficiency of PSD-93 attenuates PAF-induced increase in cGMP and retards PAF-mediated formation of NMDA receptor-neuronal nitric oxide synthase complex. These observations suggest that PSD-93 is involved in the NMDA receptor/nitric oxide-promoted pathological processing of neuronal damage triggered via platelet-activating factor receptor activation and PSD-93, and PAF may be closely associated with molecular mechanisms of neuronal damage in glutamate-mediated neurotoxicity (Xu et al., 2004).

In addition to glutamate, PAF mediates potentiation of stimulation-evoked catecholamine release in adrenal chromaffin cells (Morita et al., 1995). Pretreatment with BN 50739 blocks the release of catecholamine indicating that PAF and its receptors may modulate catecholamine release in adrenal medullary chromaffin cells. Lyso-PAF has no affect on catecholamine release.

9.5.5 PAF and Calcium Influx

PAF causes an elevation of cytosolic calcium in neural and nonneural cells (Kornecki and Ehrlich, 1988). Several mechanisms are associated with PAF-mediated elevation in calcium levels. These mechanisms include calcium influx through PAF-regulated calcium channels, elevation of calcium through signaling molecules (arachidonic acid) generated during PAF receptor stimulation, and release of calcium from intracellular stores in response to the intracellular second messengers such as InsP_3 generated during PAF receptor stimulation. The elevation in calcium levels is blocked by verapamil, a calcium channel blocker, PLA_2 inhibitors, and PAF receptor antagonists (Farooqui et al., 2006).

9.5.6 PAF in Neuroinflammation

Neuroinflammation is a protective mechanism that isolates the damaged brain tissue from uninjured area, destroys affected cells, and repairs the extracellular matrix (Farooqui et al. 2007b). Inflammation and oxidative stress are key processes associated with neural cell injury. Bioactive lipid mediators like PAF, oxidized phospholipids, prostaglandins, and reactive aldehydes (4-HNE, isoprostanes, isoketals) contribute significantly to the initiation, maintenance, and progression of inflammation, oxidative stress, and neural cell death (Fig. 9.10) (Farooqui et al., 2007b). PAF stimulates the inducible isoform of PLA_2 and cyclooxygenase-2 (COX-2). COX-2 is encoded by an immediate early gene and is responsible for prostaglandin synthesis in neuropathological processes (Fig. 9.8). PAF receptors are also involved in the release of PGE_2 from astrocytes. PAF is an essential component of the intricate mechanisms by which immune cells such as leukocytes are recruited to their targets (Zimmerman et al., 1996). Collective evidence suggests that PAF-mediated neuroinflammation is closely associated with short and long-term responses of cells to stimulation or neural trauma (Bazan et al., 1997). PAF promotes adhesive interactions between leukocytes and endothelial cells, leading to transendothelial migration of leukocytes, by a process of juxtacrine intercellular signaling. This process leads to activation of leukocytes and the release of reactive oxygen radicals, lipid mediators, cytokines, and enzymes. These reaction products subsequently contribute to the pathological features of various inflammatory diseases. Thus, neuroinflammation is a hallmark of all major CNS diseases. Levels of PAF and other bioactive lipid mediators are significantly increased in brain, plasma, and cerebrospinal fluid of patients with neural trauma and neurodegenerative diseases (Phillis et al., 2006). PAF has an acetyl group at the *sn*-2 position of its glycerol moiety. This acetyl group is essential for its proinflammatory activity. PAF acetylhydrolase blocks the proinflammatory effects of PAF by hydrolyzing the acetyl group. The antiinflammatory effect of PAF acetylhydrolase is accompanied by inhibition of PAF-induced chemotaxis and changes in intracellular Ca^{2+} (Kuijpers et al., 2001). All these processes are associated with neuroinflammation in brain tissue.

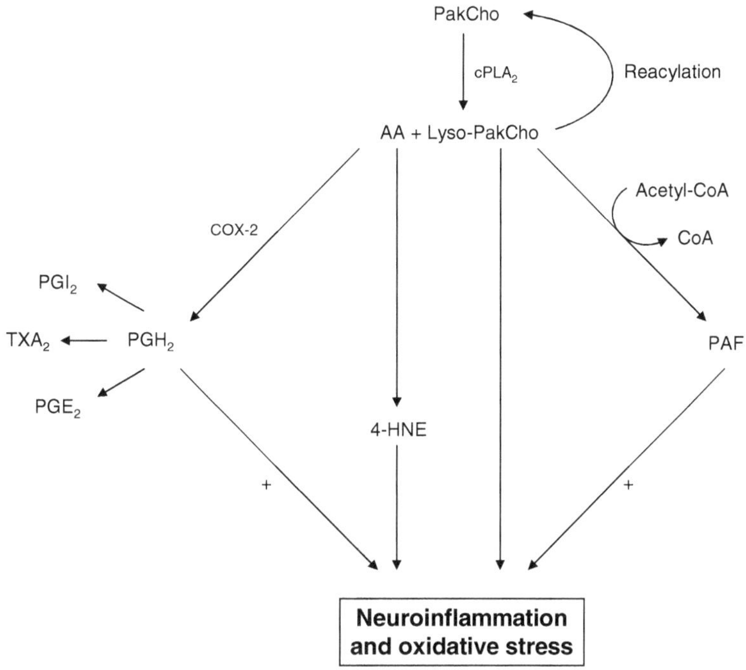

Fig. 9.10 Generation of lipid mediators closely associated with neuroinflammation. *PakCho* Alkylacyl-glycerophosphocholine, *lyso-PakCho* Alkyl-lyso-glycerphosphocholine, *cPLA*$_2$ cytosolic phospholipase A_2, *AA* arachidonic acid, *PAF* platelet-activating factor, *4-HNE* 4-ydroxynonanal, and *COX-2* cyclooxygenase-2.

9.5.7 PAF in Cerebral Blood Flow and Blood–Brain Barrier Permeability

Intracarotid infusion of PAF decreases cerebral blood flow with a concomitant increase in the global cerebral metabolic rate for oxygen (Kochanek et al., 1988; Kochanek et al., 1990). PAF administration causes a dose-dependent decrease in spinal cord blood flow. This decrease in blood flow can be blocked by a PAF receptor antagonist (Faden and Halt, 1992). In brain tissue, close interactions occur among neurons, astrocytes, and microvessels. Under normal conditions, PAF mediates these interactions and modulates cerebral blood flow. However, generation of PAF following ischemic injury results in loss of autoregulation and disturbance in cerebral blood flow (del Zoppo and Mabuchi, 2003). Collective evidence suggests that PAF plays an important role in the modulation of cerebral blood flow. The blood–brain barrier (BBB) is a complex cellular system formed by brain endothelial cells lining the cerebral microvasculature, and is an important mechanism for protecting the brain from fluctuations in plasma composition, and from circulating agents capable of disturbing neural function. The barrier also plays an important role in the homeostatic regulation of the brain microenvironment necessary for the stable and coordinated activity of neurons. Although PAF does not cross the BBB itself, it induces changes in BBB permeability (Catalán et al., 1993). PAF, when locally released after brain injury, can penetrate the BBB and mediate systemic effects, including arterial hypotension. PAF-mediated inflammatory interactions that occur at the blood–endothelium interface, involve cytokines, adhesion molecules, chemokines, and leukocytes, are critical to the pathogenesis of tissue damage in cerebral infarction. PAF is also an essential component of the intricate mechanism by which immune cells such as leukocytes are recruited to their targets (Zimmerman et al., 1996).

9.5.8 PAF in Apoptosis

Treatment glial cell cultures with PAF for 72h causes considerable cell death in astrocytes as well as oligodendrocytes, and this effect can be blocked by the PAF receptor antagonists, WEB 2170 and BN 52021 (Hostettler and Carlson, 2002). In cultured chick neurons, serum deprivation as well as staurosporine in serum-free medium results in apoptotic cell death and PAF antagonist, Ginkgolides B, protects against serum deprivation and staurosporine-induced apoptotic cell death. Immunocytochemical localization studies indicate that PAF produces the activation of caspase-3 at 24, 48, and 72h after treatment in both cell types. PAF receptors may promote enhanced clearance of apoptotic cells through phagocytosis (Zhang et al 2006). Suggestion on the involvement of caspase-3 in PAF-mediated cell death is supported by studies on PAF knockout mice (–/–). These mice are deficient in the *caspase-3* gene. Toxic effects of PAF are lost when astrocytes (–/–) are exposed to low concentrations of PAF. Oligodendrocytes derived from knockout mice (–/–) are not susceptible to PAF toxicity. Thus, PAF induces cell apoptotic death in cultured CNS glial cells and this effect is, in part, dependent on caspase-3 activation (Hostettler and Carlson, 2002).

PC12 cells do not express PAF receptor mRNA as demonstrated by Northern analysis and RT-PCR (Brewer et al., 2002). In the absence of the G-protein coupled receptor, treatment of PC12 cells with PAF produces chromatin condensation, DNA strand breaks, oligonucleosomal fragmentation, and nuclear disintegration. These are characteristic of apoptosis. Lyso-PAF does not elicit apoptotic death. Concentrations of PAF or lyso-PAF that exceeded critical micelle concentration induce physicochemical effects on plasma membrane resulting in necrosis (Brewer et al., 2002). Apoptosis but not necrosis is inhibited by the PAF antagonist BN52021 but not CV3988. Ectopic PAF receptor expression protects PC12 transfectants from ligand-induced apoptosis. PAF receptor-mediated protection is retarded by CV3988. Collectively, these studies suggest that in PC12 cells, PAF-mediated apoptosis is not dependent PAF-Rs. PAF-mediated signaling initiated by its G-protein coupled receptor is cytoprotective. Finally, pro and antiapoptotic effects of PAF on PC12 cells can be pharmacologically distinguished using two different PAF antagonists (Brewer et al., 2002).

9.5.9 PAF in Noniception

Pain is a complex process that is characterized by peripheral and central mechanisms (Svensson and Yaksh, 2002). It is generally agreed that both peripheral and central mechanisms are involved in neuropathic and inflammatory pain. Pain-like behavior in rats can be produced by peripheral administration of PGI_2, PGE_2, and PAF (Vahidy et al., 2006). Subplantar injections of PAF into the rat hindpaw increase pain sensitivity indicating that PAF and its receptors modulate the processing of inflammation-related pain. Involvement of PAF in nociceptive transmission, especially in persistent pain, is supported by studies on formalin test (Teather et al., 2002). Systemic administration of PAF antagonists, BN 52021 and BN 50730 (Fig. 9.7), decreases nociceptive behavior during the late phase of the formalin test in rats. It is stated that PAF antagonists cross BBB and block intracellular hippocampal PAF-binding sites and modulate the intensity of nociception. Administration of PAF antagonists within the hippocampus, and of using agents specific for either plasma membrane (BN 52021) or intracellular (BN 50730) PAF-binding sites, it is shown that hippocampal plasma membrane PAF receptors, but not intracellular PAF-binding sites, mediate tonic inflammatory pain processing in rats (Teather et al., 2006). Collective evidence suggests that nociception is modulated by interactions between central as well as peripheral nociceptive mechanisms and both these mechanisms involve PAF and COX-2-generated metabolites. However, the relative contribution of these metabolites toward pain intensity of neuropathic and peripheral nociception still remains unknown. Studies on the involvement of PAF in pain-related processes in the spinal cord in mice have indicated that intrathecal injection of PAF induce-tactile pain, while lyso-PAF has no effect. Tactile allodynia induced by PAF is blocked by PAF receptor antagonists, (TCV-309, WEB 2086, and BN 50739). ATP P2X receptor antagonists, pyridoxalphosphate-6-azophenyl-2′,4′-disulfonic acid and 2′,3′-*O*-(2,4,6-trinitrophenyl)adenosine 5-triphosphate, NMDA receptor antagonist, MK 801, and nitric oxide synthetase inhibitor, 7-nitroindazole, retard the PAF-mediated tactile allodynia. Collective evidence suggests that PAF is a potent mediator of tactile allodynia and thermal hyperalgesia at the level of the spinal cord. PAF-induced tactile allodynia may be mediated by ATP and NMDA and nitric oxide cascade through capsaicin-sensitive fiber (Morita et al., 2004). Studies on the involvement of PAF in pain related signaling have indicated that mice lacking PAF receptor (PAF-R−/− mice) exhibit almost normal responses to thermal and mechanical insults, but display attenuated persistent pain behaviors mediated by locally injecting formalin and capsaicin at the periphery, and visceral inflammatory pain. Although the molecular mechanism associated with these pain behaviors is not fully understood. However, mice lacking PAF-R show reduced phosphorylation of extracellular signal-related protein kinase (ERK), an important kinase for the sensitization of primary sensory and DRG neurons after formalin injection. Furthermore, U0126, a specific inhibitor of the ERK pathway suppresses the persistent pain mediated by formalin (Tsuda et al., 2007a). Activation of P2X(3) and P2X(2/3) receptors (P2X(3)R/P2X(2/3)R) in primary sensory neurons is also associated with neuropathic pain. These receptors are coupled with $cPLA_2$. Application of ATP to cultured dorsal root ganglion (DRG) neurons results in activation of $cPLA_2$ (Tsuda et al 2007b). The activation is due to increases in the phosphorylation of Ser^{505} in $cPLA_2$. The phosphorylation of $cPLA_2$ promotes the translocation of phosphorylated enzyme to the plasma membrane. The ATP-mediated $cPLA_2$ activation is blocked by a selective antagonist of P2X(3)R/P2X(2/3)R and by a selective inhibitor of $cPLA_2$. In vivo studies have indicated that in the DRG, the number of $cPLA_2$-activated neurons increases after peripheral nerve injury but not after peripheral inflammation caused by complete Freund's adjuvant (Tsuda et al 2007b). Pharmacological inhibition of P2X(3)R/P2X(2/3)R reverses the nerve injury-mediated $cPLA_2$ activation in DRG neurons. Moreover, the administration of $cPLA_2$ inhibitor near the DRG prevents nerve injury-mediated tactile allodynia, a process closely associated with neuropathic pain. Collectively, these studies suggest that PAF and PAF-R and $cPLA_2$ play an important role in both persistent pain and sensitization of primary sensory neurons after tissue injury (Tsuda et al., 2007a, b).

9.5.10 PAF in Immune Response

In pig kidney–human blood xenoperfusion model, PAF mediates allergic reaction and has been proposed to contribute to the pathogenesis of xenograft rejection (Cruzado et al., 1998). The administration of PAF antagonist, BN 52021, exerts a protective effect not only by attenuating the acute inflammatory response, but also by blocking vascular microthrombi formation. It is well known that PAF receptors are found in human monocytes, neutrophil, and B lymphocytes cell line. PAF plays an important role in early B cell activation, enhances IgG and IgA secretion, and is an important B cell immunomodulator, which can interact with other leukocyte cell mediators. Although IgE receptors have not been identified, but PAF has been shown to increase IgE binding, IgE-dependent adherence, and cytotoxicity of normal human eosinophils (Moqbel et al., 1990). Activation of neurophils by PAF increases phagocytosis in a calcium-dependent manner. Collectively, these studies suggest that PAF may contribute to allergic reaction and immune challenge such as infection.

9.6 Conclusion

PAF is a potent phospholipid mediator that plays various roles in neuronal function and brain development. PAF acts through specific receptors. These receptors are widespread in different brain regions and are present on the neuronal and glial cell surface as well as in intracellular membrane compartments. A variety of stimuli, including those producing inflammation, promote the synthesis and release of PAF from various types of neural and nonneural cells. PAF-Rs are linked through G proteins to several intracellular signal transduction pathways, including activation of phospholipases, activation of GTPase, activation of polyphosphatidylinositide turnover, calcium mobilization, activation of various protein kinases, and modulation of transcription of immediate early genes. All the above signal transduction processes and expressions of growth factors are blocked by PAF antagonists. PAF and its receptors modulate the expression of nerve growth factors in astrocyte and vascular endothelial growth factor and heparin-binding factor in nonneural cells. PAF modulate cerebral blood flow and plays a role in neuronal differentiation. PAF mediates interactions between neurons and glial cells during development as well as adult life. PAF and PAF receptors act as a signal for cellular communication within brain and between the brain and the immune and inflammatory systems. Stimulation of PAF receptors plays important roles in modulating neuronal plasticity, memory formation, modulation of neuroinflammation and nociceptive responses, and apoptotic cell death during neuronal injury. Recent advances in molecular biology procedures have not only resulted in cloning of PAF receptor cDNAs and genes, but also in generation of PAF receptor mutant animals, i.e., PAF receptor-over-expressing mouse and PAF receptor-deficient mouse. These mutant mice provide a novel and specific approach for identifying the pathophysiological and physiological functions of PAF.

References

Adachi T., Aoki J., Manya H., Asai H., Arai H. and Inoue K. (1997). PAF analogues capable of inhibiting PAF acetylhydrolase activity suppress migration of isolated rat cerebellar granule cells. Neurosci. Lett. 235:133–136.

Aihara M., Ishii S., Kume K., and Shimizu T. (2000). Interaction between neurone and microglia mediated by platelet-activating factor. Genes Cells 5:397–406.

Bazan N. G., Squinto S. P., Braquet P., Panetta T., and Marcheselli V. L. (1991). Platelet-activating factor and polyunsaturated fatty acids in cerebral ischemia or convulsions: Intracellular PAF-binding sites and activation of a fos/jun/AP-1 transcriptional signaling system. Lipids 26:1236–1242.

Bazan N. G., Fletcher B. S., Herschman H. R., and Mukherjee P. K. (1994). Platelet-activating factor and retinoic acid synergistically activate the inducible prostaglandin synthase gene. Proc. Natl. Acad. Sci. USA 91:5252–5256.

Bazan N. G., Packard M. G., Teather L., and Allan G. (1997). Bioactive lipids in excitatory neurotransmission and neuronal plasticity. Neurochem. Int. 30:225–231.

Bennett S.A., Chen J., Pappas B.A., Roberts D.C., and Tenniswood M. (1998). Platelet activating factor receptor expression is associated with neuronal apoptosis in an in vivo model of excitotoxicity. Cell Death Differ. 5: 867–875.

Bito H., Nakamura M., Honda Z., Izumi T., Iwatsubo T., Seyama Y., Ogura A., Kudo Y., and Shimizu T. (1992). Platelet-activating factor (PAF) receptor in rat brain: PAF mobilizes intracellular Ca^{2+} in hippocampal neurons. Neuron 9:285–294.

Brewer, C., Bonin, F., Bullock, P., Nault, M.C., Morin, J., Imbeault, S., Shen, T.Y., Franks, D.J., and Bennet, S.A. (2002). Platelet activating factor-induced apoptosis is inhibited by ectopic

expression of the platelet activating factor G-protein coupled receptor. J. Neurochem. 82:1502–1511
Brodie C. (1995). Platelet activating factor induces nerve growth factor production by rat astrocytes. Neurosci. Lett. 186:5–8.
Catalán R. E., Martínez A. M., Aragonés M. D., Garde E., and Díaz G. (1993). Platelet-activating factor stimulates protein kinase C translocation in cerebral microvessels. Biochem. Biophys. Res. Commun. 192:446–451.
Chao W. and Olson M. S. (1993). Platelet-activating factor: Receptors and signal transduction. Biochem. J. 292:617–629.
Clark, G.D., Zorumski, C.F., McNeil, R.S., Happel, L.T., Ovella, T., McGuire, S., Bix, G.J., and Swann, J.W. (2000). Neuronal platelet-activating factor receptor signal transduction involves a pertussis toxin-sensitive G-protein. Neurochem. Res. 25:603–611.
Cruzado J.M., Torras J., Riera M., Lloberas N., Herrero I., Condom E., Martorell J., Alsina J., and Grinyo J.M. (1998). Effect of a platelet-activating factor (PAF) receptor antagonist on hyperacute xenograft rejection; Evaluation in a pig kidney-human blood xenoperfusion model. Clin. Exp. Immunol. 113:136–144.
DeCoster M. A., Mukherjee P. K., Davis R. J., and Bazan N. G. (1998). Platelet-activating factor is a downstream messenger of kainate-induced activation of mitogen-activated protein kinases in primary hippocampal neurons. J. Neurosci. Res. 53:297–303.
del Zoppo G.J. and Mabuchi T. (2003). Cerebral microvessel responses to focal ischemia. J. Cereb. Blood Flow Metab. 23:879–894.
Faden A. I. and Halt P. (1992). Platelet-activating factor reduces spinal cord blood flow and causes behavioral deficits after intrathecal administration in rats through a specific receptor mechanism. J. Pharmacol. Exp. Ther. 261:1064–1070.
Farooqui A. A. and Horrocks L. A. (1994). Excitotoxicity and neurological disorders: Involvement of membrane phospholipids. Int. Rev. Neurobiol. 36:267–323.
Farooqui A. A., Horrocks L. A., and Farooqui T. (2007a). Interactions between neural membrane glycerophospholipid and sphingolipid mediators: A recipe for neural cell survival or suicide. J. Neurosci. Res. 85:1834–1850.
Farooqui A. A., Horrocks L. A., and Farooqui T. (2007b). Modulation of inflammation in brain: A matter of fat. J. Neurochem. 101:577–599.
Farooqui A. A., Ong W. Y., and Horrocks L. A. (2006). Inhibitors of brain phospholipase A_2 activity: Their neuropharmacological effects and therapeutic importance for the treatment of neurologic disorders. Pharmacol. Rev. 58:591–620.
Furnkranz A. and Leitinger N. (2004). Regulation of inflammatory responses by oxidized phospholipids structure–function relationships. Curr. Pharmaceut. Design. 10: 915–921.
Han X.B., Liu X., Hsueh W., De Plaen I.G. (2004) Macrophage inflammatory protein-2 mediates the bowel injury induced by platelet-activating factor. Am. J. Physiol. Gastrointest. Liver Physiol. 287:G1220–1226.
Harris E.W. and Cotman C.W. (1986). Long-term potentiation of guinea pig mossy fiber responses is not blocked by *N*-methyl D-aspartate antagonists. Neurosci. Lett. 70:132–137.
Honda Z., Nakamura M., Miki I., Minami M., Watanabe T., Shimizu T. (1991). Cloning by functional expression of platelet-activating factor receptor from guinea-pig lung. Nature 349:342–346.
Honda Z., Ishii S., and Shimizu T. (2002). Platelet-activating factor receptor. J. Biochem. 131:773–779.
Hosford D. J., Domingo M. T., Chabrier P. E., and Braquet P. (1990). Ginkgolides and platelet-activating factor binding sites. Method Enzymol. 187:433–446.
Hostettler M. E. and Carlson S. L. (2002). PAF antagonist treatment reduces pro-inflammatory cytokine mRNA after spinal cord injury. NeuroReport. 13:21–24.
Ishii S., Matsuda Y., Nakamura M., Waga I., Kume K., Izumi T., and Shimizu T. (1996). A murine platelet-activating factor receptor gene: cloning, chromosomal localization and up-regulation of expression by lipopolysaccharide in peritoneal resident macrophages. Biochem. J. 314:671–678.

Ishii S. and Shimizu T. (2000). Platelet-activating factor (PAF) receptor and genetically engineered PAF receptor mutant mice. Prog. Lipid Res. 39:41–82.

Izumi T. and Shimizu T. (1995). Platelet-activating factor receptor: Gene expression and signal transduction. Biochim. Biophys. Acta Lipids Lipid Metab. 1259:317–333.

Izquierdo I., Fin C., Schmitz P.K., Da Silva R.C., Jerusalinsky D., Quillfeldt J.A., Ferreira M.B., Medina J.H. and Bazan N.G. (1995). Memory enhancement by intrahippocampal, intraamygdala, or intraentorhinal infusion of platelet-activating factor measured in an inhibitory avoidance task. Proc Natl Acad Sci USA. 92:5047–5051.

Junier M.P., Tiberghien C., Rougeot C., Fafeur V., and Dray F. (1988). Inhibitory effect of platelet-activating factor (PAF) on luteinizing hormone-releasing hormone and somatostatin release from rat median eminence *in vitro* correlated with the characterization of specific PAF receptor sites in rat hypothalamus. Endocrinology. 123:72–80.

Kato K., Clark G. D., Bazan N. G., and Zorumski C. F. (1994). Platelet-activating factor as a potential retrograde messenger in CA1 hippocampal long-term potentiation. Nature 367:175–179.

Ko H.M., Seo K.H., Han S.J., Ahn K.Y., Choi I.H., Koh G.Y., Lee H.K., Ra M.S., and Im S.Y. (2002). Nuclear factor kappaB dependency of platelet-activating factor-induced angiogenesis. Cancer Res. 62:1809–1814.

Ko H.M., Park Y.M., Jung B., Kim H.A., Choi J.H., Park S.J., Lee H.K., and Im S.Y. (2005). Involvement of matrix metalloproteinase-9 in platelet-activating factor-induced angiogenesis. FEBS Lett. 679:2369–2375.

Ko H., Jung H.H., Seo K.H., Kang Y.R., Seo K.H., Kang Y.R., Kim H.A., Park S.J., Lee H.K., and Im S.Y. (2006). Platelet-activating factor-induced NF-kappaB activation enhances VEGF expression through a decrease in p53 activity. FEBS Lett. 580:3006–3012.

Kochanek P. M., Melick J. A., Schoettle R. J., Magargee M. J., Evans R. W., and Nemoto E. M. (1990). Endogenous platelet activating factor does not modulate blood flow and metabolism in normal rat brain. Stroke. 21:459–462.

Kochanek P. M., Nemoto E. M., Melick J. A., Evans R. W., and Burke D. F. (1988). Cerebrovascular and cerebrometabolic effects of intracarotid infused platelet-activating factor in rats. J. Cereb. Blood Flow Metab. 8:546–551.

Kornecki E. and Ehrlich Y. H. (1988). Neuroregulatory and neuropathological actions of the ether-phospholipid platelet-activating factor. Science 240:1792–1794.

Kornecki E. and Ehrlich Y. H. (1991). Calcium ion mobilization in neuronal cells induced by PAF. Lipids 26:1243–1246.

Kornecki E., Wieraszko A., Chan J. C., and Ehrlich Y. H. (1996). Platelet activating factor (PAF) in memory formation: Role as a retrograde messenger in long-term potentiation. J. Lipid Mediat. Cell Signal. 14:115–126.

Kuijpers T. W., Van den Berg J. M., Tool A. T. J., and Roos D. (2001). The impact of platelet-activating factor (PAF)-like mediators on the functional activity of neutrophils: Anti-inflammatory effects of human PAF-acetylhydrolase. Clin. Exp. Immunol. 123:412–420.

Kunievsky B., Bazan N. G., and Yavin E. (1992). Generation of arachidonic acid and diacylglycerol second messengers from polyphosphoinositides in ischemic fetal brain. J. Neurochem. 59:1812–1819.

Leitinger N. (2003). Oxidized phospholipids as modulators of inflammation in atherosclerosis. Curr. Opin. Lipidol. 14:421–430.

Lo Nigro C., Chong S.S., Smith A.C.M., Dobyns W.B., Carrozzo R., and Ledbetter D.H. (1997). Point mutations and an intragenic deletion in LIS1, the lissencephaly causative gene in isolated lissencephaly sequence and Miller-Dieker syndrome. Hum. Mol. Genet. 6:157–164.

Maclennan K. M., Smith P. F., and Darlington C. L. (1996). Platelet-activating factor in the CNS. Prog. Neurobiol. 50:585–596.

Marcheselli V. L. and Bazan N. G. (1994). Platelet-activating factor is a messenger in the electroconvulsive shock-induced transcriptional activation of c-*fos* and *zif*-268 in hippocampus. J. Neurosci. Res. 37:54–61.

Marcheselli V. L., Rossowska M. J., Domingo M. T., Braquet P., and Bazan N. G. (1990). Distinct platelet-activating factor binding sites in synaptic endings and in intracellular membranes of rat cerebral cortex. J. Biol. Chem. 265:9140–9145.

Melnikova V. O., Mourad-Zeidan A. A., Lev D. C., and Bar-Eli M. (2006). Platelet-activating factor mediates MMP-2 expression and activation via phosphorylation of cAMP-response element-binding protein and contributes to melanoma metastasis. J Biol Chem. 281:2911–2922.

Miller B., Sarantis M., Traynelis S. F., and Attwell D. (1992). Potentiation of NMDA receptor currents by arachidonic acid. Nature 355:722–725.

Montrucchio G., Alloatti G., and Camussi G. (2000). Role of platelet-activating factor in cardiovascular pathophysiology.Physiol. Rev. 80:1669–1699.

Moqbel R., Walsh G. M., Nagakura T., MacDonald A. J., Wardlaw A. J., Iikura Y., Kay A. B. (1990). The effect of platelet-activating factor on IgE binding to, and IgE-dependent biological properties of, human eosinophils. Immunology. 70:251–257.

Mori M., Aihara M., Kume K., Hamanoue M., Kohsaka S., and Shimizu T. (1996). Localization of platelet-activating factor receptor in the rat brain. Adv. Exp. Med. Biol. 407:357–363:357–363.

Morita K., Suemitsu T., Uchiyama Y., Miyasako T., and Dohi T. (1995). Platelet-activating factor mediated potentiation of stimulation- evoked catecholamine release and the rise in intracellular free Ca^{2+} concentration in adrenal chromaffin cells. J. Lipid Mediat. Cell Signal. 11:219–230.

Morita K., Morioka W., Abdin J., Kitayama S., Nakata Y., and Dohi T. (2004). Development of tactile allodynia and thermal hyperalgesia by intrathecally administered platelet-activating factor in mice. Pain 111:351–359.

Nogami K., Hirashima Y., Endo S., and Takaku A. (1997). Involvement of platelet-activating factor (PAF) in glutamate neurotoxicity in rat neuronal cultures. Brain Res. 754:72–78.

Olney J. W., Fuller T., and de Gubareff T. (1979). Acute dendrotoxic changes in the hippocampus of kainate treated rats. Brain Res. 176:91–100.

Ottino P., He J., Axelrod T. W., Bazan H. E. (2005). PAF-induced furin and MT1-MMP expression is independent of MMP-2 activation in corneal myofibroblasts. Invest. Ophthalmol. Vis. Sci. 46:487–496.

Packard M. G., Teather L. A., and Bazan N. G. (1996). Effects of intrastriatal injections of platelet-activating factor and the PAF antagonist BN 52021 on memory. Neurobiol. Learn. Mem. 66:176–182.

Pan Z., Kravchenko V. V., Ye R. D. (1995). Platelet-activating factor stimulates transcription of the heparin-binding epidermal growth factor-like growth factor in monocytes. Correlation with an increased kappa B binding activity. J Biol Chem. 270:7787–7790.

Panwala C. M., Jones J. C., and Viney J. L. (1998). A novel model of inflammatory bowel disease: Mice deficient for the multiple drug resistance gene, mdr1a, spontaneously develop colitis. J. Immunol. 161:5733–5744.

Pettorossi V. E., and Grassi S. (2001). Different contributions of platelet-activating factor and nitric oxide in long-term potentiation of the rat medial vestibular nuclei. Acta Otolaryngol Suppl. 545:160–165.

Phillis J. W., Horrocks L. A., and Farooqui A. A. (2006). Cyclooxygenases, lipoxygenases, and epoxygenases in CNS: Their role and involvement in neurological disorders. Brain Res. Rev. 52:201–243.

Prescott S. M., McIntyre T. M., Zimmerman G. A., and Stafforini D. M. (2002). Sol Sherry lecture in thrombosis – Molecular events in acute inflammation. Arterioscler. Thromb. Vasc. Biol. 22:727–733.

Qu X. W., Wang H., Rozenfeld R. A., Huang W., and Hsueh W. (1999). Type I nitric oxide synthase (NOS) is the predominant NOS in rat small intestine. Regulation by platelet-activating factor. Biochim. Biophys. Acta. 1451:211–217.

Raggers R. J., Vogels I., and Van Meer G. (2001). Multidrug-resistance P-glycoprotein (MDR1) secretes platelet-activating factor. Biochem. J. 357:859–865.

Reiner O., Carrozzo R., Shen Y., Wehnert M., Faustinella F., Dobyns W. B., Caskey C. T., and Ledbetter D. H. (1993). Isolation of a Miller-Dieker lissencephaly gene containing G protein beta-subunit-like repeats. Nature 364:717–721.

Snyder F. (1995). Platelet-activating factor: The biosynthetic and catabolic enzymes. Biochem. J. 305:689–705.

Squinto S. P., Block A. L., Braquet P., and Bazan N. G. (1989). Platelet-activating factor stimulates a fos/jun/AP-1 transcriptional signaling system in human neuroblastoma cells. J. Neurosci. Res. 24:558–566.

Svensson C. I. and Yaksh T. L. (2002). The spinal phospholipase-cyclooxygenase-prostanoid cascade in nociceptive processing. Annu. Rev. Pharmacol. Toxicol. 42:553–583.

Taheri F. and Bazan H. E. (2007). Platelet-activating factor overturns the transcriptional repressor disposition of Sp1 in the expression of MMP-9 in human corneal epithelial cells. Invest. Ophthalmol. Vis. Sci. 48:1931–1941.

Teather L. A., Afonso V. M., and Wurtman R. J. (2006). Inhibition of platelet-activating factor receptors in hippocampal plasma membranes attenuates the inflammatory nociceptive response in rats. Brain Res. 1097:230–233.

Teather L. A., Magnusson J. E., Chow C. M., and Wurtman R. J. (2002). Environmental conditions influence hippocampus-dependent behaviours and brain levels of amyloid precursor protein in rats. Eur. J. Neurosci. 16:2405–2415.

Tokuoka S. M., Ishii S., Kawamura N., Satoh M., Shimada A., Sasaki S., Hirotsune S., Wynshaw-Boris A., and Shimizu T. (2003). Involvement of platelet-activating factor and LIS1 in neuronal migration. Eur. J. Neurosci. 18:563–570.

Tsuda M., Ishii S., Masuda T., Hasegawa S., Nakamura K., Nagata K., Yamashita T., Furue H., Tozaki-Saitoh H., Yoshimura M., Koizumi S., Shimizu T., and Inoue K. (2007a). Reduced pain behaviors and extracellular signal-related protein kinase activation in primary sensory neurons by peripheral tissue injury in mice lacking platelet-activating factor receptor. J Neurochem. 102:1658–1668.

Tsuda M., Hasegawa S., and Inoue K. (2007b). P2X receptors-mediated cytosolic phospholipase A(2) activation in primary afferent sensory neurons contributes to neuropathic pain. J Neurochem.103:1408–1416.

Vahidy W.H., Ong W.Y., Farooqui A.A., and Yeo J.-F. (2006). Pronociceptive effects of central nervous lysophospholipids in a mouse model of orofacial pain. Exp. Brain Res. 174:781–785.

Wang J.H., and Sun G.Y. (2000). Platelet activating factor (PAF) antagonists on cytokine induction of iNOS and $sPLA_2$ in immortalized astrocytes (DITNC). Neurochem. Res. 25:613–619.

Williams J. H., Errington M. L., Lynch M. A., and Bliss T. V. P. (1989). Arachidonic acid induces a long-term activity dependent enhancement of synaptic transmission in the hippocampus. Nature. 341:739–742.

Xu Y., Zhang B. S., Hua Z. C., Johns R. A., Bredt D. S., and Tao Y. X. (2004). Targeted disruption of *PSD-93* gene reduces platelet-activating factor-induced neurotoxicity in cultured cortical neurons. Exp Neurol. 189:16–24.

Yoshida H., Imaizumi T., Tanji K., Matsumiya T., Sakaki H., Kimura D., Cui X. F., Kumagai M., Tamo W., Shibata T., Hatakeyama M., Sato Y., and Satoh K. (2002). Platelet-activating factor enhances the expression of vascular endothelial growth factor in normal human astrocytes. Brain Res. 944:65–72.

Yoshida H., Imaizumi T., Tanji K., Sakaki H., Metoki N., Hatakeyama M., Yamashita K., Ishikawa A., Taima K., Sato Y., Kimura H., and Satoh K. (2005). Platelet-activating factor enhances the expression of nerve growth factor in normal human astrocytes under hypoxia. Mol. Brain Res. 133:95–101.

Zhang Q., Seltmann H., Zouboulis C. C., and Travers J. B. (2006). Activation of platelet-activating factor receptor in SZ95 sebocytes results in inflammatory cytokine and prostaglandin E2 production. Exp. Dermatol. 15:769–774.

Zimmerman G. A., Elstad M. R., Lorant D. E., McIntyre T. M., Prescott S. M., Topham M. K., Weyrich A. S., and Whatley R. E. (1996). Platelet-activating factor (PAF): Signalling and adhesion in cell–cell interactions. Adv. Exp. Med. Biol. 416:297–304.

Chapter 10
Involvement of Platelet-Activating Factor in Neurological Disorders

10.1 Introduction

Platelet-activating factor (PAF) is a potent proinflammatory lipid mediator that is not stored, but synthesized by activated neural cells (neurons, astrocytes, oligodendrocytes, and microglial cells) as well as by nonneural cells (platelets, inflammatory, and endothelial cells) on demand by remodeling and de novo synthesis pathways. PAF receptors (PAF-Rs) are widely distributed in different brain regions and are present on the cell surface as well as in intracellular membrane compartments. In normal brain, levels of PAF are low, but levels of lyso-PAF are quite high. Thus levels of PAF in the hippocampus are higher than in cerebellum and cortex. These observations suggest that PAF is present in its inactive form in the brain tissue (Tiberghien et al., 1991). Concentration of PAF decreases with age (Tokumura et al., 1992). Under normal conditions, the synthesis of PAF in brain occurs through de novo synthesis. Normally, de novo synthesis is not influenced by the external stimulus. In response to PAF-R stimulation, injury, chemo-electroconvulsion and oxidative stress, the remodeling pathway is activated in neural and nonneural cells. Treatment of neural or nonneural cells with neurotransmitters such as dopamine and acetylcholine stimulates PAF synthesis in a calcium-dependent manner (Sogos et al., 1990). PAF is also synthesized by neurons and glial cells following stimulation with glutamate. PAF synthesis requires glutamate mediated stimulation of NMDA receptors and subsequent elevation of intracellular calcium ions.

Microglia, which express functional PAF-Rs to a high level show a marked chemotactic response to PAF. Microglia derived from PAF-receptor-deficient mice do not show chemotactic response (Aihara et al., 2000). Thus, PAF functions as a key messenger in neuron–microglial interactions. PAF-Rs generate specific signals that are transduced by downstream effector and pathways, which may be specific to each brain cell type. Although the synthesis and release of PAF under pathological conditions in the brain has been recognized, the relative contribution of various neural and nonneural cell types for the synthesis of PAF remains unknown. Furthermore, target cells and brain regions for PAF action have not been fully identified.

10.2 Involvement of Platelet-Activating Factor in Neurological Disorders

Following excessive cellular activation and oxidative stress or exposure to toxins, neural cells produce oxidatively fragmented phospholipids or PAF-like lipids from membrane glycerophospholipids at an improper location, at inappropriate times, and in large quantities in brain regions where the density of PAF-Rs and activity of PAF acetyl hydrolase (PAF-AH) are very low. This may cause an increased production and accumulation of PAF and PAF-like molecules and abnormal signal transduction processes, resulting in inflammation and oxidative stress associated with neurological diseases (Table 10.1) (Farooqui and Horrocks, 2004; Farooqui et al., 2007). Thus, PAF and PAF-like molecules accumulate during ischemic and traumatic injuries. Excessive levels of PAF have been implicated in inflammatory syndrome, epileptic seizures, bacterial meningitis, multiple sclerosis, prion diseases, Miller-Dieker lissencephaly, and HIV replication associated with AIDS dementia complex (Fig. 10.1) (Feuerstein, 1996). PAF has also been implicated in the neuronal damage in Alzheimer disease (AD). The mechanisms linking neural cell injury to PAF synthesis are not fully understood. However, it is known that like nonneural cells and tissues, the stimulation of PAF-R is mediated through G proteins that are often pertussis-toxin-sensitive (Clark et al., 2000). The activation of PAF-Rs results in the mobilization of calcium through calcium channels and from intracellular stores via inositol 1,4,5-trisphosphate. At the nuclear level in nonneural cells, PAF mediates the stimulation of a DNA binding activity with specificity to the κB sequence. The p50 and p65 proteins, composing the prototypic nuclear factor kappa B (NF-κB),

Table 10.1 Involvement of platelet-activating factor (PAF) in neurological disorders.

Neurological disorder	Levels of PAF	Reference
Ischemia	Increased	Lindsberg et al., 1991
Spinal cord injury	Increased	Hostettler et al., 2002
Head injury	Increased	Faden and Tzendzalian, 1992
Cold injury	Increased	Tokutomi et al., 2001
Experimental autoimmune encephalomyelitis	Increased	Kihara et al., 2005
Migraine	Increased	Sarchielli et al., 2004
Perinatal asphyxia	Increased	Akisu et al., 2003
Subarachnoid hemorrhage	Increased, and decreased with time Hirashima et al., 1994	Hirashima et al., 1993a, b
Dyslexia	Increased	Kelley et al., 1999 Taylor et al., 2001
Multiple sclerosis	Increased	Callea et al., 1999
Heminegalencephaly	Decreased	Hirashima et al., 1999
Human immunodeficiency	Increased	Gelbard et al., 1994
Meningitis	Increased	Arditi et al., 1990
Seizures/convulsions	Increased	Kumar et al., 1988
Shock	Increased	Arimura et al., 1990
Pneumonia	Increased	Makristathis et al., 1993

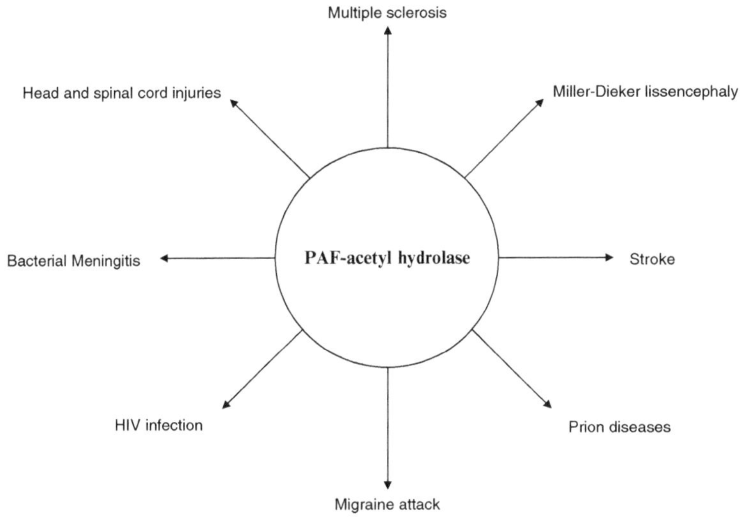

Fig. 10.1 Neurological disorders associated with platelet-activating factor and its metabolism.

have been identified as components of the DNA–protein complexes by antipeptide antibodies in gel supershift experiments (Kravchenko et al., 1995). These results suggest a potential role of PAF in the regulation of gene expression through a G-protein-coupled transcription factor activation pathway. Collectively, these studies suggest that PAF plays an important role in gene expression, inflammation, oxidative stress, and neurodegeneration.

10.2.1 PAF in Ischemia

Cerebral ischemic injury is accompanied by the stimulation of PAF synthesis and its release from neural and nonneural cell membranes. Neurons, macro- and microglial cells, monocytes, macrophages, and endothelial cells are the targets of PAF. Administration of PAF antagonists has beneficial effects in various models of cerebral ischemia (Maclennan et al., 1996). Thus, PAF-R antagonists partially prevent neural cell injury and leukocyte adhesion to endothelial cells. In hippocampus-derived cell line, HN33.11, PAF-mediated neural cell injury is accompanied by apoptosis. Based on pharmacological studies, it is stated that PAF acts through its receptors and mediates apoptosis under ischemia/postischemia-like conditions (Shi et al., 1998). PAF also contributes to leukocyte adherence and blood–brain barrier breakdown after cerebral ischemia. The expression of PAF-R is modulated by excitotoxicity (Bennett et al., 1998). While PAF-R mRNA normally occurs in neurons and microglia in rat hippocampus, their expression becomes restricted to apoptotic neurons and to glia involved in phagocytosing apoptotic debris following treatment with glutamate. PAF-R mRNA is rarely detected in surviving neurons. RT-PCR analysis indicates that ischemia-reperfusion injury reduces PAF-R mRNA levels, which return to normal levels gradually (Zhang et al., 2007). PAF binding to platelets of stroke victims correlates with both the extent of neurodegeneration and the associated neurological impairment. This may serve as an additional index in the assessment of stroke severity and clinical outcome of stroke victims (Adunsky et al., 1999). PAF-R antagonists, BN50726 and BN50739, enhance metabolic recovery during postischemic reperfusion in dogs (Gilboe et al., 1991). Pretreatment of rats with PAF antagonist Egb-761 also promotes metabolic recovery from hypoxic recovery (Karcher et al., 1984). PAF causes inhibition of Na^+, K^+-ATPase activity in rat cerebral cortex in a concentration- and time-dependent manner, and this inhibition can be prevented by the PAF antagonist PCA-4248 (Catalan et al., 1994). Cerebral ischemia induces changes in protein kinase C (PKC) and ornithine decarboxylase (ODC), and preadministration of PAF antagonist BN52021 blocks postischemic induction of PKC and ODC (Zablocka et al., 1995). These studies indicate that ischemic injury is accompanied not only by increased synthesis and release of PAF from neural and nonneural cells, but also by the downregulation of PAF-R gene expression in the perifocal regions of cerebral infarction. PAF-R antagonists can protect brain tissue from neural cell injury.

10.2.2 PAF in Head Injury and Spinal Cord Trauma

Most of the damage that occurs in brain and spinal cord tissues following trauma is due to secondary effects of glutamate-mediated toxicity, Calcium overload, and oxidative stress. Interactions between excitotoxicity and oxidative stress activate neutrophil-mediated inflammation and promote secondary damage. PAF has been implicated in head injury. Neurons, macro- and microglial cells, monocyte cell populations, macrophages, and endothelial cells of blood vessels are the targets of PAF. Immunofluorescent staining studies have shown that PAF can be detected in the rat brain after cold-induced local brain injury. Cold injury is accompanied by increased immediate-early PAF staining within the cold lesion, followed later by immunoreactivity in the ipsilateral white matter. PAF immunoreactivity can be clearly observed both in cortical neurons adjacent to the cold lesion and in the ipsilateral hippocampus, which shows delayed neuronal degeneration (Tokutomi et al., 2001). Collective evidence suggests that PAF synthesis occurs in the neuronal cells in the perilesional area of hippocampus as well as within the cold lesion site during the early stages of cold-induced brain injury. It is proposed that PAF expression may be associated with the onset and progression of further brain damage, such as delayed axotomy and delayed neuronal loss (Tokutomi et al., 2001). Brain damage caused by head injury can be partially prevented by BN52021, a PAF antagonist.

PAF also contributes to secondary damage after spinal cord trauma. The molecular mechanisms associated with PAF-mediated changes in spinal cord trauma are not fully understood. However, it is proposed that increased expression of cytokine induction may contribute to spinal cord trauma (Hostettler and Carlson, 2002). Contusion injury spinal cord trauma at T10 using the NYU impactor results in enhancement of cytokine mRNA. Thus, interleukin-1α (IL-1α) mRNA peaks at 1 h after injury while IL-1β and IL-6 mRNA levels peak at 6 h. Tumor necrosis factor alpha (TNF-α) mRNA remains undetectable. mRNA for all cytokines comes down to the baseline levels by 24 h. Treatment with PAF antagonist WEB 2170 15 min prior to injury significantly decreases mRNA levels for all three cytokines at 6 h after injury. These results suggest that PAF modulates the induction of proinflammatory cytokine after spinal cord trauma (Hostettler et al., 2002; Hostettler and Carlson, 2002).

10.2.3 PAF in Meningitis

PAF levels are significantly increased in the cerebrospinal fluid (CSF) of children with bacterial meningitis, compared with those in age-matched controls (Arditi et al., 1990). Bacterial meningitis produces pathophysiological alterations in the brain tissue. These changes may be mediated by components from bacterial cell wall such as LPS, which is known to cause cytokine induction. PAF-AH activity has also been determined in CFS samples from children with meningitis (Moon et al., 2003). PAF plays a crucial role in neuroinflammation caused by bacterial infection in the brain and brain meninges. *Streptococcus pneumoniae* is the major cause of bacterial meningitis, sepsis, and pneumonia in humans. The molecular mechanism associated with its traversal from the circulation across the blood–brain barrier into the subarachnoid space is not fully understood. One mechanism involves transcytosis through microvascular endothelial cells (Ring et al., 1998). *S. pneumoniae* targets blood–brain barrier through microvascular cells that are rich in PAF-Rs. Interactions between G-protein-coupled PAF-R and bacterial phosphorylcholine may promote enhanced adherence and invasion of bacteria in blood–brain barrier endothelial cells. These interactions are inhibited by PAF-R antagonists, and therefore, PAF antagonists may be useful for the treatment of bacterial meningitis (Cabellos et al., 1992). During infection, cPLA$_2$ (cytosolic phospholipase A$_2$)-induced PAF synthesis plays an important role in neutrophil-mediated bacterial killing (Rubin et al., 2005). Collective evidence suggests that PAF is associated with the pathogenesis of bacterial meningitis. PAF-AH activity is significantly increased (3-fold) in the CSF from patient with meningitis, compared with that in control subjects (Table 10.2). CSF PAF acetyl hydrolase is a calcium-independent enzyme that has a broad pH spectrum and is relatively heat stable. The enzyme activity is strongly inhibited by phenylmethanesulfonyl fluoride and partially inhibited by *p*-bromophenacylbromide. Iodoacetamide has no effect, but dithiothreitol moderately stimulates PAF-AH. In addition, this enzyme does not degrade phospholipid with a long chain fatty acyl group at the *sn*-2 position, but hydrolyzes PAF and oxidatively modified PtdCho (Moon et al., 2003). Thus, the biochemical profile of CSF PAF acetyl hydrolase is different from other known acetyl hydrolases.

Table 10.2 Activities of platelet-activating factor (PAF) acetyl hydrolases in neurological disorders.

Neurological disorder	Effect on PAF acetyl hydrolase	Reference
Spinal cord injury	Increased	Zhu et al., 2006
Meningitis	Increased	Kim et al., 2003
Migraine	Decreased	Sarchielli et al., 2004
L1S1 lissencephaly	Decreased	Shmueli et al., 1999
Atherosclerotic occlusive disease	Decreased	Unno et al., 2000
Subarachnoid hemorrhage	Decreased and then increased	Hirashima et al., 1994

These findings suggest that CSF PAF acetyl hydrolase activity may be a sensitive marker of the host response to central nervous system infections (Chang et al., 2002).

10.2.4 PAF in HIV Infection

Human immunodeficiency virus type 1 (HIV-1) is a neurotropic virus linked to many neurological disorders. Although the molecular mechanism associated with HIV-1 entry in brain and interactions with neuronal and nonneuronal cells to initiate and sustain neurologic dysfunction are not fully understood, overwhelming evidence suggests that majority of cells infected with HIV-1 in the brain are microglia/macrophages that are rich in PAF-Rs (Gelbard et al., 1994; Glass and Wesselingh, 2001). Microglia/macrophages are involved in immune regulation as well as in generation and release of cytotoxic neurotoxins such as PAF, eicosanoids, quinolinic acid. These toxins have been implicated in the pathogenesis of HIV-associated dementia complex (Nottet et al., 1995). Their actions on brain tissue lead to neuronal injury, glial proliferation, and myelin pallor during advance stages of the disease. PAF levels are significantly increased at 6 and 12 weeks after LP-BM5 murine leukemia virus (LP-BM5 MuLV) inoculation in cerebral cortex and hippocampal region of mice brain (Nishida et al., 1996). Significant increases in striatal and cerebellar PAF levels also occur at 12 weeks after virus inoculation. MK-801, an NMDA antagonist, significantly reduces the increased PAF levels in the cerebral cortex and hippocampus of LP-BM5 MuLV-infected mice. It is suggested that the LP-BM5 MuLV-mediated increase in brain PAF levels is associated with NMDA receptor activation, and is consistent with the hypothesis that elevated brain PAF levels contribute to the behavioral deficits observed in LP-BM5 MuLV-infected mice (Nishida et al., 1996). Increase in PAF levels in brain is also seen in guinea pigs following Pichinde virus infection (Guo et al., 1993). In addition to pathological changes in brain, the infection leads to fever, electrolyte imbalance, and cardiopulmonary dysfunction. Suggestions on the accumulation of PAF in HIV-1-associated dementia are supported by reports that indicate the elevation of PAF levels in CSF (Gelbard et al., 1994).

In HIV-1-associated dementia, the accumulation of PAF not only mediates neurotoxicity but also promotes dendrite injury following elevated synaptic activity. The molecular mechanism underlying dendritic injury remains unknown. This process is closely related to the replication of HIV (Bellizzi et al., 2005). Treatment of hippocampal slices with stable PAF analog enhances calcium influx and caspase-dependent dendritic beading, which is retarded by PAF antagonists. Randomized, double-blind, placebo-controlled trials in HIV-infected patients have indicated that PAF antagonist lexipafant promotes improvement in cognitive function and neurophysiological performance, especially verbal memory (Schifitto et al., 1999). HIV-1 Tat protein, which mediates chemotaxis and recruitment of monocytes, also induces the PAF synthesis. This process plays a critical role in triggering the events involved in the migratory response of monocytes (Del Sorbo et al., 1999). Collectively, these studies indicate that PAF promotes neural injury in HIV-1-associated dementia complex.

10.2.5 PAF in Prion Diseases

The molecular mechanism associated with the pathogenesis of prion diseases is not fully understood. However, the conversion of the cellular prion protein (PrP^{C}) into β-sheet-rich disease-related isoforms (PrP^{Sc}) and the accumulation of PrP^{Sc} may lead to neurodegeneration (Bate et al., 2004a, b). PrP^{C} interacts with cholesterol- and glycosphingolipid-rich lipid rafts through association of its glycosylphosphatidylinositol anchor with saturated raft lipids and through interaction of its N-terminal region with an as yet unidentified raft associated molecule (Cundell et al., 1995). Squalestatin, a squalene synthase inhibitor, not only reduces the cholesterol content of cells, but also retards the accumulation of PrP^{Sc} in three prion-infected cell lines (ScN2a, SMB, and ScGT1 cells) (Bate et al., 2004c). Treatment of ScN2a cells with squalestatin also protects against microglial-cell-mediated cell death. Treatment of neurons with squalestatin results in a redistribution of PrP^{C} away from Triton X-100 insoluble lipid rafts in a dose-dependent manner and is partially reversed by cholesterol. These studies indicate that cholesterol may play a pivotal role in controlling PrP^{Sc} formation, and in the activation of signaling pathways associated with PrP^{Sc}-induced neuronal death.

Pretreatment of neurons with PAF antagonists and PLA_2 inhibitors makes them resistant to PrP^{Sc} peptide or amyloid-β1–42 neurotoxicity (Bate et al., 2004a, b). Although the molecular mechanism underlying this process is not fully understood, PAF antagonists downregulate the activation of caspase-3, a marker of apoptosis, and the synthesis of prostaglandin E_2. These markers are closely associated with neuronal death in prion and AD (Farooqui et al., 2007). Nanomolar concentrations of the ginkgolides protect neurons against sPrP106- or amyloid-β1–42-mediated neuronal cell death (Bate et al., 2004d). The ginkgolides also block the PAF-induced neurotoxicity and downregulate the production of prostaglandin E_2 in response to PAF, amyloid-β1–42, or sPrP106 (Bate et al., 2004c). Thus, PAF antagonists, ginkgolides, can be used as potential therapeutic agent for the treatment of prion or AD diseases.

10.2.6 PAF in Multiple Sclerosis

High-pressure liquid chromatography tandem mass spectrometry of plasma and CSF indicates that PAF levels are significantly higher in multiple sclerosis (MS) patients than in healthy control subjects (Callea et al., 1999). These PAF levels correlate with the MRI number of gadolinium enhancing lesions, which are markers of blood–brain barrier injury. Based on these findings, it is proposed that PAF mediates blood–brain barrier injury in the early stages of MS, rather than a marker of its progression and severity (Callea et al., 1999).

Experimental autoimmune encephalomyelitis (EAE) is an important potential animal model for MS. Studies on PAF metabolism in brain and spinal cord indicate that PAF synthesis and PAF-R mRNA expression in the spinal cord correlate with the EAE symptoms (Kihara et al., 2005). PAF-R-knockout mice exhibit lower incidence and less severe symptoms in the chronic phase of EAE than do wildtype mice. Compared to wildtype mice, PAF-R-knockout mice show downregulation of mRNA for IL-6 and chemokine and their receptors in the spinal cord. Moreover, spinal cords from PAF-R-knockout mice in chronic phase of EAE show substantial reduction in the severity of inflammation and demyelination. PAF-R-knockout mice macrophages have reduced phagocytic activity and produce less TNF-α. These results suggest that PAF plays a dual role in EAE pathology. PAF is not only involved in the induction phase, but also modulates the chronic phase through the T cells' activation (Kihara et al., 2005).

10.2.7 PAF in Miller-Dieker Lissencephaly

A mutation of PAF-AH in man causes a devastating neurodevelopmental syndrome, Miller-Dieker lissencephaly. Lissencephalic cerebral cortex is thick and formed without its usual folds, indicating that PAF-AH plays a crucial role in brain development. Miller-Dieker syndrome is characterized by seizures and severe mental retardation (Hattori, 1994). Miller-Dieker syndrome causing *LIS1* gene has been mapped to chromosome 17p 13.3. It contains β-transducin-like repeats and is associated with neuronal migration and axonal growth (Chong et al., 1997). In the in vitro models of lissencephaly, PAF agonists and antagonists can alter migration of cerebral granule and hippocampal cells, indicating that PAF may act as a neuronal migration stop signal (Adachi et al., 1997; Bix and Clark, 1998). The overexpression of LIS1 protein rescues the migration defect in $LIS1^{+/-}$ neurons (Tanaka et al., 2004). *LIS1* is predominantly localized in centrosome, and after disruption of microtubules, it is redistributed to the perinuclear region. *LIS1* is also associated with microtubules extending from the perinuclear "cage" to the centrosome (Tanaka et al., 2004). *LIS1*$^{+/-}$ neurons display increased separation between the nucleus and the preceding centrosome during migration. As stated here, centrosome coupling defects can be rescued by LIS1 overexpression. LIS1 complexes with dynein, a motor protein that converts the chemical energy contained in ATP into the mechanical energy of movement. Dynein transports various cellular cargo by "walking" along cytoskeletal tubules toward the minus end of the microtubule, which is usually oriented toward the cell center. Substantial evidence indicates that LIS1 functions in a highly conserved pathway with dynein to regulate neuronal migration and other motile events. Interactions of LIS1 with dynein mediate centrosome coupling during migration. The defective coupling between LIS1 and dynein may contribute to migration defects in lissencephaly. It is suggested that LIS1 participates in cytoplasmic dynein-mediated nuclear migration (Tai et al., 2002; Tanaka et al., 2004), and PAF levels and PAF-AH activity may be closely associated with the pathogenesis of Miller-Dieker lissencephaly.

10.2.8 PAF in Migraine Attacks

Migraine is an episodic syndrome consisting of a variety of clinical features caused by dysfunction of the sympathetic nervous system. Increase in PAF levels is observed in the internal jugular venous blood of migraine patients. PAF levels are decreased at the end of the migraine attack, reaching levels significantly lower than those measured before migraine attack (Sarchielli et al., 2004). PAF-AH activity shows an opposite trend with higher value before significantly lower value after the migraine attack (Sarchielli et al., 2004). The increased production of PAF may also account for persistent platelet activation during migraine crises, even in the presence of an increased generation of nitric oxide end products that are known to counteract and limit platelet activation. The potential sources of PAF production and neuroinflammation in migraine attack are the endothelial cells from cerebral vessels, stimulated by trigeminal neuropeptides, platelets themselves, and mast cells (Sarchielli et al., 2004). The relative contribution of these cells in PAF metabolism during migraine attack remains unknown.

10.2.9 PAF in Kainic-Acid-Mediated Neurodegeneration

As stated earlier, the localization of PAF-R mRNA is altered during kainic-acid-induced neural cell injury. PAF-R mRNA is normally expressed by neurons and microglia in rat hippocampus following kainic acid treatment. In kainic-acid-mediated neurotoxicity, the expression of PAF becomes restricted to apoptotic neurons and to glia involved in phagocytosing apoptotic debris.

PAF-R mRNA, is rarely detected in surviving neurons (Bennett et al., 1998). Kainate injections not only affect PAF-R mRNA, but also reduce PAF-AH activity as early as 30 min following initiation of seizures. PAF-AH activity returned to and surpasses control values 1 week following initiation of seizures. Expression of LIS1 (the gene mutated in lissencephaly) in the dentate gyrus modulates PAF-AH activity (Shmueli et al., 1999). Reduction in LIS1 protein levels found in lissencephaly patients may render animals more susceptible to seizures. PAF, along with excitatory amino acids, is also involved in modulating long-term potentiation and neuronal survival (Bazan et al., 1997). Excitatory amino acids and PAF increase mitogen-activated protein (MAP) kinases in neurons and glial cells. Studies on comparison between PAF and kainate result in activation of MAP kinases in primary hippocampal neurons in vitro (De Coster et al., 1998). Extracellular signal-regulated kinase, c-Jun *N*-terminal kinase, and p38 kinases are also activated by kainate or PAF in hippocampal neurons and nonneural cells (Margues et al., 2002). The activation of kinases is inhibited by the KA receptor antagonists CNQX and PAF-R antagonist BN50730. The PAF-R antagonist BN50730 also prevents kainate activation. CNQX has no effect on PAF activation of the kinases, indicating that PAF is downstream of kainate activation (De Coster et al., 1998). Coapplication of PAF and kainate has a less than additive activation, indicating similar mechanisms of activation by the two agonists. Both CNQX and BN50730 inhibit kainate-mediated neurotoxicity in neural cell cultures. These results indicate that PAF and kainate activate similar kinase pathways. PAF acts downstream of the kainate receptors, and excessive stimulation of PAF-Rs contributes to neurodegeneration in PAF-mediated neurotoxicity (De Coster et al., 1998).

10.3 Involvement of PAF in Nonneural Injuries

PAF is involved in several pathological processes, including allergy, inflammation, shock, and trauma. These processes affect the cardiovascular, cerebral, respiratory, renal, hepatic, gastrointestinal, and reproductive systems. The identification of PAF as a pathological mediator has been reported either by injecting it into animals and measuring the intensity of inflammation or by the ability of PAF antagonists to block PAF-mediated responses (Maclennan et al., 1996). Recent studies on PAF-R-deficient and PAF-R-overexpressing mice strongly support the involvement and role of PAF in allergy, inflammation, shock, and trauma (Nagase et al., 1997, 1999; Ishii et al., 1997). Increased levels of PAF have been observed in human plasma during asthma attacks (Tsukioka et al., 1996). During thrombosis, PAF accumulates in the atherosclerotic plaques of patients with advanced coronary artery disease (Mueller et al., 1995). Levels of urinary PAF are significantly higher in patients with membranous nephropathy and are positively correlated with proteinuria (Noris et al., 1993). In addition, PAF is also involved in anaphylaxis, endotoxic shock, psoriasis, and dermatitis (Ishii and Shimizu, 2000).

10.4 Consequences of Altered PAF Acetyl Hydrolase in Cardiovascular System

PAF-AH is a proatherogenic enzyme and its deficiency is associated with increased risk for vascular diseases and stroke. A G^{994} (M allele) $\rightarrow$ T (m allele) mutation in the plasma PAF-AH gene, which results in a $Val^{279} \rightarrow$ Phe substitution in the mature protein, leads to the loss of catalytic activity (Yamada et al., 2000). A study of Val^{279} mutation in Japanese adult population with asthma indicates that plasma PAF-AH deficiency due to $V^{279}F$ mutation is not essential to the pathophysiology of asthma in the Japanese adult population. Human subjects with complete PAF-AH deficiency are unable to release F_2-isoprostanes from esterified glycerophospholipids, and have low circulating and/or urinary levels of isoprostanes. These subjects have a high risk of developing atherosclerosis, compared to control subjects having normal PAF-AH activity (Yamada et al., 2000).

Deficiency of PAF-AH is associated with a number of diseases. The most common inactivating mutation, $V^{279}F$, is found in >30% of randomly surveyed Japanese subjects (4% homozygous, 27% heterozygous). The prevalence of the mutant allele is significantly greater in patients with asthma, stroke, myocardial infarction, brain hemorrhage, and nonfamilial cardiomyopathy. Administration of recombinant plasma PAF-AH attenuates inflammation and oxidative stress in a number of animal models. In addition, the recombinant PAF-AH may have pharmacologic potential in human inflammatory disease as well. These observations point toward new approaches for controlling inflammation (Tjoelker and Stafforini, 2000).

The overexpression of PAF-AH in balloon-injured carotid arteries results in anti-inflammatory, antithrombotic, and antiproliferative effects. Similarly, following postangioplasty restenosis local adenovirus-mediated transfer of PAF-AH also produces significant downregulation of neointima formation in balloon-denuded rabbit aortas and may be useful for the prevention of restenosis after arterial manipulations (Turunen et al., 2005). Moreover, adenovirus-mediated transfer of PAF-AH to atherosclerosis-prone $ApoE^{-/-}$ mice reduces the extent of atherosclerosis lesion formation (Quarck et al., 2001). Collectively, these studies suggest that PAF-AH modulates vascular inflammation and this may be beneficial against oxidative stress and inflammatory processes.

PAF-AH activity is also altered in many other nonneural diseases, including asthma (Grissom et al., 2003), systemic lupus erythematosus (Tetta et al., 1990; Cederholm et al., 2004), juvenile rheumatoid arthritis (Tselepis et al., 1999), multiple organ failure (Partrick et al., 1997), acute myocardial infarction (Serebruany et al., 1998), sepsis (Graham et al., 1994), and Crohn's disease (Kald et al., 1996; Mueller et al., 2007) (Table 10.3).

Table 10.3 Activities of platelet-activating factor (PAF) acetyl hydrolases in visceral diseases.

Disease	Effect on PAF acetyl hydrolase	Reference
Asthma	Increased	Grissom et al., 2003
Systemic lupus erythematosus	Decreased	Camussi et al., 1990
Juvenile rheumatoid arthritis	Decreased	Tselepis et al., 1999
Multiple organ failure	Decreased	Partrick et al., 1997
Acute myocardial infarction	Decreased	Serebruany et al., 1998
Sepsis	Decreased	Graham et al., 1994
Crohn's disease	Decreased	Kald et al., 1996

10.5 Molecular Mechanism of PAF-Mediated Neural Injury

As stated in Chap. 9, PAF-Rs are linked to the activation of PLA_2, phospholipase C, and phospholipase D. Interactions of PAF with its receptors not only enhance the turnover of various phospholipids, including PtdCho, PlsEtn, and PtdIns, but also mobilizes calcium (Clark et al., 2000; Honda et al., 2002). This results in arachidonic acid release, synthesis of eicosanoids, and generation of diacylglycerol and inositol 1,4,5-trisphosphate, causing elevation in intracellular calcium concentration (Fig. 10.2). This elevation in calcium is known to cause neuronal growth cone collapse (Rehder et al., 1992). Activation of guanylate cyclase produces cGMP, and phosphorylation of various proteins through the activation of various kinases, including MAP kinases, phosphatidylinositol 3-kinase, and tyrosine kinases (Chao and Olson, 1993; Honda et al., 2002). Excessive production of the above-mentioned lipid mediators and phosphorylation of neural protein may cause neural injury.

PAF and reactive oxygen species generated through the oxidation of arachidonic acid interact with NF-κB (nuclear factor kappa B), a transcription factor that occurs in inactivated form in the cytoplasm attached to its inhibitory protein, IκB (Kravchenko et al., 1995; Farooqui et al., 2007). Upon stimulation IκB is rapidly phosphorylated, ubiquinated, and then degraded by proteasomes, resulting in the release and subsequent nuclear translocation of active NF-κB (Yamamoto and Gaynor, 2004). In the nucleus NF-κB mediates the transcription of many genes implicated in inflammatory and immune responses (Fig. 10.2). These genes include cyclooxygenase 2 (COX-2), intracellular adhesion molecule 1 (ICAM-1), vascular adhesion molecule 1 (VCAM-1), E-selectin, TNF-α, IL-1β, IL-6, $sPLA_2$, inducible nitric oxide synthase (iNOS), and matrix metalloproteinases (MMPs). The PAF-induced expression of the above-mentioned cytokines is inhibited by p65 antisense or antioxidants (Ko et al., 2002). A significant inhibition of the inflammatory and angiogenic effect of PAF can be achieved by anti-VEGF (vascular endothelial growth factor) antibodies or soluble vascular endothelial growth factor receptors such as KDR (kinase insert domain receptor) and flt-1 (fms-related tyrosine kinase 1), but not by antibodies against TNF-α and IL-1α. These results indicate that PAF enhances inflammation and angiogenesis through inducing NF-κB activation, which in turn promotes the production of other inflammatory and angiogenic factors.

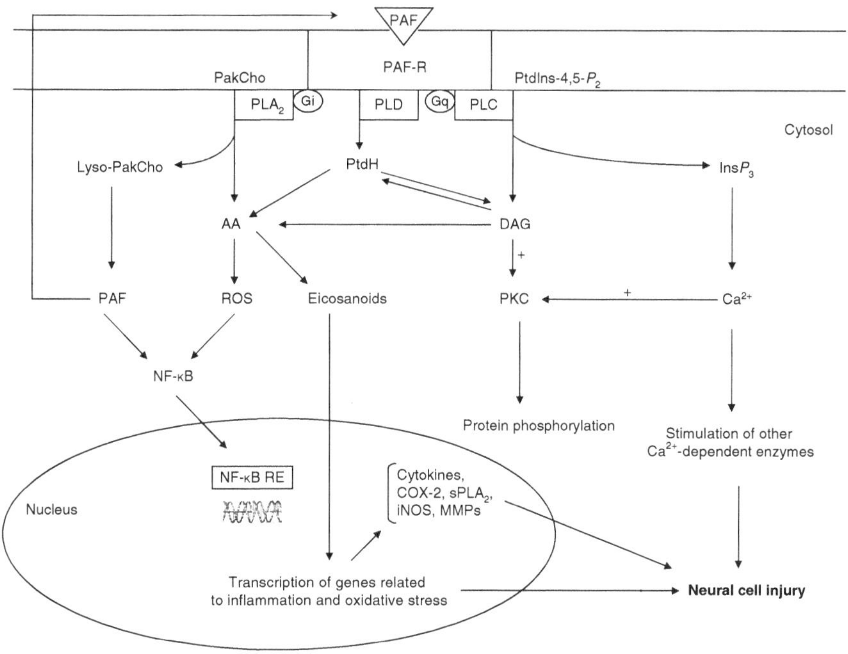

Fig. 10.2 Signal transduction processes associated with PAF-mediated neural cell injury. *cPLA*$_2$ cytosolic phospholipase A$_2$, *PLC* phospholipase C, *PLD* phospholipase D, *PakCho* phosphatidylcholine containing ether bond at the *sn*-1 position, lyso-*PakCho* lyso-phosphatidylcholine, *AA* arachidonic acid, *PAF* platelet-activating factor, *PAF-R* platelet-activating receptor, *PtdH* phosphatidic acid, *DAG* diacylglycerol, *PtdIns-4,5-P*$_2$ phosphatidylinositol 4,5-bisphosphate, InsP_3 inositol 1,4,5-trisphosphate, *PKC* protein kinase C, G_i and G_q G proteins, *ROS* reactive oxygen species, *NF-κB* nuclear factor κB, *NF-κB-RE* nuclear factor κB-response element, *COX-2* cyclooxygenase 2, *iNOS* inducible nitric oxide synthase, *MMPs* matrix metalloproteinase, and *sPLA*$_2$ secretory phospholipase A$_2$.

Reactive-oxygen-species-mediated NF-κB activation involves NADPH oxidase, which is an important component of the innate immune response against toxic agents (metabolic as well as microbial), and is involved in shaping the cellular response to a variety of physiological and pathological signals (Anrather et al., 2006; Frey et al., 2006; Rubin et al., 2005; Miller et al., 2006). In neutrophils, macrophages NADPH oxidase modulates PAF synthesis via the remodeling pathway. Stimulation of neural and nonneural cells results in translocation of cPLA$_2$ to the plasma membranes, where it interacts with NADPH oxidase (Shmelzer et al., 2003). The interaction between these two enzymes provides the molecular basis for arachidonic acid release by cPLA$_2$ and generation of reactive species to activate the NADPH oxidase (Shmelzer et al., 2003). The ability of cPLA$_2$ to modulate superoxide production and generation of eicosanoid indicates its importance in inflammatory processes. Meanwhile, in endothelial cells lining, the local cerebral blood vessels are stimulated to produce adhesion molecules, causing the migration of peripheral circulating leukocytes into the brain tissue, an event that amplifies inflammatory signaling cascades. Collective evidence suggests that PAF-mediated stimulation of NF-κB and increased expression of cytokines are closely associated with neural cell injury.

10.6 Clinical Application of PAF Antagonists for the Treatment of Neurological Disorders

It is well known that PAF-R antagonists elicit neuroprotection in a variety of neurological disorders and their animal models. Thus, administration of PAF antagonists has beneficial effects in various models of cerebral ischemia and head and spinal cord injuries. It also improves the neurological score of patients with MS, meningitis, and AIDS dementia complex (Serradji et al., 2004; Birkle et al., 1988; Brochet et al., 1995). PAF is a transcriptional activator of the COX-2 gene. BN50730, a potent PAF antagonist, inhibits COX-2 induction. COX-2 transcription and protein expression are upregulated in the hippocampus in kainic acid neurotoxicity. There is a selectively elevated induction of COX-2 by kainic acid preceding neuronal cell death. BN50730 injections prevent seizure-mediated COX-2 induction. Collective evidence suggests that PAF is a dual modulator of neural function and becomes an endogenous neurotoxin when overproduced (Bazan, 1998).

The molecular mechanisms by which PAF antagonists produce neuroprotection are not clear at present. However, PAF-mediated neural cell injury is accompanied not only by the modulation of genes expression for cytokines, PLA_2, and COX-2, and an increase in intracellular calcium, but also by the release of neurotransmitters, and modulation of their receptor activity. It is likely that PAF antagonists protect against neural damage by blocking calcium mobilization. PAF amplifies excitotoxicity and PAF antagonists block excitotoxicity (Clark et al., 1992). PAF and its antagonists interact with other neurotransmitters and their receptors and modulate oxidative stress and neuroinflammation. This view is supported by several observations. PAF synthesis is stimulated by acetycholine and inhibited by acetylcholine receptor antagonist, atropine (Sogos et al., 1990). Specific PAF antagonist Y-24180 interacts with low affinity to benzodiazepine receptors in synaptosomal membranes (Takehara et al., 1990). Adrenocorticotrophic hormone (ACTH) fragments are known to protect against neural cell injury, and treatment with PAF also results in a decrease in ACTH levels (Blasquez et al., 1990). Collectively, these studies suggest that PAF antagonists have promising therapeutic effects in experimental models of neurological disorders and may eventually prove valuable for clinical trials along with other therapeutic agents. In brain PAF antagonists act not only by blocking inflammation and oxidative stress, but also by producing procoagulant effects such as decrease in blood–brain barrier permeability and edema.

10.7 Conclusion

PAF is a potent lipid mediator synthesized in brain by neural and nonneural cells. PAF-Rs have been demonstrated in the brain tissue. PAF stimulates intracellular Ca^{2+} mobilization, activates phospholipases A_2, C, and D, and enhances polyphosphoinositide turnover through G proteins that are coupled to PAF-Rs. PAF has potent actions on cerebral vessels and cerebral metabolism when administered in vivo. Levels of PAF are elevated in neurological and visceral disorders, including brain trauma, seizures, stroke, MS, viral and bacterial infections, and in a variety of other conditions such as asthma, thrombosis, toxic shock, and dermatitis. PAF is also involved in the pathophysiology of AIDS dementia and Miller-Dieker lissencephaly. Administration of PAF antagonists slows down the progression of these disorders. Thus, the development of new nontoxic PAF antagonists would result in better treatment of visceral and neurological disorders. Collectively, these studies support the view that PAF is an important mediator in the pathophysiology of brain injury.

References

Arditi M., Manogue K. R., Caplan M., and Yogev R. (1990). Cerebrospinal fluid cachectin/tumor necrosis factor-α and platelet-activating factor concentrations and severity of bacterial meningitis in children. J. Infect. Dis. 162:139–147.

Adachi T., Aoki J., Manya H., Asou H., Arai H., and Inoue K. (1997). PAF analogues capable of inhibiting PAF acetylhydrolase activity suppress migration of isolated rat cerebellar granule cells. Neurosci. Lett. 235:133–136.

Adunsky A., Hershkowitz M., Atar E., Bakoun M., and Poreh A. (1999). Infarct volume, neurological severity and PAF binding to platelets of patients with acute cerebral ischemic stroke. Neurol. Res. 21:645–648.

Aihara M., Ishii S., Kume K., and Shimizu T. (2000). Interaction between neurone and microglia mediated by platelet-activating factor. Genes Cells. 5:397–406.

Akisu M., Huseyinov A., Yalaz M., Cetin H., Kultursay N. (2003). Selective head cooling with hypothermia suppresses the generation of platelet-activating factor in cerebrospinal fluid of newborn infants with perinatal asphyxia. Prostaglandins Leukot. Essent. Fatty Acids. 69:45–50.

Anrather J., Racchumi G., and Iadecola C. (2006). NF-κB regulates phagocytic NADPH oxidase by inducing the expression of gp91^{phox}. J. Biol. Chem. 281:5657–5667.

Arimura A., Nagata M., Watanabe A., Nakamura K., Takeuchi M., and Harada M. (1990). Production of active and passive anaphylactic shock in the WBB6F1 mouse, a mast cell-deficient strain. Experientia. 46:739–742.

Bate C., Salmona M., and Williams A. (2004a). The role of platelet activating factor in prion and amyloid-beta neurotoxicity. NeuroReport. 15:509–513.

Bate C., Reid S., and Williams A. (2004b). Phospholipase A_2 inhibitors or platelet-activating factor antagonists prevent prion replication. J. Biol. Chem. 279:36405–36411.

Bate C., Salmona M., Diomede L., and Williams A. (2004c). Squalestatin cures prion-infected neurons and protects against prion neurotoxicity. J. Biol. Chem. 279:14983–14990.

Bate C., Salmona M., and Williams A. (2004d). Ginkgolide B inhibits the neurotoxicity of prions or amyloid-beta1–42. J. Neuroinflamm. 1:4.

Bazan N. G. (1998). The neuromessenger platelet-activating factor in plasticity and neurodegeneration. Prog. Brain Res. 118:281–291.

Bazan N. G., Packard M. G., Teather L., and Allan G. (1997). Bioactive lipids in excitatory neurotransmission and neuronal plasticity. Neurochem. Int. 30:225–231.

Bellizzi M. J., Lu S. M., Masliah E., and Gelbard H. A. (2005). Synaptic activity becomes excitotoxic in neurons exposed to elevated levels of platelet-activating factor. J. Clin. Invest. 115:3185–3192.

Bennett S. A., Chen J., Pappas B. A., Roberts D. C., and Tenniswood M. (1998). Platelet activating factor receptor expression is associated with neuronal apoptosis in an in vivo model of excitotoxicity. Cell Death Differ. 5:867–875.

Birkle D. L., Kurian P., Braquet P., and Bazan N. G. (1988). Platelet-activating factor antagonist BN52021 decreases accumulation of free polyunsaturated fatty acid in mouse brain during ischemia and electroconvulsive shock. J. Neurochem. 51:1900–1905.

Bix G. J., and Clark G. D. (1998). Platelet-activating factor receptor stimulation disrupts neuronal migration in vitro. J. Neurosci. 18:307–318.

Blasquez C., Jegou S., Delarue C., Delbrnde C., Bunel D. T., Blasquez P., and Vaudry H. (1990). Effect of platelet-activating factor on hypothalamic and hypophyseal pro-opiomelanocortin-related peptides and hypothalamo-pituitary-adrenal axis in the rat. Eur. J. Pharmacol. 177:145–153.

Brochet B., Guinot P., Orgogozo J. M., Confavreux C., Rumbach L., Lavergne V., and The Ginkgolide Study Group in Multiple Sclerosis. (1995). Double blind placebo controlled multicentre study of ginkgolide B in treatment of acute exacerbations of multiple sclerosis. J. Neurol. Neurosurg. Psychiatry. 58:360–362.

Cabellos C., MacIntyre D. E., Forrest M., Burroughs M., Prasad S., and Tuomanen E. (1992). Differing roles for platelet-activating factor during inflammation of the lung and subarachnoid space. The special case of *Streptococcus pneumoniae*. J. Clin. Invest. 90:612–618.

Callea L., Arese M., Orlandini A., Bargnani C., Priori A., and Bussolino F. (1999). Platelet activating factor is elevated in cerebral spinal fluid and plasma of patients with relapsing-remitting multiple sclerosis. J. Neuroimmunol. 94:212–221.

Camussi G., Tetta C., and Baglioni C. (1990). Antiflammins inhibit synthesis of platelet-activating factor and intradermal inflammatory reactions. Adv. Exp. Med. Biol. 279:161–172.

Catalan R. E., Martinez A. M., Aragones M. D., Fernandez I., Miguel B. G., Calcerrada M. C., and Perez M. J. (1994). Platelet-activating factor inhibits Na^{+}, K^{+} ATPase activity in rat brain. Neurosci. Res. 19:241–244.

Cederholm A., Svenungsson E., Stengel D., Fei G. Z., Pockley A. G., Ninio E., and Frostegard J. (2004). Platelet-activating factor-acetylhydrolase and other novel risk and protective factors for cardiovascular disease in systemic lupus erythematosus. Arthritis Rheum. 50:2869–2876.

Chang H. W., Kwon S., Kim H., Lee K., Kim M., Moon T., and Baek S. (2002). Platelet-activating factor acetylhydrolase activity in cerebrospinal fluid of children with acute systemic or neurological illness. Ann. Neurol. 51:760–763.

Chao W., and Olson M. S. (1993). Platelet-activating factor: Receptors and signal transduction. Biochem. J. 292:617–629.

Chong S. S., Pack S. D., Roschke A. V., Tanigami A., Carrozzo R., Smith A. C., Dobyns W. B., and Ledbetter D. H. (1997). A revision of the lissencephaly and Miller-Dieker syndrome critical regions in chromosome 17p13.3. Hum. Mol. Genet. 6:147–155.

Clark G. D., Happel L. T., Zorumski C. F., and Bazan N. G. (1992). Enhancement of hippocampal excitatory synaptic transmission by platelet-activating factor. Neuron. 9:1211–1216.

Clark G. D., Zorumski C. F., McNeil R. S., Happel L. T., Ovella T., McGuire S., Bix G. J., and Swann J. W. (2000). Neuronal platelet-activating factor receptor signal transduction involves a pertussis toxin-sensitive G-protein. Neurochem. Res. 25:603–611.

Cundell D. R., Gerard N. P., Gerard C., Idanpaan H. I., and Tuomanen E. I. (1995). *Streptococcus pneumoniae* anchor to activated human cells by the receptor for platelet-activating factor. Nature. 377:435–438.

De Coster M. A., Mukherjee P. K., Davis R. J., and Bazan N. G. (1998). Platelet-activating factor is a downstream messenger of kainate-induced activation of mitogen-activated protein kinases in primary hippocampal neurons. J. Neurosci. Res. 53:297–303.

Del Sorbo L., DeMartino A., Biancone L., Bussolati B., Conaldi P. G., Tonioli A., and Cammassi G. (1999). The synthesis of platelet-activating factor modulates chemotaxis of monocytes induced by HIV-1 Tat. Eur. J Immunol. 29:1513–1521.

Faden A. I., and Tzendzalian P. A. (1992). Platelet-activating factor antagonists limit glycine changes and behavioral deficits after brain trauma. Am. J. Physiol. 263:R909–R914.

Farooqui A. A., and Horrocks L. A. (2004). Plasmalogens, and other ether lipids. In *Ether Lipids*, A. Nicolaou and G. Kokotos (Eds.). Oily, Bridgwater, England, pp. 107–134.

Farooqui A. A., Horrocks L. A., and Farooqui T. (2007). Modulation of inflammation in brain: A matter of fat. J. Neurochem. 101:577–599.

Feuerstein G. Z. (1996). Platelet-activating factor: A case for its role in CNS function and brain injury. J. Lipid Mediat. Cell Signal. 14:109–114.

Frey R. S., Gao X., Javaid K., Siddiqui S. S., Rahman A., and Malik A. B. (2006). Phosphatidylinositol 3-kinase γ signaling through protein kinase Cζ induces NADPH oxidase-mediated oxidant generation and NF-κB activation in endothelial cells. J. Biol. Chem. 281:16128–16138.

Gelbard H. A., Nottet H. S., Swindell S., Jett M., Dzenko K. A., Genis P., White P., Wang L., Choi Y. B., and Zhang D. (1994). Platelet-activating factor: A candidate human immunodeficiency virus type 1-induced neurotoxin. J. Virol. 68:4628–4635.

Gilboe D. D., Kintner D., Fitzpatrick J. H., Emoto S. E., Esanu A., Braquet P. G., and Bazan N. G. (1991). Recovery of post-ischemic brain metabolism and function following treatment with a free radical scavenger and platelet-activating factor antagonists. J. Neurochem. 56:311–319.

Glass J. D., and Wesselingh S. L. (2001). Microglia in HIV-associated neurological diseases. Microsc. Res. Tech. 54:95–105.

Graham R. M., Strahan M. E., Norman K. W., Watkins D. N., Strum M. J., and Taylor R. R. (1994). Platelet and plasma platelet-activating factor in sepsis and myocardial infarction. J. Lipid Mediat. Cell Signal. 9:167–182.

Grissom C. K., Orme J. F., Jr., Richer L. D., McIntyre T. M., Zimmerman G. A., and Elstad M. R. (2003). Platelet-activating factor acetylhydrolase is increased in lung lavage fluid from patients with acute respiratory distress syndrome. Crit. Care Med. 31:770–775.

Guo Z. M., Qian C., Peters C. J., and Liu C. T. (1993). Changes in platelet-activating factor, catecholamine, and serotonin concentrations in brain, cerebrospinal fluid, and plasma of pichinde virus-infected guinea pigs. Lab. Anim. Sci. 43:569–574.

Hattori M., Adachi H., Tsijimoto M., Arai H., Inoue K. (1994). Miller-Dieker lissencephaly gene encodes a subunit of brain platelet-activating factor acetylhydrolase. Nature. 370:216–218.

Hirashima Y., Kato R., Endo S., Takaku A., Karasawa K., and Nojima S. (1993a). Immunofluorescent localization of platelet-activating factor (PAF) in the rat. Histochem. J. 25:830–833.

Hirashima Y., Endo S., Otsuji T., Karasawa K., Nojima S., and Takaku A. (1993b). Platelet-activating factor and cerebral vasospasm following subarachnoid hemorrhage. J. Neurosurg. 78:592–597.

Hirashima Y., Endo S., Ohmori T., Kato R., and Takaku A. (1994). Platelet-activating factor (PAF) concentration and PAF acetylhydrolase activity in cerebrospinal fluid of patients with subarachnoid hemorrhage. J. Neurosurg. 80:31–36.

Hirashima Y., Endo S., KaraSawa K., Sato N., YokoYama K., Kurimoto M., Ikeda H., Setaka M., and Takaku A. (1999). Deficient platelet-activating factor and related enzymes in hemimegalencephaly. Childs Nerv. Syst. 15:98–101.

Honda Z., Ishii S., and Shimizu T. (2002). Platelet-activating factor receptor. J. Biochem. 131:773–779.

Hostettler M. E., and Carlson S. L. (2002). PAF antagonist treatment reduces pro-inflammatory cytokine mRNA after spinal cord injury. Neuroreport. 13:21–24.

Hostettler M. E., Knapp P. E., and Carlson S. L. (2002). Platelet-activating factor induces cell death in cultured astrocytes and oligodendrocytes: Involvement of caspase-3. Glia. 38:228–239.

Ishii I., Izumi T., Ui M., and Shimizu T. (1997). High and low affinity mutants of platelet-activating factor receptor. Adv. Exp. Med. Biol. 433:249–253.

Ishii S., and Shimizu T. (2000). Platelet-activating factor (PAF) receptor and genetically engineered PAF receptor mutant mice. Prog. Lipid Res. 39:41–82.

Kald B., Smedh K., Olaison G., Sjodahl R., and Tagesson C. (1996). Platelet-activating factor acetylhydrolase activity in intestinal mucosa and plasma of patients with Crohn's disease. Digestion. 57:472–477.

Karcher L., Zagermann P., and Krieglstein J. (1984). Effect of an extract of *Ginkgo biloba* on rat brain energy metabolism in hypoxia. Naunyn Schmiedebergs Arch. Pharmakol. 327:31–35.

Kelley D. S., Taylor P. C., Nelson G. J., Schmidt P. C., Ferretti A., Erickson K. L., Yu R., Chandra R. K., and Mackey B. E. (1999). Docosahexaenoic acid ingestion inhibits natural killer cell activity and production of inflammatory mediators in young healthy men. Lipids. 34:317–324.

Kihara Y., Ishii S., Kita Y., Toda A., Shimada A., and Shimizu T. (2005). Dual phase regulation of experimental allergic encephalomyelitis by platelet-activating factor. J. Exp. Med. 202:853–863.

Kim H. J., Lee S. R., and Moon K. D. (2003). Ether fraction of methanol extracts of *Gastrodia elata*, medicinal herb protects against neuronal cell damage after transient global ischemia in gerbils. Phytother. Res. 17:909–912.

Ko H. M., Seo K. H., Han S. J., Ahn K. Y., Choi I. H., Koh G. Y., Lee H. K., Ra M. S., and Im S. Y. (2002). Nuclear factor kappa B dependency of platelet-activating factor-induced angiogenesis. Cancer Res. 62:1809–1814.

Kravchenko V. V., Pan Z., Han J., Herbert J. M., Ulevitch R. J., and Ye R. D. (1995). Platelet-activating factor induces NF-kappa B activation through a G protein-coupled pathway. J. Biol. Chem. 270:14928–14934.

Kumar R., Harvey S. A., Kester M., Hanahan D. J., and Olson M. S. (1988). Production and effects of platelet-activating factor in the rat brain. Biochim. Biophys. Acta. 963:375–383.

Lindsberg P. J., Hallenbeck J. M., and Feuerstein G. (1991). Platelet-activating factor in stroke and brain injury. Ann. Neurol. 30:117–129.

Maclennan K. M., Smith P. F., and Darlington C. L. (1996). Platelet-activating factor in the CNS. Prog. Neurobiol. 50:585–596.

Makristathis A., Stauffer F., Feistauer S. M., Georgopoulos A. (1993). Bacteria induce release of platelet-activating factor (PAF) from polymorphonuclear neutrophil granulocytes: Possible role for PAF in pathogenesis of experimentally induced bacterial pneumonia. Infect. Immun. 61:1996–2002.

Manya H., Aoki J., Kato H., Ishii J., Hino S., Arai H., and Inoue K. (1999). Biochemical characterization of various catalytic complexes of the brain platelet-activating factor acetylhydrolase. J. Biol. Chem. 274:31827–31832.

Margues S. A., Dy L. C., Southall M. D., Yi O., Smietana E., Kapur R., Margues M., Travers J. B., and Spandau D. F. (2002). The platelet-activating factor receptor activates the extracellular signal-regulated kinase mitogen-activated protein kinase and induces proliferation of epidermal cells through an epidermal growth factor-receptor-dependent pathway. J. Pharmacol. Exp. Ther. 300:1026–1035.

Miller A. A., Drummond G. R., and Sobey C. G. (2006). Novel isoforms of NADPH-oxidase in cerebral vascular control. Pharmacol. Ther. 111:928–948.

Moon T. C., Kim M. S., Lee S. J., Lee T. Y., Kwon S. H., Baek S. H., and Chang H. W. (2003). Detection and characterization of 45 kDa platelet activating factor acetylhydrolase in cerebrospinal fluid of children with meningitis. Arch. Pharm. Res. 26:554–558.

Mueller H. W., Haught C. A., McNatt J. M., Chi K., Gaskell S. J., Johnston D. A., and Willerson J. T. (1995). Measurement of platelet-activating factor in a canine model of coronary thrombosis and in endarterectomy samples from patients with advanced coronary artery disease. Circ. Res. 77:54–63.

Mueller M. H., Geis M., Glatzle J., Kasparek M., Meile T., Jehle E. C., Kreis M. E., and Zittel T. T. (2007). Risk of fecal diversion in complicated perianal Crohn's disease. J. Gastrointest. Surg. 11:529–537.

Nagase T., Ishii S., Katayama H., Fukuchi Y., Ouchi Y., and Shimizu T. (1997). Airway responsiveness in transgenic mice overexpressing platelet-activating factor receptor. Roles of thromboxanes and leukotrienes. Am. J. Respir. Crit. Care Med. 156:1621–1627.

Nagase T., Kurihara H., Kurihara Y., Aoki-Nagase T., Nagai R., and Ouchi Y. (1999). Disruption of ET-1 gene enhances pulmonary responses to methacholine via functional mechanism in knockout mice. J. Appl. Physiol. 87:2020–2024.

Nishida K., Markey S. P., Kustova Y., Morse H. C., Skolnick P., Basile A. S., and Sei Y. (1996). Increased brain levels of platelet-activating factor in a murine acquired immune deficiency syndrome are NMDA receptor-mediated. J. Neurochem. 66:433–435.

Noris M., Benigni A., Boccardo P., Gotti E., Benfenati E., Aiello S., Todeschini M., and Remmuzzi G. (1993). Urinary excretion of platelet activating factor in patients with immune-mediated glomerulonephritis. Kidney Int. 43:426–429.

Nottet H. S., Jett M., Flanagan C. R., Zhai Q. H., Persidsky Y., Rizzino A., Bernton E. W., Genis P., Baldwin T., Schwartz J. H., LaBenz C. J., and Gendelman H. E. (1995). A regulatory role for astrocytes in HIV-1 encephalitis. An overexpression of eicosanoids, platelet-activating factor, and tumor necrosis factor-alpha by activated HIV-1-infected monocytes is attenuated by primary human astrocytes. J. Immunol. 154:3567–3581.

Partrick D. A., Moore E. E., Moore F. A., Biffl W. L., and Barnett C. C. (1997). Reduced PAF-acetylhydrolase activity is associated with postinjury multiple organ failure. Shock. 7:170–174.

Quarck R., De Geest B., Stengel D., Mertens A., Lox M., Theilmeier G., Michiels C., Raes M., Bult H., Collen D., Van Veldhoven P., Ninio E., and Holvolt P. (2001). Adenovirus-mediated gene transfer of human platelet-activating factor-acetylhydrolase prevents injury-induced neointima formation and reduces spontaneous atherosclerosis in apolipoprotein E-deficient mice. Circulation. 103:2495–2500.

Rehder V., Jensen J. R., and Kater S. B. (1992). The initial stages of neural regeneration are dependent upon intracellular calcium levels. Neuroscience. 51:565–574.

Ring A., Weiser J. N., and Tuomanen E. I. (1998). Pneumococcal trafficking across the blood–brain barrier. Molecular analysis of a novel bidirectional pathway. J. Clin. Invest. 102:347–360.

Rubin B. B., Downey G. P., Koh A., Degousee N., Ghomashchi F., Nallan L., Stefanski E., Harkin D. W., Sun C. X., Smart B. P., Lindsay T. F., Cherepanov V., Vachon E., Kelvin D., Sadilek M., Brown G. E., Yaffe M. B., Plumb J., Grinstein S., Glogauer M., and Gelb M. H. (2005). Cytosolic phospholipase A_2-α is necessary for platelet-activating factor biosynthesis, efficient neutrophil-mediated bacterial killing, and the innate immune response to pulmonary infection – cPLA$_2$-α does not regulate neutrophil NADPH oxidase activity. J. Biol. Chem. 280:7519–7529.

Sarchielli P., Alberti A., Coppola F., Baldi A., Gallai B., Floridi A., Floridi A., Capocchi G., and Gallai V. (2004). Platelet-activating factor (PAF) in internal jugular venous blood of migraine without aura patients assessed during migraine attacks. Cephalalgia. 24:623–630.

Schifitto G., Sacktor N., Marder K., McDermott M. P., McArthur J. C., Kieburtz K., Small S., Epstein L. G., and The Neurological AIDS Research Consortium. (1999). Randomized trial of the platelet-activating factor antagonist lexipafant in HIV-associated cognitive impairment. Neurology. 53:391–396.

Serebruany V. L., Gurbel P. A., Murugesan S. R., Lowry D. R., Sturm E., and Svetlov S. I. (1998). Depressed plasma platelet-activating factor acetylhydrolase in patients presenting with acute myocardial infarction. Cardiology. 90:127–130.

Serradji N., Martin M., Bensaid O., Cisternino S., Rousselle C., Dereuddre-Bosquet N., Huet J., Redeuilh C., Lamouri A., Doug L. Z., Clayette P., Scherrmann J. M., Dormont D., and

Heymans F. (2004). Structure–activity relationships in platelet-activating factor. 12. Synthesis and biological evaluation of platelet-activating factor antagonists with anti-HIV-1 activity. J. Med. Chem. 47:6410–6419.

Shi L. C., Wang H. Y., and Friedman E. (1998). Involvement of platelet-activating factor in cell death induced under ischemia/postischemia-like conditions in an immortalized hippocampal cell line. J. Neurochem. 70:1035–1044.

Shmueli O., Cahana A., and Reiner O. (1999). Platelet-activating factor (PAF) acetylhydrolase activity, LIS1 expression, and seizures. J. Neurosci. Res. 57:176–184.

Sogos V., Bussolino F., Pilia E., Torrelli S., and Gremo F. (1990). Acetylcholine-induced production of platelet-activating factor by human fetal brain cells in culture. J. Neurosci. Res. 27:706–711.

Shmelzer Z., Haddad N., Admon E., Pessach I., Leto T. L., Eitan-Hazan Z., Hershfinkel M., and Levy R. (2003). Unique targeting of cytosolic phospholipase A_2 to plasma membranes mediated by the NADPH oxidase in phagocytes. J. Cell Biol. 162:683–692.

Tai C. Y., Dujardin D. L., Faulkner N. E., and Vallee R. B. (2002). Role of dynein, dynactin, and CLIP-170 interactions in LIS1 kinetochore function. J. Cell Biol. 156:959–968.

Takehara S., Mikashima H., Muramoto Y., Terasawa M., Setoguchi M., and Tahara T. (1990). Pharmacological actions of Y-24180, a new specific antagonist of platelet activating factor (PAF). II. Interactions with PAF and benzodiazepine receptors. Prostaglandins. 40:571–583.

Tanaka T., Iimori M., Tsukatani H., Tokumura A. (1994). Platelet-aggregating effects of platelet-activating factor-like phospholipids formed by oxidation of phosphatidylcholines containing an sn-2-polyunsaturated fatty acyl group. Biochim. Biophys. Acta. 1210:202–208.

Tanaka T., Serneo F. F., Higgins C., Gambello M. J., Wynshaw-Boris A., and Gleeson J. G. (2004). Lis1 and doublecortin function with dynein to mediate coupling of the nucleus to the centrosome in neuronal migration. J. Cell Biol. 165:709–721.

Taylor K. E., Richardson A. J., and Stein J. F. (2001). Could platelet activating factor play a role in developmental dyslexia? Prostaglandins Leukot. Essent. Fatty Acids. 64:173–180.

Tetta C., Bussolino F., Modena V., Montrucchio G., Segoloni G., Pescarmona G., and Camussi G. (1990). Release of platelet-activating factor in systemic lupus erythematosus. Int. Arch. Allergy Appl. Immunol. 91:244–256.

Tiberghien C., Laurent L., Junier M. P., and Dray F. (1991). A competitive receptor binding assay for platelet-activating factor (PAF): Quantification of PAF in rat brain. J. Lipid Mediat. 3:249–266.

Tjoelker L. W., and Stafforini D. M. (2000). Platelet-activating factor acetylhydrolases in health and disease. Biochim. Biophys. Acta. 1488:102–123.

Tokumura A., Tsutsumi T., and Tsukatani H. (1992). Transbilayer movement and metabolic fate of ether-linked phosphatidic acid (1-*O*-octadecyl-2-acetyl-sn-glycerol 3-phosphate) in guinea pig peritoneal polymorphonuclear leukocytes. J. Biol. Chem. 267:7275–7283.

Tokutomi T., Maruiwa H., Hirohata M., Miyagi T., and Shigemori M. (2001). Production of platelet-activating factor by neuronal cells in the rat brain with cold injury. Neurol. Res. 23:605–611.

Tselepis A. D., Elisaf M., Besis S., Karabina S. A., Chapman M. J., and Siamopoulou A. (1999). Association of the inflammatory state in active juvenile rheumatoid arthritis with hypo-high-density lipoproteinemia and reduced lipoprotein-associated platelet-activating factor acetylhydrolase activity. Arthritis Rheum. 42:373–383.

Tsukioka K., Matsuzaki M., Nakamata M., Kayahara H., and Nakagawa T. (1996). Increased plasma level of platelet-activating factor (PAF) and decreased serum PAF acetylhydrolase (PAFAH) activity in adults with bronchial asthma. J. Invest. Allergol. Clin. Immunol. 6:22–29.

Turunen P., Puhakka H., Rutanen J., Hiltunen M. O., Heikura T., Gruchala M., Yla-Herttuala S. (2005). Intravascular adenovirus-mediated lipoprotein-associated phospholipase A_2 gene transfer reduces neointima formation in balloon-denuded rabbit aorta. Atherosclerosis. 179:27–33.

Unno N., Nakamura T., Kaneko H., Uchiyama T., Yamamoto N., Sugatani J., Miwa M., and Nakamura S. (2000). Plasma platelet-activating factor acetylhydrolase deficiency is associated with atherosclerotic occlusive disease in Japan. J. Vasc. Surg. 32:263–267.

Yamada Y., Yoshida H., Ichihara S., Imaizumi T., Satoh K., and Yotaka M. (2000). Correlations between plasma platelet-activating factor acetylhydrolase (PAF-AH) activity and PAF-AH genotype, age, and atherosclerosis in a Japanese population. Atherosclerosis. 150:209–216.

Yamamoto Y., and Gaynor R. B. (2004). IkappaB kinases: Key regulators of the NF-kappa B pathway. Trends Biochem. Sci. 29:72–79.

Zablocka B., Lukasiuk K., Lazarewicz J. W., and Domanska-Janik S. (1995). Modulation of ischemic signal by antagonists of *N*-methyl-D-aspartate, nitric oxide synthase, and platelet-activating factor in gerbil hippocampus. J. Neurosci. Res. 40:233–240.

Zhang X., Pan X. L., Liu X. T., Wang S., and Wang L. J. (2007). Down-regulation of platelet-activating factor receptor gene expression during focal reversible cerebral ischemia in rats. Neurochem. Res. 32:451–456.

Zhu J., Brackett N. L., Aballa T. C., Lynne C. M., Witt M. A., Kort H. I., and Roudebush W. E. (2006). High seminal platelet-activating factor acetylhydrolase activity in men with spinal cord injury. J. Androl. 27:429–433.

Chapter 11
Biochemical Effects of Nonphysiological Antitumor Ether Lipids

11.1 Introduction

Antitumor ether lipids (AEL) are structural analogs of PAF, but lack the readily hydrolyzable ester substituent at the *sn*-2 position of glycerol moiety (Fig. 11.1). They contain a long carbon chain at the *sn*-1 and a short chain at the *sn*-2 position. At the *sn*-3 position, phosphocholine is the head group. Examples of common AEL are Edelfosine (1-*O*-octadecyl-2-*O*-methyl-*rac*-glycero-3-phosphocholine, Et-18-OCH_3); Ilmofosine; Miltefosine (hexadecylphosphocholine, HePC); ilmofosine (BM 41.440, 1-hexadecylthio-2-methoxymethyl-rac-glycero-3-phosphocholine); and SR1 62–834 ((+-)-2-(Hydroxy[tetrahydro-2-(octadecyloxy) methylfuran-2-yl] methoxyl phosphinyloxy)-*N*,*N*,*N*-trimethylethaniminium hydroxide). HePC lacks the glycerol backbone. Alkyl-phosphocholine and alkyl-glycerophosphocholine derivatives as well as aza-substituted alkyl-glycerophosphocholines have also been synthesized. These derivatives include 1-methoxy-2-*N,N*-methyl-octadecylamino-propyloxyphosphorylcholine (BN 52205), 1-methoxy-3-*N,N*-methyl-octadecylamino-propyloxyphosphorylcholine (BN 52207), 1-*N,N*-methyl-octadecylamino-2-methoxy-propyloxyphosphorylcholine (BN 52211) (Fig. 11.2). Besides the above-mentioned anticancer compounds, glycosidated phospholipids such as 2-glucophosphatidylcholine (1-stearoyl-2-*O*-α-D-glucopyranoside-*sn*-glycero-3-phosphocholine) and Glc-PAF (1-*O*-octadecyl-2-*O*-α-D-glucopyranosyl-*sn*-2-glycero-3-phosphatidylcholine) and other alkylphosphocholines have been synthesized (Fig. 11.3). All these compounds inhibit cell proliferation in HaCaT cells (Fischer et al., 2006). The substitution of myo-inositol, in place of α-D-glucose, in the *sn*-2 position of the glycerol moiety leads to two diastereomeric 1-*O*-octadecyl-2-*O*-(2-(myo-inositolyl)-ethyl)-*sn*-glycero-3-(*R/S*)-phosphatidylcholines (Ino-C2-PAF). The inositol-containing PAF enhances the antiproliferative capacity (IC_{50} = 1.8 μM) and reduces the cytotoxicity relative to Glc-PAF (LC_{50} = 15 μM). AEL are taken up in the tumor cell more rapidly than normal cells. Several mechanisms have been proposed for AEL uptake. They include passive diffusion (Kelley et al 1993), internalization through endocytosis (Bazill and Dexter, 1990), and active uptake through a carrier (Hanson et al., 2003; Perez-Victoria et al., 2003; Ménez et al., 2007). The uptake

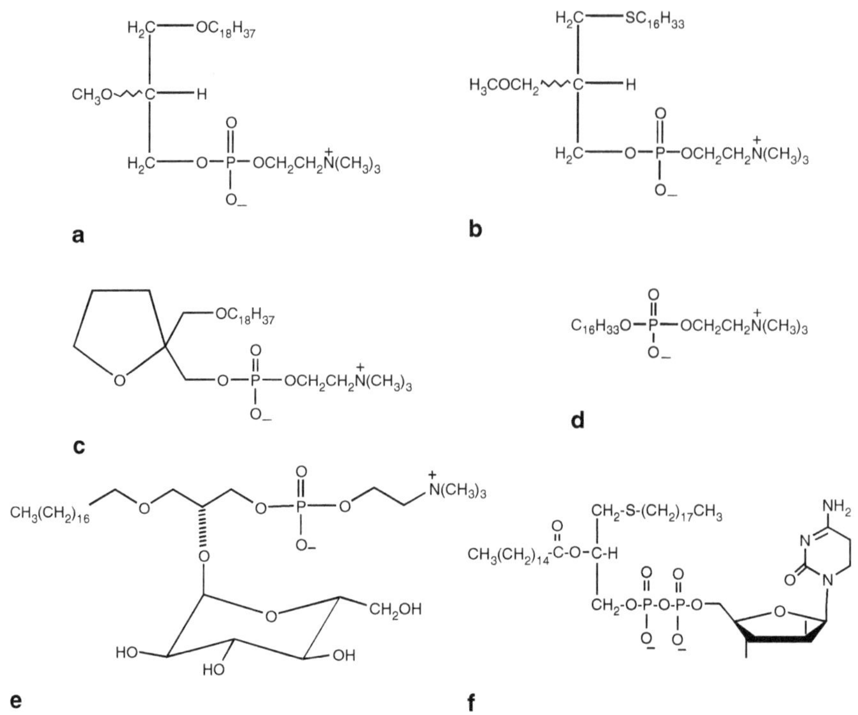

Fig. 11.1 Chemical structures of the antitumor ether lipids: Edelfosine (ET 18-OCH_3) (**a**), Ilmofosine (BM 41.440) (**b**), SR1 62–834 (**c**), Miltefosine (HePC) (**d**), Glc-PAF (**e**), and ara-CDP-DL-PTBA (1-β-D-arabinofuranosylcytosine-5′-diphosphate-L-1,2-dipalmitin) (**f**).

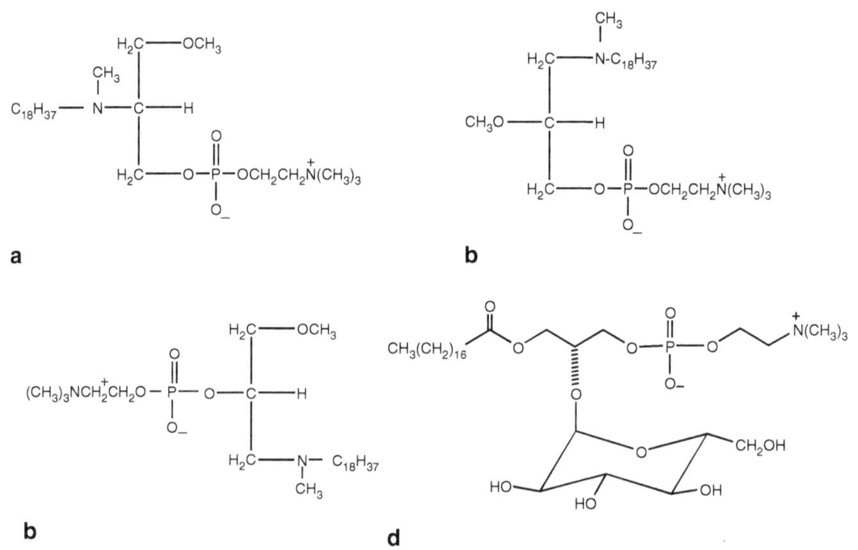

Fig. 11.2 Chemical structures of the aza-phospholipids BN52205 (**a**), BN 52207 (**b**), BN 52211 (**c**), and Glc-PtdCho (**d**).

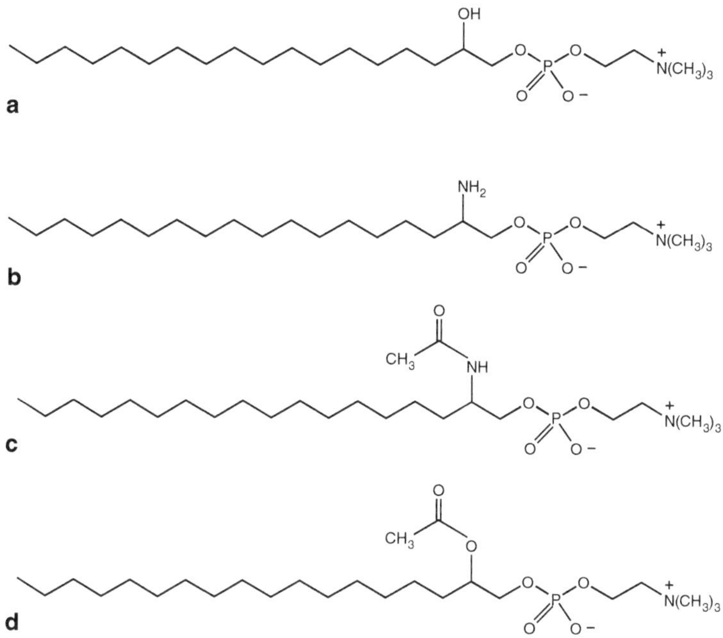

Fig. 11.3 Chemical structures of alkylphosphocholine: *R*- or *S*-1-*O*-phosphocholine-2-hydroxy-octadecane (**a**), *R*- or *S*-1-*O*-phosphocholine-2-amino-octadecane (**b**), *R*- or *S*-1-*O*-phosphocholine-2-*N*-acetyl-octadecane (**c**), *R*- or *S*-1-*O*-phosphocholine-2-*O*-acetyl-octadecane (**d**).

of AEL is an energy-dependent process because cellular ATP depletion through the inhibition of oxidative phosphorylation and glycolytic pathway reduce the uptake and accumulation of AEL in Caco-2 cells (Ménez et al., 2007). AEL are highly resistant to the normal metabolic reactions, and are metabolized very slowly by normal mammalian tissues. The AEL are antineoplastic agents that have remarkable antiproliferative activity not only on tumor cell lines but also on some tumors in vivo (Berkovic, 1998). They have very little effect on normal neural and nonneural cells. The antitumor properties of AEL include activation of macrophages, reduction of tumor cell invasion in vitro, inhibition of tumor metastases, inhibition of tumor development and shrinkage of tumors, differentiation of tumor cells, and selective inhibition of tumor cell proliferation (Arthur and Bittman, 1998). AEL exert minimal hematologic toxicity. In contrast, the majority of conventional anticancer drugs cause severe side effects because of bone marrow suppression. AELs differ from other cancer drugs in two ways. First, being nonphysiological compounds, they are highly resistant to normal metabolic reactions and are very slowly hydrolyzed. Second, unlike most conventional cancer chemotherapeutic agents, AEL do not interact directly with the cell's DNA, but act at the cell membrane. This suggests that instead of giving rise to mutations, AEL may block the growth of different types of cancer cells through interactions with enzyme-mediated signal transduction processes. An important drawback of AEL therapy is that some AEL cause dose-dependent haemolysis when administered intravenously (Mollinedo, 2007).

11.2 Effect of AEL on Enzymes Involved in Signal Transduction

The molecular mechanisms associated with cytotoxic actions of AEL are poorly understood. However, there is an agreement that cellular uptake and accumulation of AEL are crucial for antitumor activity. Unlike conventional chemotherapeutic drugs, as stated earlier, AEL do not target DNA. They act at the level of cellular membranes where they not only inhibit mitogenic and survival signaling pathways but also activate the stress-activated protein kinase/c-Jun NH_2-terminal kinase pathway (Ruiter et al., 1999). AEL also modulate a number of enzymes associated with signal transduction processes such as phospholipases, protein and lipid kinases, acyltransferases, sialyltransferases, and cyclooxygenases (Table 11.1). In leukemic HL-60 cells, AEL (Et-18-OCH_3) promotes selective apoptotic changes in ultrastructural morphology, including blebbing, chromatin condensation, nuclear membrane

Table 11.1 Effect of AEL on enzymic activities associated with signal transduction processes in cancer cells.

Antitumor ether lipid	Enzyme	Effect on activity	Reference
Et-18-OCH_3	Protein Kinase C	Decreased	Helfman et al., 1983; Conesa-Zamora et al., 2005
Et-18-OCH_3	Mitogen-activated protein kinase	Decreased	Ruiter et al., 1999; Samadder et al., 2003
Et-18-OCH_3	PtdIns 3-kinase-Akt/PKB	Decreased	Ruiter et al., 2003
Et-18-OCH_3	c-Jun NH(2)-terminal protein kinase	Increased	Samadder et al., 2004
Et-18-OCH_3	Lysophosphocholine acyltransferase	Decreased	Herrmann and Neumann, 1986
Et-18-OCH_3	Sialyltransferase	Decreased	Bador et al., 1983
Et-18-OCH_3	Na^+, K^+-ATPase	Decreased	Zheng et al., 1990
Et-18-OCH_3	Ca^{2+}-ATPase	Increased	Grosman, 2001
Et-18-OCH_3	Caspase-3	Increased	Gajate et al., 2000
Et-18-OCH_3	Nitric oxide synthase	Increased	Cardile et al., 1996
Et-18-OCH_3	Cyclooxygenase-2	Increased	Na et al., 2005
Et-18-OCH_3	Phospholipase C	Decreased	Powis et al., 1992
He-PC	Phospholipase D	Increased	Lucas et al., 2001
He-PC	Phospholipase A_2	Decreased	Berkovic et al., 1997
He-PC	CTP-phosopho-choline cytidylyl-transferase	Decreased	Jiménez-López et al., 2004
He-PC	PtdEtn methyl-transferase	Decreased	Jiménez-López et al., 2004
He-PC	PtdEtn methyl-transferase	Decreased	Jiménez-López et al., 2004
Glc-PAF	Transglutaminase	Increased	Fischer et al., 2006

breakdown, extensive vacuolation, and release of cytochome c from mitochondria in tumor cells (Gajate et al., 2000). The overexpression of Bcl-X(L) prevents the cytochrome c release and apoptosis. ET-18-OCH_3 also mediates disruption of the mitochondrial transmembrane potential (DeltaPsi$_m$) followed by production of reactive oxygen species (ROS) and DNA fragmentation in leukemic cells. Et-18-OCH_3 also induces caspase-3 activation in human leukemic cells (Gajate et al., 2000).

Collective evidence suggests that AEL behaves as a potent and highly selective antitumor drug able to induce selective apoptotic pathway in tumor cells, but not in nonmalignant cells. AEL also inhibit the invasion of malignant cells into normal tissue (Schallier et al., 1991).

11.2.1 Effects of AEL on Phospholipases A_2, C, and D

Treatment of cultured cell with AEL alters the activities of phospholipases A_2 (PLA_2), PLC, and PLD (Fig. 11.4). Thus, in human leukemia cell line U937 cells, HePC stimulates TNF-α-mediated activation of PLA_2 (Berkovic et al., 1997). In untreated U937 cells, TNF-α has no effect on PLA_2 activity. HePC inhibits PLA_2 activation in U937 that have been differentiated with DMSO. Molecular mechanism associated with these processes is not fully understood. As it is well known that under certain conditions PKC regulates PLA_2 activity, attempts have been made to explain PLA_2 activation through PKC-mediated mechanism. Treatment of U937 cells with TNF-α, TPA, and specific PKC inhibitors indicates a significant decrease in TPA-induced PLA_2 activation by B II and H7, while TNF-α-induced

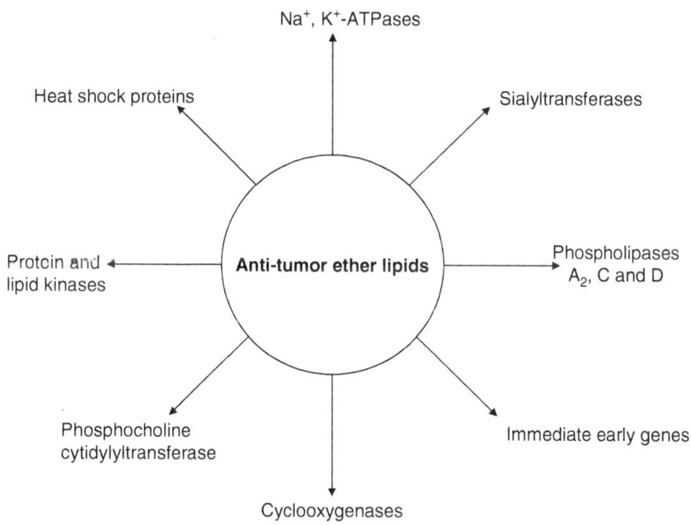

Fig. 11.4 Effects of AEL on enzymic acivities associated with signal transduction.

enzyme activation remains unchanged suggesting that HePC inhibits PLA_2 activity directly (Berkovic et al., 1997).

In Swiss 3T3 fibroblast and BG1 ovarian adenocarcinoma cells ether lipid analog, Et-18-OCH_3 (1-octadecyl-2-methyl-rac-glycero-3-phosphocholine) inhibits cytosolic phosphoinositide selective phospholipase C (PtdIns-PLC) in a dose-dependent manner with an IC50 of 9.6 μM (Powis et al., 1992). The noncytotoxic analog, 1-*O*-alkyl-2-hydroxy-*sn*-glycero-3-phosphocholine, also inhibits PtdIns-PLC when incorporated into the [^{3}H]PtdIns-4,5-P_2 substrate micelles. CP10, another cytotoxic ether lipid is a potent inhibitor of PtdIns-PLC than Et-18-OCH_3. Similarly, lyso-PAF, which is a nontoxic phospholipid, also inhibits enzymic activity. Another AEL, HePC, retards the bombesin-mediated formation of Ins(1,4,5) P_3 and the concomitant mobilization of intracellular Ca^{2+}. Dose-response curves for the inhibition of Ins(1,4,5) P_3 formation and Ca^{2+} mobilization reveal 50% inhibitory concentrations of 2 or 5 μM, respectively (Uberall et al., 1991). Polyphosphorylated phosphoinositides accumulate in HePC-treated cells indicating that the depression of Ins(1,4,5) P_3 generation is not caused by the inhibition of phosphoinositide kinases. The addition of HePC to purified PtdIns-PLC-δ also inhibits the enzymic activity. Inspite of the above evidence, it is suggested that ether lipid analogs at cytotoxic concentrations are potent inhibitors of PtdIns-PLC. However, the relationship of this inhibition with growth inhibitory activity of ether lipid analogs still remains elusive, and more studies are required on this important topic (Powis et al., 1992)(Uberall et al., 1991; Pawelczyk and Lowenstein, 1993).

In contrast to PLA_2 and PLC, HePC stimulates PLD activity through PKC-dependent or independent mechanisms (Lucas et al., 2001; Wieder et al 1996). The action of PLD on PtdCho results in the formation of phosphatidic acid (PtdH), an intracellular signaling molecule that is hydrolyzed into diacylglycerol by an enzyme called phosphatidic acid phosphohydrolase. This reaction not only results in termination of PtdH signal, but also generates diacylglycerol that activates protein kinase C (PKC). Both PLD1 and PLD2 isoenzymes are sensitive to HePC activation. HePC-stimulated PLD activity is strongly inhibited by cellular pretreatment with phorbol dibutyrate, and 59% diminished by pretreatment of cells with staurosporine, a PKC inhibitor. Furthermore, the chronic exposure of the cells to HePC prevents PLD activation by either phorbol esters or HePC itself with no effect on total cellular PLD levels. This is reflected in a strong inhibition of PLD activity. At present, the relationship of PLD to the anticancer properties of HePC is not clear. However, it is proposed that the inhibitory effects of HePC on PLD may be related to its antitumoral action (Lucas et al., 2001).

11.2.2 Effects of AEL on Protein and Lipid Kinases

AEL modulate activities of protein and lipid kinases in dose and time-dependent manner (Table 11.1). Some studies report inhibition of PKC activity, while others indicate stimulation. Et-18-OCH_3, BM 41.440, and HePC have been reported to inhibit PKC activity by competing for PtdSer-binding site (Zheng et al., 1990; Uberall et al., 1991). In contrast, the activity of the membrane-bound protein kinase C is increased in HL-60 cells treated with Et-18-OCH_3 compared with untreated HL-60 cells (Heesebeen et al., 1991). These results are due to the manner in which AEL are presented to the enzyme. Inhibition is seen when Et-18-OCH_3 is presented as mixed micelles with substrate, whereas activation of enzymic activity is observed when AEL is incorporated into cellular membranes prior to preparation of membrane fractions for the determination of PKC activity (Arthur and Bittman, 1998) Although earlier studies have indicated that inhibition of PKC by ether-linked lipids is not correlated with their antineoplastic activity on WEHI-3B and R6X-B15 cells, recent studies provide some support to the view that inhibitory effect of AEL on PKC may be associated with antineoplastic activity of various homologs of Et-18-OCH_3 (Salari et al., 1992; Conesa-Zamora et al., 2005). Antineoplastic ether lipid such as Et-18-OCH_3 inhibits activities of PKC isozymes in various cancer cell lines. Thus, Et-18-OCH_3 inhibits PKCα activity as the concentration is increased up to 30 mol% of the total lipid, above which PKCα activity is increased. The molecular mechanism associated with this process is not fully understood. However, detailed investigations indicate that the methoxy group located at the *sn*-2 position of the glycerol of Et-18-OCH_3 is essential for both the initial inhibitory effect and the subsequent activation of PKCα. In contrast, variations of substituents at the *sn*-1 position with ether or ester bonds do not play a role in determining the activity of the enzyme. On PKCζ, Et-18-OCH_3 has a triphasic effect, activating the enzyme at low concentrations, inhibiting it at slightly higher concentrations, and then activating it again at higher concentrations (Conesa-Zamora et al., 2005). In this case, the presence of the methoxy group linked to the *sn*-2 position of glycerol moiety and the type of bond linking substitutions to the *sn*-1 position have been reported to be important for activating the enzymic activity. Thus homologues with ester bonds as lyso-PtdCho and 1-palmitoyl-2-arachidonyl-*sn*-PtdCho induce the initial activation step that is similar to Et-18-OCH_3. Substitution of the phosphocholine group of Et-18-OCH_3 by phosphoserine led to a greater activation of PKCα. This effect is probably mediated through interactions between phosphoserine and Ca^{2+}-phospholipid-binding site (Conesa-Zamora et al., 2005). On the basis of various studies, it is proposed that AEL interfere with the binding of DAG/PtdSer or TPA (12-*O*-tetradecanoylphorbol-13-acetate) to PKC isozymes. Thus, AEL produce selective inhibition or activation of PKC isozymes in various types of tumor cell line cultures. Collective evidence suggests that the selective inhibition of tumor cell growth by ET-18-OCH_3 may be due to altered signal transduction mechanisms involving the inhibition of PKC.

PKC modulates NF-κB, an ubiquitous transcription factor that plays a pivotal role in the regulation of many genes associated with inflammatory responses and cell growth or apoptosis. In unstimulated cells, NF-κB is present in the cytosol in an inactive complex with the inhibitor I-κB. Phosphorylation of I-κB by PKC results in dissociation of NF-κ B inactive complex and the release of NF-κB, which migrates to the nucleus to initiate transcription (Lozano et al., 1994) (Fig. 11.5). Tumorogenesis is associated with oxidative stress caused by the generation of reactive oxygen

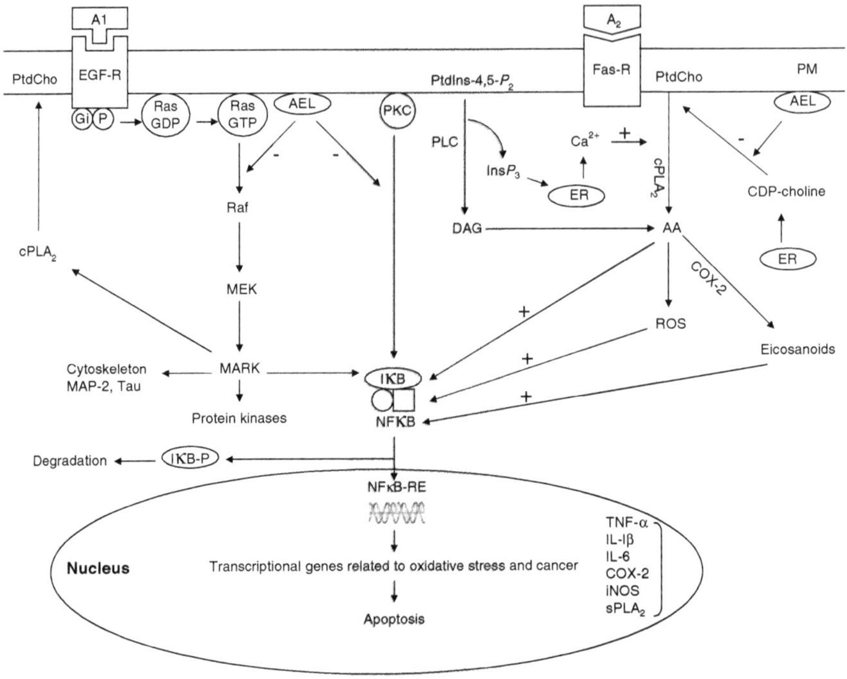

Fig. 11.5 Effects of AEL on enzymic activities associated with signal transduction processes. A1 and A2 (agonists); epidermal growth factor receptor (EGF-R); antitumor ether lipid (AEL); Fas-receptor (Fas-R); cytosolic phospholipase A_2 ($cPLA_2$); phospholipase C (PLC); protein kinase C (PKC); phosphatidylcholine (PtdCho); lyso-phosphatidyl choline (lyso-PtdCho); arachidonic acid (AA); diacylglycerol (DAG); inositol-1,4,5-trisphosphate (Ins-P_3); reactive oxygen species (ROS); mitogen-activated protein kinase (MAPK); nuclear factor-κB (NF-κB); nuclear factor-κB-response element (NF-κB-RE); inhibitory subunit of NFκB (IκB); cyclooxygenase-2 (COX-2); inducible nitric oxide synthase (iNOS); secretory phospholipase A_2 ($sPLA_2$,); tumor necrosis factor-α (TNF-α); interleukin-1β (IL-1β); interleukin-6 (IL-6); *positive sign* (+) indicates stimulation and *negative sign* (–) indicates inhibition.

species (ROS), which also stimulate the dissociation of NF-κB inactive complex and the release of NF-κB. NF-κB controls the expression of a large array of genes (TNF-α, IL-1β, IL-6, $sPLA_2$, COX-2, inducible nitric oxide synthase, and matrix metalloproteinases) involved in immune function, inflammation, oxidative stress, and tumorogenesis. It is also reported that upon stimulation, $cPLA_2$ is recruited to the plasma membranes where it interacts with NADPH oxidase (Shmelzer et al., 2003). The interaction between these two enzymes provides the molecular basis for arachidonic acid release by $cPLA_2$ and generation of ROS to activate the NADPH oxidase (Shmelzer et al., 2003). The ability of $cPLA_2$ to modulate superoxide production and generation of eicosanoid indicates its importance in inflammatory reactions, and oxidative stress involved in tumorogenesis. In 293.27.2 human kidney cells, transcription factor NF-κB is not only modulated by PKC but also by TPA, ROS, and cytokines (Daniel et al., 1995). Inhibition of TPA-induced NF-κB

Table 11.2 Effect of AEL on various cellular targets associated with cancer.

Antitumor ether lipid	Cellular target	Effect	Reference
Et-18-OCH_3	Fas/CD95	Increased	Mollinedo et al., 2004
Et-18-OCH_3	c-fos and Zif 268	Increased	Dell'Albani et al., 1993
Et-18-OCH_3	NF-κB	Decreased	Daniel et al., 1995
Et-18-OCH_3	Raf-1/Ras	Decreased	Samadder, et al., 2003
Et-18-OCH_3	HSP 70	Increased	Botzler et al., 1999

activation depends upon preincubation with Et-18-OCH_3 and levels of antioxidants in the cells (Table 11.2).

Growth factor receptors such as epidermal growth factor receptors have intrinsic protein-tyrosine kinase activity. During signal transduction process, EGF receptor recruits Sos, a bifunctional guanine nucleotide exchange factor that interacts with a small guanine nucleotide-binding protein is called Ras. In GDP, bound inactive form Ras is associated with the plasma membrane and becomes active through the exchange of GDP for GTP (Arthur and Bittman, 1998). Ras proteins also have a rather weak intrinsic GTPase activity that converts bound GTP back to GDP, thereby changing Ras to the inactive state. The active GTP-bound Ras promotes the translocation and activation of Raf, which phosphorylates and activate a dual-specificity kinase MEK that in turn activates MARK. Activated MARK not only phosphorylates cytoskeletal elements and transcription factor NF-κB, but also a number of other transcription factors, cytoskeletal proteins, and enzymes such as c-Myc, c-jun, c-fos, TAL1, ATF2, Tau, MAP-2, $cPLA_2$, and tyrosine kinase (Arthur and Bittman, 1998). Et-18-OCH_3 modulates EGF receptor by inhibiting its internalization and modulating signal transduction pathways that transduce growth signals through mitogen-activated protein kinase (MAPK) cascade in various types of tumor cells (Zhou et al., 1996). The treatment of MCF-7 cells with Et-18-OCH_3 inhibits cell proliferation. Et-18-OCH_3 also inhibits the sustained phosphorylation of MAPK resulting in a decrease in the magnitude and duration of activation of MAPK in cells stimulated with serum or EGF. Et-18-OCH_3 has no effect on the binding of EGF to its receptors, their activation, or p21ras activation. However, activation of the MAPK pathway subsequent to growth factor stimulation requires the recruitment of Raf-1 from the cytosol to the membrane. Et-18-OCH_3 decreases the level of membrane-associated Raf-1 relative to untreated control cells. As Et-18-OCH_3 does not inhibit the activities of the kinases in the cascade, the reduced Raf-1 levels may cause further reduction in the magnitude and duration of MAPK activity (Samadder et al., 2003). In this system, Et-18-OCH_3 associates specifically with Raf-1 in the cytosol and interferes in the interaction of Raf-1 with activated GTP-bound Ras, thereby reducing the levels that are translocated to the membrane for activation (Fig. 11.4). Thus, Raf-1 may be a molecular target of Et-18-OCH_3 (Samadder et al., 2003). Collective evidence suggests that there is a link between Et-18-OCH_3 accumulation, inhibition of cell proliferation, Raf association with the membranes, and MAPK activation in MCF-7 cells. It is proposed that inhibition of the MAPK cascade by Et-18-OCH_3 may be mediated by Raf-1 activation. This may constitute an important mechanism by which Et-18-OCH_3 inhibits cell proliferation (Zhou et al., 1996; Samadder et al., 2003).

The stimulation of the stress-activated protein kinase/c-Jun NH_2-terminal kinase (SAPK/JNK) pathway is essential for radiation-mediated apoptosis in certain types of tumor cell lines. The effect of ALP is tested in combination with ionizing radiation on MAPK/SAPK signaling pathway and its association with apoptosis. Three ALP, Et-18-CH_3, HePC, and a novel compound octadecyl-(1,1-dimethyl-piperidinio-4-yl)-phosphate (D-21266) induce time and dose-dependent apoptosis in the human leukemia cell lines U937 and Jurkat T but not in normal vascular endothelial cells (Ruiter et al., 1999). In combination with radiation, Et-18-CH_3 and HePC strongly promote the induction of apoptosis in both leukemic cell lines. Thus, Et-18-CH_3 and HePC not only block MAPK activation, but like radiation stimulate the SAPK/JNK cascade within minutes. A dominant-negative mutant of c-Jun prevents radiation and ALP-induced apoptosis, indicating a requirement for the SAPK/JNK pathway. These results support the view that ALPs and ionizing radiation promote apoptotic effect by modulating the balance between the mitogenic, antiapoptotic MAPK, and the apoptotic SAPK/JNK pathways (Ruiter et al., 1999).

Phosphatidylinositol-3-kinase (PtdIns-3-kinase) is an enzyme found associated with many growth factor receptor protein tyrosine kinases and oncogene protein tyrosine kinases. It plays an important role in mitogenesis and the malignant transformation of cells. Et-18-OCH_3 inhibits PtdIns-3-kinase activity in Swiss mouse 3T3 fibroblast and bovine brain with an IC50 value of 35 μM (Berggren et al., 1993). The inhibition of PtdIns-3-kinase by Et-18-OCH_3 is noncompetitive with ATP. Other antitumor ether lipid analogs also inhibit PtdIns-3-kinase activity, but the extent of inhibition is weaker than Et-18-OCH_3 (IC50 of 96 μM). These results suggest that inhibition of PtdIns-3-kinase may contribute to the antiproliferative activity of the antitumor ether lipid analogs. Similarly, in epithelial carcinoma cell lines A431 and HeLa cell cultures, AEL also inhibit the phosphatidylinositol 3-kinase (PtdIns 3-Kinase)-Akt/PKB survival pathway in dose and time-dependent manner. This inhibition of the PtdIns 3-Kinase-Akt/PKB pathway by wortmannin or ALPs may be linked to the activation of proapoptotic SAPK/JNK pathway. It is stated that these processes may significantly contribute to inhibition of proliferation and induction of apoptotic cell death (Ruiter et al., 2003).

11.2.3 Effect of AEL on Cellular Receptors

It is well known that mitogens interact with extracellular receptors and initiate a cascade of intracellular signaling associated with cellular proliferation. As AEL interfere with the binding of mitogens with receptors, there have been many studies on AEL-mediated alterations in receptor linked signal transduction processes (Kosano and Takatani, 1988; Arthur and Bittman, 1998). In MCF-7 and ZR-75-1 cell lines, Et-18-OCH_3 downregulates the number of epidermal growth factor receptors (EGF) (Kosano and Takatani, 1988), but has no effect on the affinity of the ligand for the EGF receptor. It is also shown that labeled EGF is taken up by the cells in a temperature-dependent manner and Et-18-OCH_3 blocks the internalization of EGF receptor only in Et-18-OCH_3-sensitive human breast cancer cell lines. It is proposed that the inhibition of the internalization process for EGF may be one of the modes of antitumoral action of Et-18-OCH_3 (Kosano and Takatani, 1989; Kosano et al., 1990). Et-18-OCH_3 also modulates the uptake of estrogen, the secretion of transforming growth factor-α (TGF- α), and the content of progesterone receptors in the hormone-dependent breast cancer cell line, MCF-7. This AEL decreases the uptake of labeled estradiol by MCF-7 in a dose-dependent manner, and this decrease occurs prior to the onset of the inhibitory action of Et-18-OCH_3 on MCF-7 growth. Scatchard analysis indicates that Et-18-OCH_3 downregulates the number of estrogen receptors in MCF-7 without affecting their affinity. Both the secretion of TGF-α from MCF-7 into the conditioned medium and the progesterone receptor content of MCF-7 are decreased by 48 h treatment. The estradiol uptake, the TGF-α secretion, and the progesterone receptor content are not affected by platelet-activating factor, lyso-PAF, and palmitoyl-lysoPtdCho. These results suggest that the reduction of estrogen receptor level mediated by Et-18-OCH_3 cause decrease in both the secretion of TGF-α and the content of progesterone receptor in MCF-7. In addition, these effects are specific to Et-18-OCH_3. It is proposed that the effects of Et-18-OCH_3 may lead, at least partly, to its antitumor action in hormone-dependent breast cancer cell lines. Et-18-OCH_3 also upregulates both CD71 and CD11b, but have no effect on the expression of CD82 in U-937 cells (Pushkareva et al., 2000).

In U937, HL60, and KG-1 leukemic cells, AEL inhibits granulocyte-macrophage colony-stimulating factor (GM-CSF) binding to its receptor in a dose-dependent manner (Nicola, 1989; Shoji et al., 1994). Thus, Et-18-OCH_3 significantly decreases the total uptake, surface binding, and internalization of GM-CSF in a dose-dependent manner (Shoji et al., 1994). The internalization of GM-CSF is more profoundly inhibited by Et-18-OCH_3 than GM-CSF activity or its surface binding. TPA also prevents GM-CSF binding to its receptor. Inhibition of GM-CSF binding by a combination of Et-18-OCH_3 and TPA is less than additive and Et-18-OCH_3 partially blocks TPA-mediated PKC depletion in the cytosol and translocation to the particulate fractions. It is suggested that the inhibition of GM-CSF binding by Et-18-OCH_3 is partly due to disruption of the plasma membrane and that the inhibition of GM-CSF binding by TPA is due to activation of PKC.

AEL also increase the number of transferring receptors and the affinity of transferring for its receptors in human breast cancer cell lines, MCF-7, ZR-75-1 and BT-20, and mouse fibroblast, Balb/c 3T3 (Kosano and Takatani, 1990; Kosano et al., 1990). This increase is specific for Et-18-OCH_3 and is not duplicated by lyso-PtdCho, lyso-PAF, or PAF. The relationship between increased expression of transferring receptors and inhibition of growth by Et-18-OCH_3 is not understood. However, Et-18-OCH_3-induced modulation of transferring receptor-mediated signal transduction process may contribute to cytotoxicity in human breast cancer cell lines. Collectively, these studies suggest that interactions between membrane receptors and Et-18-OCH_3

may cause membrane perturbations, which may either result in alterations in the conformation of receptor or loss of agonist-binding sites on the membrane surface (Shoji et al., 1994). Factor that determines and modulates the specificity of the above membrane receptors and Et-18-OCH_3 binding still remains unknown. Thus, AEL modulate many receptors that are associated with cellular growth in tumor cell lines.

11.2.4 Other Effects of AEL on Cellular Metabolism

Biochemically, AEL also block de novo phosphatidylcholine synthesis by inhibiting CTP: phosphocholine cytidylyltransferase and promoting its translocation to membranes (Tronchere et al., 1991; van der Luit et al., 2002). AEL have no effect on choline kinase or choline phosphotransferase activities. AEL exert a strong inhibitory effect on membranous or solubilized asialomucin-sialyltransferase activity (Bador et al 1983) (Table 11.1). The inhibitory effect of AEL depends not only on the presence of the choline moiety at *sn*-3 position of the glycerol moiety but also on the presence of long ether-linked aliphatic side chain at the *sn*-1 position. The absence of any large substituent at the *sn*-2 position also dictates the degree of inhibition. For example, 1-octadecyl-2-*O*-methyl-glycero-3-phosphorylcholine acts as a mixed-type inhibitor. This inhibition of sialyltransferase activity may contribute an additional factor to the destructive effects of AEL on tumor cells (Bador et al., 1983).

Et-18-O-CH_3 also induces COX-2 expression and apoptotic cell death in H-ras transformed human breast epithelial cells (MCF10A-ras) (Na et al., 2005). The addition of a selective COX-2 inhibitor SC-58635 and COX-2 gene knock down with the siRNA prevents Et-18-O-CH_3-mediated apoptosis suggesting that COX-2 induction by Et-18-OCH_3 is probably linked to apoptotic cell death. Et-18-O-CH_3 also promotes the transcriptional activities of cyclic AMP response element, which is a key regulator of COX-2 expression. Et-18-O-CH_3 treatment produces the elevated release of 15d-PGJ_2 and DNA binding and transcriptional activity of PPARγ. On the basis of these findings, it is suggested that Et-18-O-CH_3 mediates COX-2 expression and production of 15d-PGJ_2, and this process may be associated with apoptosis in MCF10A-ras cells (Na et al., 2005).

Incubation of HL60 and K562 cells with AEL under noncytotoxic conditions induces significant membrane fluidization, which is related to the membrane cholesterol levels. HL60 cells that are sensitive to the cytotoxic action of AEL have lower basal cholesterol content. The loading of cholesterol in HL60 cells results in the decrease of Na^+, K^+-ATPase activity compared with that of untreated cells (Diomede et al., 1992). In contrast, cholesterol-deprived K562 cells have twice the Na^+, K^+-ATPase activity of unmodified K562 cells. AEL treatment stimulates Na^+, K^+, and Mg^{2+}-ATPase activities at very low concentration. This stimulation is greater in cells that are rich in cholesterol, such as K562 cells and cholesterol-enriched HL60 cells. In contrast, Na^+, K^+-ATPase in both cell lines is inhibited by AEL at high concentration regardless of the cholesterol content. Mg^{2+}-ATPase activity is neither related to cell cholesterol content nor affected by AEL (Diomede et al., 1992).

Noncytotoxic concentrations of Et-18-OCH_3 not only induce increase in the amount of cell membrane-bound hsp70 on leukemic K562 cells, but also in freshly isolated bone marrow of chronic myelogeneous leukemia patients (Botzler et al., 1999). Et-18-OCH_3 has no effect on peripheral blood lymphocytes or CD34 + hematopoietic progenitor cells of healthy human individuals. The increased hsp70 membrane expression on leukemic K562 cells results in a significantly increased sensitivity to lysis mediated by natural killer cells (Table 11.2). In contrast, the lysis of peripheral blood lymphocytes and CD34 + progenitor cells that lack expression of hsp70 on their plasma membrane is not influenced by AEL treatment suggesting that AEL-mediated expression of hsp70 may be related to tumor-selective immune responses in chronic myelogeneous leukemia patients (Botzler et al., 1999).

Et-18-OCH_3 induces a rapid and transient increase of c-fos and zif/268 mRNA level in astroglial cells (Table 11.2). Pretreatment of astroglial cells with the PAF antagonist BN50730 prior to the addition of Et-18-OCH_3 completely blocks the activation of the immediate early genes. These results support the view that astroglial cells are a cellular target for Et-18-OCH_3 and like macrophages respond to this methoxy-analog. Similar observation has also been made on the expression of c-fos and c-jun in human leukemic cells (Mollinedo et al., 1994).

AEL are active against HIV (Carballeira, 2002). They can be used along with azidothymidine (AZT) to inhibit the replication of HIV. AEL also possess antileishmanial activity (Unger et al., 1998; Rakotomanga et al., 2007) with a cure rate of 88–100%. The use of AEL has also been recommended for multiple sclerosis, autoimmune and inflammatory diseases, psoriasis, and viral infections (Mollinedo, 2007). Thus besides cancer, AEL can also be used for the treatment of AIDS, leishmaniosis, multiple sclerosis, and autoimmune diseases.

11.3 Molecular Mechanism and Site of Action of AEL

Although the primary target of AEL is plasma membranes, molecular mechanisms of cytotoxic action of AEL remain unclear. As stated earlier AEL do not target the DNA, but act at the level of cellular membranes. They interact with lipid rafts of tumor cells, followed by coaggregation with Fas death receptor (APO-1 or CD95) and recruitment of apoptotic molecules into Fas-enriched rafts (Gajate et al., 2004; Mollinedo et al., 2004; Mollinedo, 2007). AEL sensitivity depends on its uptake and Fas expression, regardless of the presence of other major death receptors, such as tumor necrosis factor (TNF) receptor 1 or TNF-related apoptosis-inducing ligand R2/DR5, in the target cell. Partial deletion of the Fas intracellular domain blocks apoptosis. Unlike normal lymphocytes, leukemic T cells incorporate AEL into rafts coaggregating with Fas and undergo apoptosis. Fas-associated death domain protein, procaspase-8, procaspase-10, c-Jun amino-terminal kinase, and Bid are recruited into rafts, linking Fas and mitochondrial signaling routes (Gajate and Molinedo, 2005; Gajate and Mollinedo, 2007). Blocking Fas/FasL interaction partially inhibit AEL-induced apoptotic cell death. Actin-linking proteins, ezrin, moesin, RhoA, and RhoGDI are conveyed into Fas-enriched rafts in AEL-treated leukemic cells. Disruption of lipid rafts and interference with actin cytoskeleton blocks Fas clustering and apoptotic cell death (Gajate and Mollinedo, 2007). Although clustering of rafts is necessary, but not sufficient for Et-18-OCH_3-mediated cell death, Fas is required for triggering the apoptotic cell death. Et-18-OCH_3-mediated apoptosis does not require sphingomyelinase activation. Normal cells, including human and rat hepatocytes, do not incorporate Et-18-OCH_3 and are spared from AEL-mediated apoptotic cell death (Gajate et al 2004; Mollinedo et al., 2004).

11.4 Conclusion

AEL interact with cellular membranes. They not only promote immunomodulation, but also modulate many signal transduction pathways associated with inhibition of cell proliferation and apoptotic cell death in tumor cells. AEL modulate phospholipases, protein and lipid kinases, cycloxygenases, and sialyltransferases. Many of these enzymes are simultaneously involved in activation of gene transcription through NF-κB. These genes are closely associated with oxidative stress and tumorogenesis. As AEL do not act through receptors, the specificity of their effects on signal transduction processes and gene transcription remains illusive. Many authors claim that the actions of AEL are nonspecific and may vary not only from one AEL to another, but also from one cell type to another. Effect of AEL also depends upon the way in which AEL has been presented to the enzyme. Inhibition is observed when AEL is presented as mixed micelles, whereas stimulation is reported when AEL incorporated into cell membrane prior to determination of enzymic activity. AEL enhance host defense mechanisms against tumors. Their major antitumor action lies in a direct effect on cancer cells. Even if the AEL effects are nonspecific, they may involve downstream signaling pathways associated with tumor formation (such as CD95) at the plasma membrane of target tumor cells. Recent progress has provided good evidence on the molecular mechanism associated with Et-18-OCH_3-mediated apoptotic cell death. Et-18-OCH_3 selectively incorporates into tumor cell membranes and causes cell death by intracellular activation of the cell death receptor Fas/CD95. This intracellular Fas/CD95 activation is a novel mechanism of action for an antitumor drug and represents a new way to target tumor cells in cancer chemotherapy.

References

Arthur G. and Bittman R. (1998). The inhibition of cell signaling pathways by antitumor ether lipids. Biochim. Biophys. Acta Lipids Lipid Metab. 1390:85–102.

Bador H., Morelis R., and Louisot P. (1983). Biochemical evidence for the role of alkyllysophospholipids on liver sialyltransferase. Int. J. Biochem. 15:1137–1142.

Bazill G. W. and Dexter T. M. (1990). Role of endocytosis in the action of ether lipids on WEHI-3B, HL60, and FDCP-mix A4 cells. Cancer Res. 50:7505–7512.

Berggren M. I., Gallegos A., Dressler L. A., Modest E. J., and Powis G. (1993). Inhibition of the signalling enzyme phosphatidylinositol-3-kinase by antitumor ether lipid analogues. Cancer Res. 53:4297–4302.

Berkovic D. (1998). Cytotoxic etherphospholipid analogues. Gen. Pharmacol. 31:511–517.

Berkovic D., Luders S., Goeckenjan M., Hiddemann W., and Fleer E. A. (1997). Differential regulation of phospholipase A_2 in human leukemia cells by the etherphospholipid analogue hexadecylphosphocholine. Biochem. Pharmacol. 53:1725–1733.

Botzler C., Ellwart J., Gunther W., Eissner G., and Multhoff G. (1999). Synergistic effects of heat and ET-18-OCH_3 on membrane expression of hsp70 and lysis of leukemic K562 cells. Exp Hematol. 27:470–478.

Carballeira N. M. (2002). New advances in the chemistry of methoxylated lipids. Prog. Lipid Res. 41:437–456.

Cardile V., Jiang X., Renis M., and Bindoni M. (1996). Effects of ether lipid 1-*O*-octadecyl-2-methoxy-rac-glycero-3-phosphocholine and its analogs PAF and CPAF on the release of nitric oxide in primary cultures of rat astrocytes. Brain Res. 715:98–103.

Conesa-Zamora P., Mollinedo F., Corbalán-García S., and Gómez-Fernández J. C. (2005). A comparative study of the effect of the antineoplastic ether lipid 1-*O*-octadecyl-2-*O*-methyl-glycero-3-phosphocholine and some homologous compounds on PKCα and PKCε. Biochim. Biophys. Acta Mol. Cell Biol. Lipids 1687:110–119.

Daniel L.W., Civoli F., Rogers M.A., Smitherman P.K., Raju P.A., Roederer M. (1995). ET-18-OCH_3 inhibits nuclear factor-kappa B activation by 12-*O*-tetradecanoylphorbol-13-acetate but not by tumor necrosis factor-alpha or interleukin 1 alpha. Cancer Res. 55:4844–4849.

Dell'Albani P., Condorelli D.F., Mudo G., Amico C., Bindoni M., and Belluardo N. (1993). Platelet-activating factor and its methoxy-analogue ET-18-OCH_3 stimulate immediate early gene expression in rat astroglial cultures. Neurochem. Int. 22:567–574.

Diomede L., Bianchi R., Modest E.J., Piovani B., Bubba F., and Salmona M. (1992). Modulation of ATPase activity by cholesterol and synthetic ether lipids in leukemic cells. Biochem Pharmacol. 43:803–807.

Fischer A., Muller D., Zimmermann-Kordmann M., Kleuser B., Mickeleit M., Laabs S., Lowe W., Cantagrel F., Reutter W., and Danker K. (2006). The ether lipid inositol-C2-PAF is a potent inhibitor of cell proliferation in HaCaT cells. Chembiochem 7:441–449.

Gajate C., Del Canto-Janez E., Acuna Au, Veldman R.J., and Molinedo F. (2004). Intracellular triggering of Fas aggregation and recruitment of apoptotic molecules into Fas-enriched rafts in selective tumor cell apoptosis.J. Exp. Med. 200:353–365.

Gajate C., Santos-Beneit A.M., Macho A., Lazaro M., Hernandez-De Rojas A., Modolell M., Munoz E., and Mollinedo F. (2000). Involvement of mitochondria and caspase-3 in ET-18-OCH_3-induced apoptosis of human leukemic cells. Int. J. Cancer 86:208–218.

Gajate C. and Molinedo F. (2005). Cytoskeleton-mediated death receptor and ligand concentration in lipid rafts forms apoptosis-promoting clusters in cancer chemotherapy. J. Biol. Chem. 280:11641–11647.

Gajate C. and Mollinedo F. (2007). Edelfosine and perifosine induce selective apoptosis in multiple myeloma by recruitment of death receptors and downstream signaling molecules into lipid rafts. Blood 109:711–719.

Grosman N. (2001). Similar effects of ether phospholipids, PAF and lyso-PAF on the Ca^{2+}-ATPase activity of rat brain synaptosomes and leukocyte membranes. Int. Immunopharmacol. 1:1321–1329.

Hanson P.K., Malone L., Birchmore J., and Nichols J. (2003). Lem3p is essential for the uptake and potency of alkylphosphocholine drugs, edelfosine and miltefosine. J. Biol. Chem. 278:36041–36050.

Helfman D.M., Barnes K.C., Kinkade J.M., Jr., Vogler W.R., Shoji M., and Kuo J.F. (1983). Phospholipid-sensitive Ca^{2+}-dependent protein phosphorylation system in various types of leukemic cells from human patients and in human leukemic cell lines HL60 and K562, and its inhibition by alkyl-lysophospholipid. Cancer Res. 43:2955–2961.

Heesebeen E.C., Verdonck L.F., Hermans S.W.G., Van Heugten H.G., Staal G.E.J., and Rijksen G. (1991). Alkyllysophospholipid ET-18-OCH_3 acts as an activator of protein kinase C in HL-60 cells. FEBS Lett. 290:231–234.

Herrmann D.B.J. and Neumann H.A. (1986). Cytotoxic ether phospholipids. Different affinities to lysophosphocholine acyltransferases in sensitive and resistant cells. J. Biol. Chem. 261:7742–7747.

Jiménez-López J. M., Carrasco M. P., Segovia J. L., and Marco C. (2004). Hexadecylphosphocholine inhibits phosphatidylcholine synthesis via both the methylation of phosphatidylethanolamine and CDP-choline pathways in HepG2 cells. Int. J. Biochem. Cell Biol. 36:153–161.

Kelley E.E., Modest E.J., and Burns C.P. (1993). Unidirectional membrane uptake of the ether lipid antineoplastic agent edelfosine by L1210 cells. Biochem. Pharmacol. 45:2435–2439.

Kosano H. and Takatani O. (1988). Reduction of epidermal growth factor binding in human breast cancer cell lines by an alkyl-lysophospholipid. Cancer Res. 48:6033–6036.

Kosano H. and Takatani O. (1989). Inhibition by an alkyl-lysophospholipid of the uptake of epidermal growth factor in human breast cancer cell lines in relation to epidermal growth factor internalization. Cancer Res. 49:2868–2870.

Kosano H. and Takatani O. (1990). Increase of transferrin binding induced by an alkyl-lysophospholipid in breast cancer cells. J. Lipid Mediat. 2:117–121.

Kosano H., Yasutomo Y., Kugai N., Nagata N., Inagaki H., Tanaka S., and Takatani O. (1990). Inhibition of estradiol uptake and transforming growth factor alpha secretion in human breast cancer cell line MCF-7 by an alkyl-lysophospholipid. Cancer Res. 50:3172–3175.

Lozano J., Berra E., Municio M.M., Diaz-Meco M.T., Domiguez I., Sanz L., and Moscat J. (1994). Protein kinase C zeta isoform is critical for kappa B-dependent promoter activation by sphingomyelinase. J. Biol. Chem. 269:19200–19202.

Lucas L., Hernandez-Alcoceba R., Penalva V., and Lacal J. C. (2001). Modulation of phospholipase D by hexadecylphosphorylcholine: a putative novel mechanism for its antitumoral activity. Oncogene 20:1110–1117.

Ménez C., Buyse M., Farinotti R., and Barratt G. (2007). Inward translocation of the phospholipid analogue miltefosine across Caco-2 cell membranes exhibits characteristics of a carrier-mediated process. Lipids 42:229–240.

Mollinedo F. (2007). Antitumour ether liquids: Proapoptotic agents with multiple therapeutic indications. Expert Opin. Ther. Patents 17:385–405.

Mollinedo F., Gajate C., Martin-Santamaria S., and Gago F. (2004). ET-18-OCH3 (Edelfosine): A selective antitumour lipid targeting apoptosis through intracellular activation of Fas/CD95 death receptor. Curr. Medicinal Chem. 11:3163–3184.

Mollinedo F., Gajate C., and Modolell M. (1994). The ether lipid 1-octadecyl-2-methyl-*rac*-glycero-3-phosphocholine induces expression of *fos* and *jun* proto-oncogenes and activates AP-1 transcription factor in human leukaemic cells. Biochem. J. 302:325–329.

Mollinedo F. and Gajate C. (2006). Fas/CD95 death receptor and lipid rafts: New targets for apoptosis-directed cancer therapy. Drug. Resist. Updat. 9:51–73.

Na H.K., Inoue H., Surh Y.J. (2005). ET-18-O-CH_3-induced apoptosis is causally linked to COX-2 upregulation in H-ras transformed human breast epithelial cells. FEBS Lett. 579:6279–6287.

Nicola N.A. (1989). Hemopoietic cell growth factors and their receptors. Annu. Rev. Biochem. 58:45–77.

Pawelczyk T. and Lowenstein J.M. (1993). Inhibition of phospholipase C delta by hexadecylphosphorylcholine and lysophospholipids with antitumor activity. Biochem. Pharmacol. 45:493–497.

Rerez-Victoria F.J., Gamarro F., Ouellette M., and Castanys S. (2003). Functional cloning of the miltefosine transporter. A novel P-type phospholipid translocase from *Leishmania* involved in drug resistance. J. Biol. Chem. 278:49965–49971.

Powis G., Seewald M.J., Gratas C., Melder D., Riebow J., and Modest E.J. (1992). Selective inhibition of phosphatidylinositol phospholipase C by cytotoxic ether lipid analogues. Cancer Res. 52:2835–2840.

Pushkareva M.Y., Wannberg S.L., Janoff A.S., and Mayhew E. (2000). Increased cell-surface receptor expression on U-937 cells induced by 1-*O*-octadecyl-2-*O*-methyl-sn-glycero-3-phosphocholine. Cancer Immunol. Immunother. 48:569–578.

Rakotomanga M., Blanc S., Gaudin K., Chaminade P., and Loiseau P.A. (2007). Miltefosine affects lipid metabolism in *Leishmania donovani* promastigotes. Antimicrob. Agents Chemother. 51:1425–1430.

Ruiter G.A., Zerp S.F., Bartelink H., van Blitterswijk W.J., and Verheij M. (1999). Alkyl-lysophospholipids activate the SAPK/JNK pathway and enhance radiation-induced apoptosis. Cancer Res. 59:2457–2463.

Ruiter G.A., Zerp S.F., Bartelink H., van Blitterswijk W.J., and Verheij M. (2003). Anti-cancer alkyl-lysophospholipids inhibit the phosphatidylinositol 3-kinase-Akt/PKB survival pathway. Anticancer Drugs 14:167–173.

Salari H., Dryden P., Dovenport R., Howard S., Jones K., and Bittman R. (1992). Inhibition of protein kinase C by ether-linked lipids is not correlated with their antineoplastic activity on WEHI-3B and R6X-B15 cells. Biochim. Biophys. Acta 1134: 81–88.

Samadder P., Richards C., Bittman R., Bhullar R.P., and Arthur G. (2003). The antitumor ether lipid 1-Q-octadecyl-2-*O*-methyl-rac-glycerophosphocholine (ET-18-OCH_3) inhibits the association between Ras and Raf-1. Anticancer Res. 3B:2291–2295.

Samadder P., Bittman R., Byun H.S., and Arthur G. (2004). Synthesis and use of novel ether phospholipid enantiomers to probe the molecular basis of the antitumor effects of alkyllysophospholipids: correlation of differential activation of c-Jun NH_2-terminal protein kinase with antiproliferative effects in neuronal tumor cells. J Med Chem. 47: 2710–2713.

Schallier D.C., Bruyneel E.A., Storme G.A., Hilgard P., Mareel M.M. (1991). Antiinvasive activity of hexadecylphosphocholine in vitro. Anticancer Res. 11:1285–1292.

Shoji M., Fukuhara T., Winton E.F., Berdel W.E., and Vogler W.R. (1994). Different mechanisms of inhibition by alkyl-lysophospholipid and phorbol ester of granulocyte-macrophage colony-stimulating factor binding to human leukemic cell lines. Exp. Hematol. 22:13–18.

Shmelzer Z., Haddad N., Admon E., Pessach I., Leto T. L., Eitan-Hazan Z., Hershfinkel M., and Levy R. (2003). Unique targeting of cytosolic phospholipase A_2 to plasma membranes mediated by the NADPH oxidase in phagocytes. J. Cell. Biol. 162:683–692.

Tronchere H., Terce F., Record M., Ribbes G., and Chap H. (1991). Modulation of CTP:phosphocholine cytidylyltransferase translocation by oleic acid and the antitumoral alkylphospholipid in HL-60 cells. Biochem. Biophys. Res. Commun. 176:157–165.

Uberall F., Oberhuber H., Maly K., Zaknun J., Demuth L., and Grunicke H.H. (1991). Hexadecylphosphocholine inhibits inositol phosphate formation and protein kinase C activity. Cancer Res. 51:807–812.

Unger C., Maniera T., Kaufmann-Kolle P., and Eibl H. (1998). In vivo antileishmanial activity of hexadecylphosphocholine and other alkylphosphocholines. Drugs Today 34(Suppl. F):133–140.

van der Luit A.H., Budde M., Ruurs P., Verheij M., and Van Blitterswijk W.J. (2002). Alkyl-lysophospholipid accumulates in lipid rafts and induces apoptosis via raft-dependent endocytosis and inhibition of phosphatidylcholine synthesis. J. Biol. Chem. 277:39541–39547.

Wieder T., Zhang Z., Geilen C.E., Orfaros A.E., Giuliano A.E., and Cabot M.C. (1996). The antitumor phospholipid analog, hexadecylphosphocholine, activates cellular phospholipase D. Cancer Lett. 100:71–79.

Zheng B., Oishi K., Shoji M., Eible H., Berdel W.E., Hajdu J., Vogler W.R., and Kou J.F. (1990). Inhibition of protein kinase C, (sodium plus potassium)-activated adenosine triphosphatase, and sodium pump by synthetic phospholipid analogues. Cancer Res. 50:3025–3031.

Zhou X., Lu X., Richard C., Xiong W., Litchfield D.W., Bittman R., and Arthur G. (1996). 1-*O*-octadecyl-2-*O*-methyl-glycerophosphocholine inhibits the transduction of growth signals via the MAPK cascade in cultured MCF-7 cells. J. Clin. Invest. 98:937–944.

Chapter 12
Perspective and Directions for Future Developments on Ether Lipids

12. 1 Introduction

Neural membranes contain glycerophospholipids, sphingolipids, cholesterol, and proteins. These lipids are asymmetrically distributed between the two leaflets of lipid bilayers (Ikeda et al., 2006; Yamaji-Hasegawa and Tsujimoto, 2006). Glycerophospholipids and sphingolipids contribute to the lipid asymmetry, while cholesterol and sphingolipids form lipid microdomains or lipid rafts. Glycerophospholipids are made up of glycerol backbone, fatty acids, phosphoric acid, and nitrogenous base. Depending on the substituent at the *sn*-1 position of glycerol moiety, glycerophospholipids are classified into two groups. One group is represented by glycerophospholipids that contain ester bond at the *sn*-1 position, and the other group is represented by glycerophospholipids that contain ether bond at the *sn*-1 position. Ester bond containing glycerophospholipids include phosphatidylcholine (PtdCho), phosphatidylethanolamine (PtdEtn), phosphatidylserine (PtdSer), and phosphatidylinositol (PtdIns), whereas ether bond containing glycerophospholipids include plasmalogens, platelet-activating factor (PAF) and its analogs (Farooqui and Horrocks, 2001). PtdCho is mainly located in the outer leaflet, whereas PtdSer, PtdEtn, and PtdIns are mainly located in the inner leaflet (Farooqui and Horrocks, 2007; Farooqui and Horrocks, 2008). Among ether lipids, choline plasmalogen (PlsCho) is located in the outer leaflet, whereas ethanolamine plasmalogen is mainly associated with the inner leaflet.

Among ether lipids, plasmalogens contain a vinyl ether (enol ether) linkage at the *sn*-1 position with palmitic, stearic, and oleic acid [16:0, 18:0, and 18:1 (n-7 and n-9) side-chains (alk-1-enyl groups)], an ester bond linking arachidonic acid or docosahexaenoic acid or another unsaturated fatty acid at the *sn*-2 position, and a phosphoethanolamine or phosphocholine group at the *sn*-3 position of the glycerol moiety. Plasmalogens play important roles in neural cells, not only in signal transduction processes, but also in membrane fusion and antioxidant activity. PAF contains an *O*-alkyl ether linkage at the *sn*-1 position (fatty alcohol side chain), a short acyl chain (acetyl moiety) at the *sn*-2 position, and a phosphocholine group at the *sn*-3 position of the glycerol moiety. PAF stimulates a wide range of biological responses ranging from aggregation and degranulation of platelets and neutrophils to a variety of other cellular effects such as the stimulation of chemotaxis, chemokinesis, superoxide formation, protein phosphorylation, generation of arachidonic acid and phosphoinositide metabolites, and tumor necrosis factor production (Snyder, 1995).

Sphingolipids consist of a sphingoid base, a straight-chain alcohol of 18–20 carbon atoms, which is normally attached to a long saturated fatty acid (usually palmitate) through an amide bond. Sphingosine contains a *trans* double bond between carbons C-4 and C-5. On the basis of their head group, sphingolipids include glycosphingolipids with a sugar as the head group, ceramide or *N*-acylsphingosine with no head group, and phosphosphingolipids (primarily sphingomyelin, a phosphodiester of ceramide and choline). Sphingomyelin is a major constituent of plasma membranes where it is concentrated in the outer leaflet (Vaena de Avalos et al., 2004). Sphingolipids are metabolized to ceramide and sphingosine. Exogenous application of ceramide is cytotoxic. Ceramide is metabolized to less toxic forms by glycosylation, acylation, or by catabolism to sphingosine, which is then phosphorylated to the antiapoptotic sphingosine 1-phosphate (Farooqui et al., 2007a, b; Farooqui et al., 2008).

Cholesterol, which accounts for 20–25% of the total body cholesterol in fresh brain, plays a crucial role in membrane organization, dynamics, function, and sorting. Cholesterol not only serves as a precursor for steroid hormones and regulates activities of membrane-bound enzymes (Simons and Ikonen, 2000), but also modulates endocytosis and antigen expression. Dynamic clustering of cholesterol along with sphingolipids results in the formation of specialized structures called microdomains or rafts. In neural membranes, raft formation occurs by self-association of sphingolipids via their long saturated hydrocarbon chains. Cholesterol condenses this packing by positioning between these hydrocarbon chains below the large head groups of the sphingolipids. These interactions lead to the formation of a less fluid, liquid-ordered phase, which is separate from a phosphatidylcholine-rich liquid-disordered phase (Simons and Ikonen, 2000).

Lipid rafts float within the membrane and certain groups of proteins unite within these rafts. They not only serve as mobile platforms for signal transduction, but also are associated with clustering and organizing bilayer constituents including receptors, enzymes, and ion-channels. The organization of glycerophospholipids, sphingolipids, and cholesterol in lipid rafts provides neural membranes with structural and functional integrity that facilitates appropriate interactions with integral membrane proteins. The function of the signal transduction network is to convey extracellular signals from the cell surface to the nucleus to induce a biological response at the gene level. The intensity of interactions among glycerophospholipids, sphingolipids, and cholesterol-derived lipid mediators not only modulates cellular function through signal transduction processes, but also adaptive responses (Ivanova et al., 2004). Collectively, these studies suggest that composition of lipid rafts and interactions among glycerophospholipids, sphingolipids, and cholesterol-derived lipid mediators play crucial roles in neural cell functions including signal transduction, adhesion, sorting, trafficking, and organizing bilayer constituents (Simons and Ikonen, 2000; Farooqui et al., 2000; Farooqui and Horrocks, 2007). Alterations in the lipid raft composition and increased levels of lipid mediators may be associated with various chronic human diseases such as cardiovascular disease, cancer, neuropsychiatric, and neurodegenerative diseases (Vigh et al., 2005) suggesting that neural membranes are not simply an inert physical barrier separating the inside from outside or compartments within cells, regulating passage of nutrients, gases, and specific ions, but complex, well organized, and highly specialized structures that are involved in receiving, processing, transporting, and transmitting information from the plasma membrane to the nucleus and other subcellular organelles through glycerophospholipids, sphingolipids, and cholesterol-derived lipid mediators.

12.2 Interactions Among Glycerophospholipid, Sphingolipid, and Cholesterol-Derived Lipid Mediators

Changes in glycerophospholipids, sphingolipids, and cholesterol levels cause alterations in membrane fluidity, lipid packing, and permeability. Glycerophospholipid-derived mediators include PAF, lysophospholipids, eicosanoids, and docosanoids (Farooqui and Horrocks, 2006; Farooqui et al., 2007a,b; Farooqui et al., 2008). Sphingolipid-derived lipid mediators include ceramide, ceramide 1-phosphate, sphingosine 1-phosphate (Farooqui et al., 2007a,b; Farooqui et al., 2008), and cholesterol metabolites include 7-ketocholesterol and 24-hydroxycholesterol (He et al., 2006). These lipid mediators modulate signal transduction, contribute to cellular differentiation, promote regeneration, and facilitate neuronal and glial cell integrity. Multiple studies suggest that cross talk among glycerophospholipid, sphingolipid, and cholesterol-derived lipid mediators occurs in neural cells (Kihara and Igarashi, 2004). Controlled generation and coordinated signaling through glycerophospholipids, sphingolipids, and cholesterol-derived lipid mediators has been proposed to play an important role in neural cell survival and neurodegeneration. Under normal conditions, the intensity of coordinated signaling among glycerophospholipids, sphingolipids, and cholesterol-derived lipid mediators not only varies significantly from one neural cell type to another, but also with respect to the nature of stimulus and its dosage and/or duration of treatment. Thus, an interplay or cross talk is necessary between lipid mediators derived from glycerophospholipids, sphingolipids, and cholesterol for neural cell proliferation, cell mobility, neurite retraction, and survival (Farooqui et al., 2007a; Farooqui et al., 2007b; Farooqui et al., 2008). High levels of these metabolites and abnormal signaling cause oxidative stress, inflammation, membrane blebbing, and other neurochemical and morphological changes that promote neural cell death. It should be noted here that neural cell injury and death is not the result of one well-defined signaling cascade, but the consequence of extensive cross talk between several neurochemical and molecular events at different cellular and subcellular levels. Collective evidence suggests that cross talk or interplay among mediators derived from sphingolipids, glycerophospholipids, and cholesterol is not only necessary for maintaining the functional lipid asymmetry of lipid bilayers in plasma membranes, but also necessary for modulating the intensity of signal transduction process associated with neural cell survival and neurodegeneration (Farooqui et al., 2000; Koletzko et al., 2001; Kihara and Igarashi, 2004).

12.3 Interactions Between Ether Lipid and Sphingolipid-Derived Lipid Mediators

Alterations in levels of ether lipids (plasmalogens and PAF), sphingolipids (ceramide and sphingosine), and cholesterol occur in neurological disorders (Farooqui and Horrocks, 2007; Farooqui and Horrocks, 2008). Changes in the above-mentioned neural membrane components are accompanied by increase in levels of lipid mediators such as PAF, eicosanoids, and docosanoids (Farooqui et al., 2007b), ceramide, ceramide 1-phosphate, sphingosine 1-phosphate (Farooqui et al., 2007a), and 7-ketocholesterol and 24-hydroxycholesterol (Yu et al., 2000; Farooqui et al., 2001; Park et al., 2000). Low levels of these lipid mediators play important roles in signal transduction, neural cell differentiation, and cell survival, whereas high levels are associated with neural cell cycle arrest, apoptosis, and neurodegeneration (Farooqui et al., 2001; Farooqui et al., 2007a; Farooqui et al., 2001; Park et al., 2000; Reiss et al., 2004).

Interplay or cross talk between PlsEtn-PLA_2-generated second messenger (arachidonic acid) and sphingomyelinase-generated second messenger (ceramide) occurs at several sites in cytokine-mediated signal transduction process (Fig. 12.1). These cytokines are generated and released as a result of neural cell injury. In neural and nonneural cells, cytokines stimulate both sphingomyelinase and PlsEtn-PLA_2 activities in a dose and time-dependent manner. PlsEtn-PLA_2 and sphingomyelinase inhibitors (Latorre et al., 2003; Farooqui et al., 2007a,) can block this stimulation. Arachidonic acid, a product of PlsEtn-PLA_2 catalyze reaction, stimulates sphingomyelinase (Robinson et al., 1997), and ceramide, a product of sphingomyelinase reaction, enhances PlsEtn-PLA_2 activity (Latorre et al., 2003), suggesting a cross talk between receptor-mediated ether lipid and sphingomyelin signaling pathways. Phosphorylated sphingolipid metabolites, sphingosine 1-phosphate and ceramide 1-phosphate, are required for the activation and translocation of cyclooxygenase-2 and $cPLA_2$ (Chalfant and Spiegel, 2005; Nodai et al., 2007; Stahelin et al., 2007), indicating that these metabolites of sphingolipid metabolism may act in concert to regulate generation of inflammatory mediators from neural membrane plasmalogens. Quinacrine (a nonspecific inhibitor of PlsEtn-PLA_2) and 1-*O*-octadecyl-2-methyl-*rac*-glycerol-3-phosphocholine (a CoA-independent transacylase inhibitor) also inhibit ceramide mediated-stimulation of PlsEtn-PLA_2 (Latorre et al., 2003). These observations support the view that a cross talk occurs between PlsEtn degradation and sphingomyelin catabolism through the generation of arachidonic acid and ceramide (Farooqui and Horrocks, 2001; Farooqui et al., 2000; Farooqui and Horrocks, 2007).

Ethanolamine plasmalogens act as an antioxidant against cholesterol oxidation (Farooqui and Horrocks, 2004; Maeba and Ueta, 2004). Other cholesterol metabolites

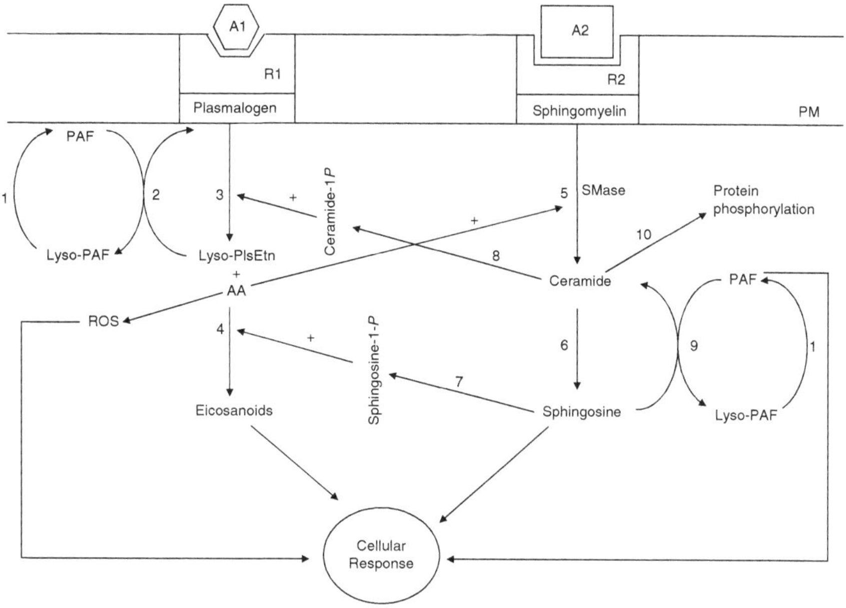

Fig. 12.1 Interactions among plasmalogen, PAF, and sphingolipid-derived lipid mediators. Acetyl CoA/lyso-PAF acetyltransferase (1); PAF/lysoplasmalogen transacetylase (2); plasmalogen-selective phospholipase A_2 (3); cyclooxygenase-2 (4); sphingomyelinase (5); ceramidase (6); sphingosine kinase (7); ceramide kinase (8); PAF/sphingosine transacetylase (9); protein kinaser (10); A1 and A2 (agonists); R1 and R2 (receptors); arachidonic acid (AA); reactive oxygen species (ROS); platelet-activating factor (PAF); lyso-platelet-activating factor (lyso-PAF); and *positive sign* (+) indicate stimulation.

such as 24-hydroxycholesterol and 7-ketocholesterol are also potent inducers of apoptotic cell death (Lizard et al., 2000; Kölsch et al., 1999). In addition, 7β-hydroxycholesterol and 7-ketocholesterol facilitate interleukin-1β secretion during apoptosis. Marked increases in lipid mediators derived from ether lipid, sphingolipids, and cholesterol occur in brain tissue (Table 12.1). The accumulation of these lipid mediators along with changes in the cellular redox status and lack of energy generation may be associated with neural cell injury and cell death in acute neural trauma (ischemia, epilepsy, head injury, and spinal cord trauma) and neurodegenerative diseases, such as Alzheimer disease (AD), Parkinson disease (PD), multiple sclerosis (MS), Creutzfeldt-Jakob disease (CJD), and AIDS dementia complex (Farooqui and Horrocks, 2006; Phillis et al., 2006; Farooqui et al., 2006; Farooqui and Horrocks, 2007; Farooqui et al., 2007a; Farooqui et al., 2008).

PAF stimulates the breakdown of sphingomyelin and the release of ceramide (Lang et al., 2005). In erythrocytes, this process is associated with apoptotic cell death. The molecular mechanism involved in apoptotic cell death in this system is not fully understood. However, PAF may activate an erythrocyte sphingomyelinase, and the generated ceramide leads to the activation of scramblase with subsequent

Table 12.1 Alterations in levels of glycerophospholipid, sphingolipid, and cholesterol-derived lipid mediators in ischemia and Alzheimer disease.

Lipid mediator	Ischemia	Alzheimer disease	References
Arachidonic acid	Increased	Increased	Farooqui et al., 2001; Farooqui and Horrocks, 2007
Eicosanoids	Increased	Increased	Farooqui et al., 2001; Farooqui and Horrocks, 2007
Ceramide	Increased	Increased	Guan et al., 2006; Yu et al., 2000
7-β-Hydroxy-cholesterol	–	–	Chang et al., 1998
7-Keto-cholesterol	–	–	Chang et al., 1998
24-Hydroxy-cholesterol	No effect	Increased	Bogdanovic et al., 2001; Reiss et al., 2004

phosphatidylserine exposure, a process associated with apoptotic cell death (Lang et al., 2005). The exposure of phosphatidylserine on cellular surface is determined by annexin-binding. The increase in annexin-binding is not only downregulated in genetic knockout of PAF receptors by the PAF receptor antagonist, but also by the inhibition of sphingomyelinase with urea (Lang et al., 2005). Sphingosine 1-phosphate, which is generated by the action of ceramidase on ceramide, stimulates PAF synthesis in endothelial cells in dose and time-dependent manner (Bernatchez et al., 2003). This response is associated with endothelial cell migration. It is attenuated by inhibiting p38 mitogen-activated protein kinase (MAPK), $cPLA_2$, $sPLA_2$ activities. It is suggested that p38 MAPK activation by sphingosine 1-phosphate promotes the conversion of membrane phospholipids into PAF through the combined activation of $cPLA_2$ and $sPLA_2$. Pretreatment of bovine aortic endothelial cells with extracellular PAF receptor antagonists, BN52021 and CV3988, reduces the cellular migration induced by sphingosine 1-phosphate (Bernatchez et al., 2003). Thus PAF, a phospholipid-derived lipid mediator, modulates the generation of ceramide and sphingosine 1-phosphate. Similarly, ceramide and sphingosine 1-phosphate, the lipid mediators of sphingolipid metabolism, modulate PAF synthesis. This interplay between PAF and sphingosine 1-phosphate constitutes cross talk between phospholipid and sphingolipid metabolism.

The transfer of the acetyl group from PAF in a CoA-independent manner to lysoplasmalogen and sphingosine is catalyzed by PAF/lysoplasmalogen transacetylase and PAF/sphingosine transacetylase activities, respectively (Lee et al., 1996; Karasawa et al., 1999; Bae et al., 2000). Purification of these enzymic acitivities from rat kidney mitochondrial and microsomal membranes indicates that one enzyme catalyzes transacetylation of the acetyl group from PAF to lysoplasmalogen forming plasmalogen analogs of PAF and to sphingosine producing *N*-acetylsphingosine (C_2-ceramide). Ceramide, sphingosylphosphocholine, stearylamine, sphingosine 1-phosphate, or sphingomyelin are not substrates, whereas sphinganine has a limited capacity to accept the acetyl group from PAF. This observation suggests that PAF-dependent transacetylase acts as a link between ether lipid and sphingolipid metabolism. In addition, this enzyme acts as a PAF-acetyl hydrolase in the absence of lipid acceptor molecules (Lee et al., 1996; Karasawa et al., 1999). The amino acid sequence of PAF-dependent transacetylase is similar to that of type II PAF-acetyl hydrolase (Bae et al., 2000). The monoclonal antibody to recombinant type II PAF-acetyl hydrolase cross-reacts with both cytosolic and membrane bound PAF-dependent transacetylases, suggesting that both activities reside in one protein. In contrast, recombinant PAF-AH Ib, which is a different subtype of intracellular PAF-acetyl hydrolase has no PAF-dependent transacetylase activity. Analysis of a series of site-directed mutant type II PAF-acetyl hydrolase proteins shows that PAF-dependent transacetylase activity is decreased, whereas type II PAF-acetyl hydrolase activity is not affected in C120S and G2A mutant proteins. On the basis of these studies, Cys^{120} and Gly^{2} are implicated in the reaction catalyzed by PAF-dependent transacetylase. In CHO-K1 cells transfected with type II PAF-acetyl hydrolase gene, the transfer of acetate from PAF to endogenous acceptor lipids is significantly increased in a time-dependent manner indicating that type II PAF-acetyl hydrolase can function as a PAF-dependent transacetylase in intact cells, and type II PAF-acetyhydrolase and PAF-dependent transacetylase are the same enzyme (Bae et al., 2000).

Studies on the determination of C_2-ceramide levels in Pex5(–/–) mice, a model for Zellweger syndrome, in which the synthesis of ether lipids such as PAF is impaired, indicate that in Pex5(–/–) mice, C_2-ceramide levels in various tissues do not differ significantly in tissues from control mice, suggesting that there is no link between ether lipid metabolism and sphingolipid metabolism (Van Overloop et al., 2007). This is in contrast to earlier studies (Lee et al., 1996; Karasawa et al., 1999). Thus more studies are required on this aspect of ether and sphingolipid metabolism.

12.4 Interactions Between Sphingolipid and Cholesterol-Derived Lipid Mediators

Interactions between sphingolipid and cholesterol metabolism modulate amyloid precursor protein (APP) processing. This process may be closely associated with the pathogenesis of AD (Yanagisawa, 2002; Kirsch et al., 2002; Grimm et al., 2005). Oxysterols, the lipid mediators of cholesterol metabolism, exert tight control over neural cell cholesterol trafficking by altering cholesterol influx/efflux. They not only modulate the generation of Aβ but also interact with lipid metabolites of glycerophospholipid and sphingolipid metabolism (Cuzzocrea and Salvemini, 2007; Farooqui and Horrocks, 2008). Both Aβ and APP oxidize cholesterol to form 7β-hydroxycholesterol, a proapoptotic oxysterol, which is neurotoxic at nanomolar concentrations. 7β-Hydroxycholesterol retards secretion of soluble APP from cultured rat hippocampal H19–7/IGF-IR neuronal cells and inhibits α-secretase activity, but has no effect on β-site APP-cleaving enzyme 1 activity (Nelson and Alkon, 2005). 7β-Hydroxycholesterol also inhibits protein kinase Cα activity. Oxidation of cholesterol is accompanied by the stoichiometric production of hydrogen peroxide and requires divalent copper. These results suggest that a function of APP may be to produce low levels of 7-β-hydroxycholesterol. The higher levels of 7-β-hydroxycholesterol produced by Aβ may contribute to the oxidative stress and neural cell loss observed in AD (Nelson and Alkon, 2005).

12.5 Use of Lipidomics, Proteomics, and Genomics for Characterization of Enzymes, Lipid Mediators, and Signal Transduction Process in Normal and Diseased Brain Tissues

The mapping of neural membrane structural components and their relation to neural cell stimulation under normal and pathologic situations involve a comprehensive approach to appreciate the interrelationship among various lipid mediators and their interactions with complex networks associated neurons and glial cells. Lipidomics has emerged as an important procedure not only for the comprehensive identification and full characterization of molecular species of glycerophospholipids and sphingolipids, and their metabolic regulation, but also for the determination and characterization of levels of their lipid mediators in normal brain as well as brain from patients with neurological disorders (Serhan, 2005; Piomelli, 2005; Guan et al., 2006; Wenk, 2005; Milne et al., 2006; Dennis et al., 2006; Han and Gross, 2005). Alterations in lipid mediators can be identified either by accurately quantifying their masses through lipidomics or by their comprehensive characterization by shotgun lipidomics and multidimensional mass spectrometry (Serhan, 2005; Piomelli, 2005; Adibhatla et al., 2006; Fonteh et al 2006). Collectively, these studies suggest that lipid mediator informatics is an important emerging field devoted to studies on the identification and characterization of bioactive lipid mediators and their biosynthetic profiles and pathways (Han and Gross, 2005; Lu et al., 2006; Adibhatla et al., 2006; Fonteh et al., 2006). Lipid mediator informatics and proteomics applied to the brain tissue undergoing through excitotoxic, inflammatory, and oxidative stress insults provide a powerful means of uncovering key intermediates and biomarkers associated with healthy and diseased brain tissue. Lipid mediator informatics employing liquid chromatography-ultraviolet-tandem mass spectrometry, gas chromatography-mass spectrometry, computer-based automated systems equipped with databases and novel searching algorithms, and enzyme-linked immunosorbent assay (ELISA) to evaluate and profile temporal and spatial production of mediators combined with proteomics at defined points during excitotoxic, inflammatory, and oxidative stress insult can enable researchers to identify novel unknown lipid mediators that contribute to pathogenesis of neurological disorders (Farooqui et al., 2008). It is proposed that the automated system including databases and searching algorithms is crucial for facilitating accurate analysis of lipid mediators such as eicosanoids, isoprostanes, isofurans, resolvins, neuroprotectins, and neuroprostanes (Lu et al., 2006). These lipid mediators play crucial roles in human neurological disorders associated with excitotoxicity, inflammation, and oxidative stress (Farooqui et al., 2007b; Alcon et al., 2002).

Synaptic function and signal transduction processes depend on the temporal and spatially coordinated interactions between lipid mediators with their receptors and on their interrelationships with the organization of lipid mediator network. Recent advances in shotgun lipidomics and multidimensional mass spectrometry can provide new insights into the roles of molecular species of glycerophospholipids and sphingolipids in neuronal function (Gross et al., 2005; Han and Gross, 2005; Han, 2007). The sensitivity and robustness of neurolipidomics can be improved with detection instruments without compromising the accuracy and specificity of molecular species identification.

Establishment of automatic systems including databases and accurate analyses of glycerophospholipids, sphingolipids, and cholesterol-derived lipid mediators can facilitate the identification of key biomarkers associated with neurodegenerative diseases (Serhan, 2005; Lu et al., 2006; Dennis et al., 2006). A comprehensive neurolipidomics analysis using liquid chromatography-tandem mass spectrometry prostaglandins in rat hippocampal tissue indicates that kainic acid-mediated neurodegeneration is accompanied by the generation of large amounts of $PGF_2\alpha$ and PGD_2 and smaller amounts of other prostaglandins and hydroxyeicosatetraenoic acids (Guan et al., 2006; Yoshikawa et al., 2006). This increase can be blocked by intracerebroventricular administration of kainic acid receptor antagonists. Neurolipidomics can also be used for analyzing small tissue and biological fluid samples from patients with neurodegenerative diseases (Butterfield et al., 2006; Han, 2007). Microarray analysis of tissue samples from brain regions associated with neurodegenerative diseases can provide information on candidate genes that influence oxidative stress, neuroinflammation, and neurodegeneration. This would not only help in understanding molecular mechanisms associated with the development of neurodegenerative diseases, but would also facilitate molecular diagnostics and targets for drug therapy on the basis of gene expression in brain tissue as well as CSF and blood (Facheris et al., 2004). With the help of neurolipidomics, proteomics, and genomics data, one can identify specific genes involved in regulation of biosynthesis of individual molecular species of glycerophospholipids and sphingolipids as well as genes that are related to their sorting and transport (Voelker, 2003; Lee et al., 2005; Forrester et al., 2004).

Collective evidence suggests that clinical neurolipidomics, proteomics, and genomics can provide information on lipid and protein expression profiles of clinical samples. This may facilitate in identification of disease-associated biomarkers that may be useful in gaining insight into the biology and pathology of disease processes. The discovery of lipid mediators and biomarker for diagnosis and prognosis of neurological disorders is essential for understanding the underlying mechanism and factors that are involved in modulating onset, progression, or severity associated with the pathogenesis of human neurological disorders (Han, 2007).

Another technique that can provide useful information on rates of glycerophospholipid, sphingolipid, and cholesterol metabolic pathways is quantitative autoradiography or positron emission tomography (PET) (Rapoport, 1999; Rapoport, 2005). Developing a kinetic strategy to examine rates of glycerophospholipid, sphingolipid, and cholesterol metabolic pathways can help not only in elucidating the role of these lipids, but also the rate of generation of their mediators in normal and diseased brain. This procedure requires intravenous injection of a radioactivity-labeled substrate that can incorporate into a lipid metabolic pathway. Using quantitative autoradiography or PET, one can determine tracer distribution of radioactive metabolites in different regions of brain as a function of time. From these measurements, fluxes, turnover rates, half-lives, and ATP consumption rates can be calculated and incorporation rates can be imaged (Rapoport, 2005). On the basis of the in vivo metabolism of glycerophospholipid, sphingolipid, and cholesterol-derived lipid mediators, PET can be used to image brain signaling and neuroplasticity in normal human brain and brain from patients with neurodegenerative diseases (Masters et al., 2006). Initial experiments on animal models of Alzheimer and Parkinson diseases with chronic unilateral lesions of nucleus basalis or substantia nigra indicate that PET and [^{11}C]arachidonic acid can be used with drug activation to image signal transduction (Rapoport, 2005; Esposito et al., 2007). Using inositol phospholipid metabolism as an index, proton magnetic resonance spectroscopy and PET can provide additional information in differentiating between Alzheimer disease, subcortical ischemic vascular dementia, and subjective cognitive impairment. Therefore, this method can contribute to the routine diagnosis of dementia. Psychiatric and behavioral symptoms associated with dementia or due to major psychiatric disorders cannot be related to changes in the measured proton magnetic resonance spectroscopy parameters (Kondo et al., 2002; Watanabe et al., 2002). Detailed investigations are required on the use of lipid mediators and MRI along with PET imaging to judge the severity and progression of dementia, during the course of various neurodegenerative diseases in patients and normal human subjects (Rapoport, 2001; Hampel et al., 2002; Rapoport, 2005; Esposito et al., 2007).

Only a few enzymes of plasmalogen and PAF metabolism have been purified, characterized, and cloned from the brain tissue (Hirashima et al., 1992; Shindou et al., 2007; Nomikos et al., 2003). Proteomics and molecular biological approaches such as cloning the cDNA for enzymes synthesizing and degrading plasmalogens and PAF and functionally expressing them in neurons, astrocytes, oligodendrocytes, and microglia will advance the understanding of the glycerophospholipid metabolism at cellular and subcellular levels in the brain tissue. The use of proteomics and genomics can lead to the identification and expression profile of genes involved in the modulation of enzymes associated with the synthesis and degradation of plasmalogen and PAF in neurons and glial cells (Hovland et al., 2001), and also in monitoring alterations in gene expression profiles in neurodegenerative diseases (Colangelo et al., 2002; Thomas et al., 2006).

12.6 Use of RNAi for the Treatment of Ether Lipid-Related Neurodegenerative Diseases

RNA interference (RNAi) is an evolutionarily conserved phenomenon that silences gene expression through double-stranded RNA species in a sequence-specific manner. In recent years, RNAi has emerged as a powerful technique to manipulate gene expression in the laboratory animals. Among its many potential biomedical applications, silencing of disease-linked genes stands out as a promising therapeutic strategy for many incurable diseases such as neurodegenerative diseases and cancer. In laboratory settings, RNAi is routinely used to suppress virtually any target gene in a sequence-specific manner, including genes associated with chronic diseases (Xia et al., 2005; Gonzalez-Alegre, 2007; Thakker et al., 2006). Neurodegenerative diseases associated with ether lipid metabolism are attractive targets for the development of therapeutic RNAi. Abnormalities in ether lipid associated diseases are accompanied by changes in enzyme activities involved in their metabolism, alterations in ether lipid levels, oxidative stress, and neuroinflammation. Interactions among these parameters results in the progressive loss of neurons that leads to the gradual appearance of disabling neurological symptoms and premature death. Currently available therapies that utilize enzyme inhibitors are aimed to improve the symptoms, but not to halt neurodegenerative processes (Gonzalez-Alegre, 2007). RNAi technique can be aimed to target processes involved in neural cell death and retard neurodegenerative processes. The development of RNAi as potential therapy for such neurodegenerative diseases has generated considerable interest, partly because of the success of early studies of therapeutic RNAi in rodent models for a range of neurodegenerative diseases (Xia et al., 2005). Although in vivo delivery of small interfering RNAs into neural cells remains a significant obstacle, information from the treatment of ether lipid-related diseases in animal model studies can still be used for establishing preclinical trials and planning and developing studies on RNAi therapy for neurodegenerative diseases associated with abnormal ether lipid metabolism in patients (Thakker et al., 2006).

The approach for designing inhibitors of enzymes associated with plasmalogen and PAF metabolism should be based on rapidly developing concept of signal transduction pathways in neurological disorders (Farooqui et al., 1999). Better drug delivery systems that target brain region associated with pathogenesis of neurological diseases need to be developed to protect inhibitors of plasmalogen and PAF metabolism from in vivo degradation or detoxification. Drugs delivered through these drug delivery systems must reach the site where damage has occurred and inflammatory processes are taking place (Yoshikawa et al., 1999; Andresen and Jorgensen, 2005). This would enhance their efficacy of these drugs. The effects of these drugs on genes can be monitored by microarray procedures (Colangelo et al., 2002; Bosetti et al., 2005). These studies can lead to better therapeutic agents for the treatment of neurological disorders associated with plasmalogen and PAF metabolism alterations. We hope that next 50 years would not only witness the better understanding of the role of plasmalogen and PAF molecular species in metabolic processes in the brain tissue (Farooqui et al., 2002), but would also provide new information on intracellular plasmalogen and PAF trafficking, sorting, and metabolic regulation at cellular and subcellular levels in the normal brain tissue and in brain tissue from patients with neurodegenerative diseases.

12.7 Conclusion

Ether lipids are normal constituents of neural membranes. Ether lipid-derived lipid mediators interact with lipid mediators of sphingolipid metabolism. Under normal conditions, these interactions are essential for neural cell function and survival. Under pathological conditions, generation of high levels of ether lipid-derived lipid mediator and sphingolipid-derived lipid mediators and their interactions with each other result in neuroinflammation and oxidative stress that may cause neurodegeneration in neurological disorders. Lipidomics, proteomics, and genomics are important technologies that can be used to identify not only levels of molecular species of ether lipids and sphingolipids associated with neurological disorders, but for characterization of lipid mediators and biomarker that can be used for diagnosis and prognosis of neurological disorders. These diseases include bipolar disorders and schizophrenia, and neurodegenerative diseases such as AD, PD, and Niemann-Pick diseases. Dysregulation of ether lipid and sphingolipid metabolism is also observed in cerebral ischemic (stroke) injury. Lipidomics, proteomics, and genomics can provide a molecular signature to certain pathways that are associated with the pathogenesis of the above-mentioned neurological disorders. Lipidomics, proteomics, and genomics together with RNA silencing are powerful tools to elucidate the specific roles of lipid intermediates and lipid mediators in neural cell survival and neurodegeneration. This information would open new opportunities for drug development.

References

Adibhatla R. M., Hatcher J. F., and Dempsey R. J. (2006). Lipids and lipidomics in brain injury and diseases. AAPS J. 8:E314–E321.

Alcon S., Morales S., Camello P. J., and Pozo M. J. (2002). Contribution of different phospholipases and arachidonic acid metabolites in the response of gallbladder smooth muscle to cholecystokinin. Biochem. Pharmacol. 64:1157–1167.

Andresen T. L. and Jorgensen K. (2005). Synthesis and membrane behavior of a new class of unnatural phospholipid analogs useful as phospholipase A_2 degradable liposomal drug carriers. Biochim. Biophys. Acta Biomembr. 1669:1–7.

Bae K., Longobardi L., Karasawa K., Malone B., Inoue T., Aoki J., Arai H., Inoue K., and Lee T. (2000). Platelet-activating factor (PAF)-dependent transacetylase and its relationship with PAF acetylhydrolases. J. Biol. Chem. 275:26704–26709.

Bernatchez P. N., Tremblay F., Rollin S., Neagoe P. E., and Sirois M. G. (2003). Sphingosine 1-phosphate effect on endothelial cell PAF synthesis: Role in cellular migration. J. Cell. Biochem. 90:719–731.

Bogdanovic N., Bretillon L., Lund E. G., Diczfalusy U., Lannfelt L., Winblad B., Russell D. W., and Björkhem I. (2001). On the turnover of brain cholesterol in patients with Alzheimer's disease. Abnormal induction of the cholesterol-catabolic enzyme CYP46 in glial cells. Neurosci. Lett. 314:45–48.

Bosetti F., Bell J. M., and Manickam P. (2005). Microarray analysis of rat brain gene expression after chronic administration of sodium valproate. Brain Res. Bull. 65:331–338.

Butterfield D. A., Perluigi M., and Sultana R. (2006). Oxidative stress in Alzheimer's disease brain: New insights from redox proteomics. Eur. J. Pharmacol. 545:39–50.

Chalfant C. E. and Spiegel S. (2005). Sphingosine 1-phosphate and ceramide 1-phosphate: Expanding roles in cell signaling. J. Cell Sci. 118:4605–4612.

Chang J. Y., Chavis J. A., Liu L. Z., and Drew P. D. (1998). Cholesterol oxides induce programmed cell death in microglial cells. Biochem. Biophys. Res. Commun. 249:817–821.

Colangelo V., Schurr J., Ball M. J., Pelaez R. P., Bazan N. G., and Lukiw W. J. (2002). Gene expression profiling of 12633 genes in Alzheimer hippocampal CA1: Transcription and neurotrophic factor down-regulation and up-regulation of apoptotic and pro-inflammatory signaling. J. Neurosci. Res. 70:462–473.

Cuzzocrea S. and Salvemini D. (2007). Molecular mechanisms involved in the reciprocal regulation of cyclooxygenase and nitric oxide synthase enzymes. Kidney Int. 71:290–297.

Dennis E. A., Brown H. A., Deems R. A., Glass C. K., Merrill A. H. J., Murphy R. C., Raetz C. R. H., Shaw W., Subramaniam S., Russell D. W., VanNieuwenhze M. S., White S. H., Witztum J. L., and Wooley J. (2006). The LIPID MAPS approach to lipidomics. In: Feng L. and Prestwich G. D. (eds.), *Functional Lipidomics*. CRC, Boca Raton, FL, pp. 1–15.

Esposito G., Giovacchini G., Der M., Liow J. S., Bhattacharjee A. K., Ma K., Herscovitch P., Channing M., Eckelman W. C., Hallett M., Carson R. E., and Rapoport S. I. (2007). Imaging signal transduction via arachidonic acid in the human brain during visual stimulation, by means of positron emission tomography. Neuroimage 34:1342–1351.

Facheris M., Beretta S., and Ferrarese C. (2004). Peripheral markers of oxidative stress and excitotoxicity in neurodegenerative disorders: Tools for diagnosis and therapy? J. Alzheimer's Dis. 6:177–184.

Farooqui A. A., Litsky M. L., Farooqui T., and Horrocks L. A. (1999). Inhibitors of intracellular phospholipase A_2 activity: Their neurochemical effects and therapeutical importance for neurological disorders. Brain Res. Bull. 49:139–153.

Farooqui A. A., Horrocks L. A., and Farooqui T. (2000). Glycerophospholipids in brain: Their metabolism, incorporation into membranes, functions, and involvement in neurological disorders. Chem. Phys. Lipids 106:1–29.

Farooqui A. A. and Horrocks L. A. (2001). Plasmalogens: Workhorse lipids of membranes in normal and injured neurons and glia. Neuroscientist 7:232–245.

Farooqui A. A., Ong W. Y., Lu X. R., Halliwell B., and Horrocks L. A. (2001). Neurochemical consequences of kainate-induced toxicity in brain: involvement of arachidonic acid release and prevention of toxicity by phospholipase A_2 inhibitors. Brain Res. Rev. 38:61–78.

Farooqui A. A., Farooqui T., and Horrocks L. A. (2002). Molecular species of phospholipids during brain development. Their occurrence, separation and roles. In: Skinner E. R. (ed.), *Brain Lipids and Disorders in Biological Psychiatry*. Elsevier Science B.V., Amsterdam, pp. 147–158.

Farooqui A. A. and Horrocks L. A. (2004). Plasmalogens, platelet-activating factor, and other ether lipids. In: Nicolaou A. and Kokotos G. (eds.), *Bioactive Lipids*. Oily Press, Bridgwater, England, pp. 107–134

Farooqui A. A. and Horrocks L. A. (2006). Phospholipase A_2-generated lipid mediators in the brain: The good, the bad, and the ugly. Neuroscientist 12:245–260.

Farooqui A. A., Ong W. Y., and Horrocks L. A. (2006). Inhibitors of brain phospholipase A_2 activity: Their neuropharmacological effects and therapeutic importance for the treatment of neurologic disorders. Pharmacol. Rev. 58:591–620.

Farooqui A. A. and Horrocks L. A. (2007). Glycerophospholipids in the Brain: Phospholipases A_2 in Neurological Disorders. Springer, Berlin Heidelberg, New York, pp. 1–394.

Farooqui A. A. and Horrocks L. A. (2008). Glutamate and cytokine-mediated alterations of phospholipids in head injury and spinal cord trauma. In: Banik N. (ed.), *Brain and Spinal Cord Trauma*. Handbook of Neurochemistry Lajtha, A. In Press Springer, Berlin Heidelberg, New York.

Farooqui A. A., Horrocks L. A., and Farooqui T. (2007a). Interactions between neural membrane glycerophospholipid and sphingolipid mediators: A recipe for neural cell survival or suicide. J. Neurosci. Res. 85:1834–1850.

Farooqui A. A., Horrocks L. A., and Farooqui T. (2007b). Modulation of inflammation in brain: a matter of fat. J. Neurochem. 101:577–599.

Farooqui A. A., Ong W. Y., and Horrocks L. A. (2008). Neurochemical Aspects of Excitotoxicity. Springer, Berlin Heidelberg New York.

Fonteh A.N., Harrington R.J., Huhmer A.F., Biringer R.G., Riggins J.N., Harrington M.G. (2006). Identification of disease markers in human cerebrospinal fluid using lipidomic and proteomic methods. Dis. Markers 22:39–64.

Forrester J. S., Milne S. B., Ivanova P. T., and Brown H. A. (2004). Computational lipidomics: A multiplexed analysis of dynamic changes in membrane lipid composition during signal transduction. Molec. Pharmacol. 65:813–821.

Gonzalez-Alegre P. (2007). Therapeutic RNA interference for neurodegenerative diseases: From promise to progress. Pharmacol. Ther. 114:34–55.

Grimm M. O. W., Grimm H. S., Pätzold A. J., Zinser E. G., Halonen R., Duering M., Tschäpe J. A., De Strooper B., Müller U., Shen J., and Hartmann T. (2005). Regulation of cholesterol and sphingomyelin metabolism by amyloid-β and presenilin. Nat. Cell Biol. 7:1118–1123.

Gross R. W., Jenkins C. M., Yang J. Y., Mancuso D. J., and Han X. L. (2005). Functional lipidomics: The roles of specialized lipids and lipid–protein interactions in modulating neuronal function. Prostaglandins Other Lipid Mediat. 77:52–64.

Guan X. L., He X., Ong W. Y., Yeo W. K., Shui G. H., and Wenk M. R. (2006). Non-targeted profiling of lipids during kainate-induced neuronal injury. FASEB J. 20:1152–1161.

Hampel H., Teipel S. J., Alexander G. E., Pogarell O., Rapoport S. I., and Moller H. J. (2002). In vivo imaging of region and cell type specific neocortical neurodegeneration in Alzheimer's disease – Perspectives of MRI derived corpus callosum measurement for mapping disease progression and effects of therapy. Evidence from studies with MRI, EEG and PET. J. Neural Transm. 109:837–855.

Han X. L. (2007). Neurolipidomics: Challenges and developments. Front. Biosci. 12:2601–2615.

Han X. L. and Gross R. W. (2005). Shotgun lipidomics: multidimensional MS analysis of cellular lipidomes. Expert Rev. Proteomics 2:253–264.

He X., Jenner A.M., Ong W.Y., Farooqui A.A., and Patel S.C. (2006) Lovasatatin modulates increased cholesterol and oxysterol levels and has a neuroprotective effect on rat hippocampal neurons after kainate injury. J. Neuropath. Exp. Neurol. 65:652–663.

Hirashima Y., Farooqui A. A., Mills J. S., and Horrocks L. A. (1992). Identification and purification of calcium-independent phospholipase A_2 from bovine brain cytosol. J. Neurochem. 59:708–714.

Hovland A. R., Nahreini P., Andreatta C. P., Edwards-Prasad J., and Prasad K. N. (2001). Identifying genes involved in regulating differentiation of neuroblastoma cells. J. Neurosci. Res. 64:302–310.

Ikeda M., Kihara A., and Igarashi Y. (2006). Lipid asymmetry of the eukaryotic plasma membrane: Functions and related enzymes. Biol. Pharm. Bull. 29:1542–1546.

Ivanova P. T., Milne S. B., Forrester J. S., and Brown H. A. (2004). Lipid arrays: New tools in the understanding of membrane dynamics and lipid signaling. Molec. Interventions 4:86–96.

Karasawa K., Qiu X., and Lee T. (1999). Purification and characterization from rat kidney membranes of a novel platelet-activating factor (PAF)-dependent transacetylase that catalyzes the hydrolysis of PAF, formation of PAF analogs, and C_2-ceramide. J. Biol. Chem. 274:8655–8661.

Kihara A. and Igarashi Y. (2004). Cross talk between sphingolipids and glycerophospholipids in the establishment of plasma membrane asymmetry. Mol. Biol. Cell 15:4949–4959.

Kirsch C., Eckert G. P., and Mueller W. E. (2002). Cholesterol attenuates the membrane perturbing properties of β-amyloid peptides. Amyloid 9:149–159.

Koletzko B., Agostoni C., Carlson S. E., Clandinin T., Hornstra G., Neuringer M., Uauy R., Yamashiro Y., and Willatts P. (2001). Long chain polyunsaturated fatty acids (LC-PUFA) and perinatal development. Acta Paediatr. 90:460–464.

Kölsch H., Lütjohann D., Tulke A., Björkhem I., and Rao M. L. (1999). The neurotoxic effect of 24-hydroxycholesterol on SH-SY5Y human neuroblastoma cells. Brain Res. 818:171–175.

Kondo M., Imahori Y., Mori S., and Nakajima K. (2002). Inositol phospholipid metabolism in Alzheimer's disease - A positron emission tomographic study. In: DeLaTorre J. C., Kalaria R., Nakajima K., and Nagata K. (eds.), *Alzheimer's Disease: Vascular Etiology and Pathology*. Annals of the New York Academy of Sciences New York Acad Sciences, New York, pp. 416–422.

Lang P. A., Kempe D. S., Tanneur V., Eisele K., Klarl B. A., Myssina S., Jendrossek V., Ishii S., Shimizu T., Waidmann M., Hessler G., Huber S. M., Lang F., and Wieder T. (2005). Stimulation of erythrocyte ceramide formation by platelet-activating factor. J. Cell Sci. 118:1233–1243.

Latorre E., Collado M. P., Fernández I., Aragonés M. D., and Catalán R. E. (2003). Signaling events mediating activation of brain ethanolamine plasmalogen hydrolysis by ceramide. Eur. J. Biochem. 270:36–46.

Lee S. H., Williams M. V., and Blair I. A. (2005). Targeted chiral lipidomics analysis. Prostaglandins Other Lipid Mediat. 77:141–157.

Lee T. C., Ou M. C., Shinozaki K., Malone B., and Snyder F. (1996). Biosynthesis of *N*-acetylsphingosine by platelet-activating factor: Sphingosine CoA-independent transacetylase in HL-60 cels. J. Biol. Chem. 271:209–217.

Lizard G., Miguet C., Bessède G., Monier S., Gueldry S., Neel D., and Gambert P. (2000). Impairment with various antioxidants of the loss of mitochondrial transmembrane potential and of the cytosolic release of cytochrome c occuring during 7-ketocholesterol-induced apoptosis. Free Radic. Biol. Med. 28:743–753.

Lu Y., Hong S., Gotlinger K., and Serhan C. N. (2006). Lipid mediator informatics and proteomics in inflammation-resolution. ScientificWorldJournal 6:589–614.

Maeba R. and Ueta N. (2004). A novel antioxidant action of ethanolamine plasmalogens in lowering the oxidizability of membranes. Biochem. Soc. Trans. 32:141–143.

Masters C. L., Cappai R., Barnham K. J., and Villemagne V. L. (2006). Molecular mechanisms for Alzheimer's disease: Implications for neuroimaging and therapeutics. J. Neurochem. 97:1700–1725.

Milne S., Ivanova P., Forrester J., and Brown H. A. (2006). Lipidomics: An analysis of cellular lipids by ESI-MS. Methods 39:92–103.

Nelson T. J. and Alkon D. L. (2005). Oxidation of cholesterol by amyloid precursor protein and β-amyloid peptide. J. Biol. Chem. 280:7377–7387.

Nodai A., Machida T., Izumi S., Hamaya Y., Kohno T., Igarashi Y., Iizuka K., Minami M., and Hirafuji M. (2007). Sphingosine 1-phosphate induces cyclooxygeriase-2 via Ca^{2+}-dependent, but MAPK-independent mechanism in rat vascular smooth muscle cells. Life Sci. 80:1768–1776.

Nomikos T. N., Iatrou C., and Demopoulos C. A. (2003). Acetyl-CoA:1-*O*-alkyl-*sn*-glycero-3-phosphocholine acetyltransferase (lyso-PAF AT) activity in cortical and medullary human renal tissue. Eur. J. Biochem. 270:2992–3000.

Park D. S., Obeidat A., Giovanni A., and Greene L. A. (2000). Cell cycle regulators in neuronal death evoked by excitotoxic stress: Implications for neurodegeneration and its treatment. Neurobiol. Aging 21:771–781.

Phillis J. W., Horrocks L. A., and Farooqui A. A. (2006). Cyclooxygenases, lipoxygenases, and epoxygenases in CNS: Their role and involvement in neurological disorders. Brain Res. Rev. 52:201–243.

Piomelli D. (2005). The challenge of brain lipidomics. Prostaglandins Other Lipid Mediat. 77:23–34.

Rapoport S. I. (1999). In vivo fatty acid incorporation into brain phospholipids in relation to signal transduction and membrane remodeling. Neurochem. Res. 24:1403–1415.

Rapoport S. I. (2001). In vivo fatty acid incorporation into brain phospholipids in relation to plasma availability, signal transduction and membrane remodeling. J. Mol. Neurosci. 16:243–261.

Rapoport S. I. (2005). In vivo approaches and rationale for quantifying kinetics and imaging brain lipid metabolic pathways. Prostaglandins Other Lipid Mediat. 77:185–196.
Reiss A. B., Siller K. A., Rahman M. M., Chan E. S. L., Ghiso J., and De Leon M. J. (2004). Cholesterol in neurologic disorders of the elderly: Stroke and Alzheimer's disease. Neurobiol. Aging 25:977–989.
Robinson B. S., Hii C. S. T., Poulos A., and Ferrante A. (1997). Activation of neutral sphingomyelinase in human neutrophils by polyunsaturated fatty acids. Immunology 91:274–280.
Serhan C. N. (2005). Mediator lipidomics. Prostaglandins Other Lipid Mediat. 77:4–14.
Shindou H., Hishikawa D., Nakanishiu H., Harayama T., Ishii S., Taguchi R., and Shimizu T. (2007). A single enzyme catalyzes both platelet-activating factor production and membrane biogenesis of inflammatory cells – Cloning and characterization of acetyl-CoA:lyso-PAF acetyltransferase. J. Biol. Chem. 282:6532–6539.
Simons K. and Ikonen E. (2000). How cells handle cholesterol. Science 290:1721–1726.
Snyder F. (1995). Platelet-activating factor: the biosynthetic and catabolic enzymes. Biochem. J. 305:689–705.
Stahelin R. V., Subramanian P., Vora M., Cho W., and Chalfant C. E. (2007). Ceramide-1-phosphate binds group IVA cytosolic phospholipase a_2 via a novel site in the C2 domain. J. Biol. Chem. 282:20467–20474.
Thakker D. R., Hoyer D., and Cryan J. F. (2006). Interfering with the brain: Use of RNA interference for understanding the pathophysiology of psychiatric and neurological disorders. Pharmacol. Ther. 109:413–438.
Thomas D. M., Francescutti-Verbeem D. M., and Kuhn D. M. (2006). Gene expression profile of activated microglia under conditions associated with dopamine neuronal damage. FASEB J. 20:515–517.
Vaena de Avalos S., Jones J. A., and Hannun Y. A. (2004). Ceramides. In: Nicolaou A. and Kokotos G. (eds.), *Bioactive Lipids*. The Oily Press, Bridgwater, England, pp. 135–167.
Van Overloop H., Denizot Y., Baes M., and Van Veldhoven P. P. (2007). On the presence of C_2-ceramide in mammalian tissues: possible relationship to ether phospholipids and phosphorylation by ceramide kinase. Biol. Chem. 388:315–324.
Vigh L., Escriba P. V., Sonnleitner A., Sonnleitner M., Piotto S., Maresca B., Horvath I., and Harwood J. L. (2005). The significance of lipid composition for membrane activity: New concepts and ways of assessing function. Prog. Lipid Res. 44:303–344.
Voelker D. R. (2003). New perspectives on the regulation of intermembrane glycerophospholipid traffic. J. Lipid Res. 44: 441–449.
Watanabe, T., Akiguchi, I., Yagi, H., Onishi, K., Kawasaki, T., Shiino, A., and Inubushi, T. (2002) Proton magnetic resonance spectroscopy and white matter hyperintensities on magnetic resonance imaging in patients with Alzheimer's disease. Ann. N. Y. Acad. Sci. 977:423–429.
Wenk M. R. (2005). The emerging field of lipidomics. Nat. Rev. Drug Discov. 4:594–610.
Xia X. G., Zhou H., and Xu Z. (2005). Promises and challenges in developing RNAi as a research tool and therapy for neurodegenerative diseases. Neurodegener. Dis. 2:220–231.
Yamaji-Hasegawa A. and Tsujimoto M. (2006). Asymmetric distribution of phospholipids in biomembranes. Biol. Pharm. Bull. 29:1547–1553.
Yanagisawa K. (2002). Cholesterol and pathological processes in Alzheimer's disease. J. Neurosci. Res. 70:361–366.
Yoshikawa K., Kita Y., Kishimoto K., and Shimizu T. (2006). Profiling of eicosanoid production in the rat hippocampus during kainic acid-induced seizure - Dual phase regulation and differential involvement of COX-1 and COX-2. J. Biol. Chem. 281:14663–14669.
Yoshikawa T., Sakaeda T., Sugawara T., Hirano K., and Stella V. J. (1999). A novel chemical delivery system for brain targeting. Adv. Drug Deliv. Rev. 36:255–275.
Yu Z. F., Nikolova-Karakashian M., Zhou D. H., Cheng G. J., Schuchman E. H., and Mattson M. P. (2000). Pivotal role for acidic sphingomyelinase in cerebral ischemia-induced ceramide and cytokine production, and neuronal apoptosis. J. Mol. Neurosci. 15:85–97.

Index

B

C

D

MIX
Papier aus verantwortungsvollen Quellen
Paper from responsible sources
FSC® C105338

If you have any concerns about our products,
you can contact us on
ProductSafety@springernature.com

In case Publisher is established outside the EU,
the EU authorized representative is:
Springer Nature Customer Service Center GmbH
Europaplatz 3, 69115 Heidelberg, Germany

Printed by Libri Plureos GmbH
in Hamburg, Germany